OPTOELECTRONICS AND LIGHTWAVE TECHNOLOGY

OPTOELECTRONICS AND LIGHTWAVE TECHNOLOGY

J. E. Midwinter and **Y. L. Guo**
Department of Electronic and Electrical Engineering
University College London
UK

JOHN WILEY & SONS
Chichester · New York · Brisbane · Toronto · Singapore

Other Wiley Editorial Offices

John Wiley & Sons, Inc., 605 Third Avenue,
New York, NY 10158-0012, USA

Jacaranda Wiley Ltd, G.P.O. Box 859, Brisbane,
Queensland 4001, Australia

John Wiley & Sons (Canada) Ltd, 22 Worcester Road,
Rexdale, Ontario M9W 1L1, Canada

John Wiley & Sons (SEA) Pte Ltd, 37 Jalan Pemimpin   05-04,
Block B, Union Industrial Building, Singapore 2057

Library of Congress Cataloging-in-Publication Data:

Midwinter, John E.
    Optoelectronics and lightwave technology / J.E. Midwinter, Y.L.
Guo.
       p.   cm.
    Includes bibliographical references and index.
    ISBN 0 471 92934 4
    1. Optical communications.   2. Optoelectronics.   3. Optical
fibers.   I. Guo, Y. L.   II. Title
TK5103.59.M53   1992
621.382'7 – dc20   91-29343
                                                               CIP

D
621. 36
MID

*A catalogue record for this book is
available from the British Library.*

Typeset in 10/12 pt Palacio by Dobbie Typesetting Limited, Tavistock, Devon
Printed in Great Britain by Courier International, East Kilbride

CONTENTS

3 OPTICAL FREQUENCY DIVISION MULTIPLEXING (FDM) TECHNOLOGY

4 HIGH-SPEED AND TIME DIVISION MULTIPLEXING OPTICAL TRANSMISSION

5 OPTICAL FIBERS—WAVEGUIDES AND DEVICES

6 SEMICONDUCTOR LASERS AND QUANTUM WELL DEVICES 212

7 GUIDED-WAVE OPTICAL DEVICES 265

PREFACE

Nobody can any longer doubt that optoelectronics has arrived and is revolutionizing telecommunications transmission through the medium of optical fiber. Moreover, the last decade has seen an astonishing proliferation not only of applications but of new technology that has led to increasingly sophisticated components and sub-systems allowing light to be manipulated in ways that would have been almost unimaginable in the 1970s. Much of this book is concerned with these components and the sub-systems in which they can be used. Hence coherent transmission features strongly as does wavelength and frequency division multiplexing and their associated devices. There are other areas that have been treated in very little detail, such as direct detection and systems, but these have been covered extensively by other authors.

The book arose primarily from a visit by one of the authors (YLG) to University College London for the calendar year of 1990. This provided a rare opportunity for almost uninterrupted writing and, as a result, his contribution is by far the greatest to the final volume. As Head of Department, one of us (JEM) was heavily distracted and the unfortunate corollary of this was that many topics we would both have liked to include could not be within the available time. Nevertheless, we hope and believe that the resulting volume will prove to be a valuable addition to existing publications on the subject. The presentation has been aimed at 3rd year BEng or 4th year MEng/MSc level or at the graduate engineer wishing to learn more of this fast moving field.

JEM and YLG
University College London

1

INTRODUCTION TO OPTOELECTRONICS AND LIGHTWAVE TECHNOLOGY

1.1 THE STORY OF FIBER OPTICAL COMMUNICATION SYSTEMS

1.1.1 The beginning

Identifying the 'beginning' of optical communications is difficult and the choice arrived at depends largely on viewpoint. Light has been used for communication purposes for millenia. For example, we now read of the Ancient Greeks using mirrors and sunlight, of the Chinese using fire beacons as early as 800BC followed by the Romans in Europe and the American Indians using smoke signals. However, all of these are examples of very simple, albeit effective, message systems with extremely limited information capacity. More recently, battery operated signalling lamps have been used to flash messages in Morse code between naval ships, an application that comes much closer to our present interest in high data rate largely digital fiber communication systems. Probably the most important starting point for discussing modern optical communications is with the discovery of the laser since this, more than anything else, changed the way we thought about light, changing from 'black body' type radiation, incoherent, broad band and more like electrical noise to coherent, very narrow spectral linewidth continuous EM radiation more closely akin to the emissions from radio or microwave oscillators.

1.1.2 Lasers and detectors

The first report of coherent light emission appeared in the now famous paper by Maiman published in 1960, where a collimated pulsed beam of red light was obtained from a crystal of ruby (chromium doped sapphire) that had been optically pumped by an electrical pulse through a gas discharge flash-lamp. Whilst it was hopelessly impractical for any telecommunications application, it nevertheless transformed thinking about light as an EM medium.

In 1961, another breakthrough in the story of the laser occurred, with the description of the first HeNe gas-discharge laser, now used primarily at 633 nm but also at 1150 and 3390 nm and other wavelengths. Today, these are ubiquitous and used for lecture room pointers, alignment tools, surveying instruments etc. However, the fact that they were rather quickly developed to provide continuous power in very monochromatic beams with excellent near-Gaussian beam profiles rapidly attracted interest in the possibility of using them for free space or guided wave telecommunications. But more of that later.

Shortly after this, in 1963, there were two simultaneous reports of laser action in electrically driven semiconductor devices based on GaAs. These represented another quantum jump in technology. Suddenly, lasers shrank from fearful room-filling beasts to small electronic devices mounted on transistor headers. Of course, there were some problems, such as the fact that the lifetime of the devices was often measured in seconds or minutes, rather than the decades that real systems would require. Nevertheless, coherent sources existed that operated at milliwatt powers and that promised acceptable conversion efficiencies from electrons to photons. They also emitted at wavelengths close to the GaAs band-edge at about 900 nm, a wavelength that initially seemed almost ideal for use with a variety of transmission media and with silicon detectors which had already been studied fairly extensively.

So we find that, stimulated by the existence of light sources that seemed akin to microwave radio sources, the stage was set to enquire whether light might perhaps be used for wideband communications purposes, given that the 'carrier frequency' of light was evidently very much greater than that of any radio spectrum source.

1.1.3 Free space optical communication and optical pipes

To understand what happened next, it is helpful to have an appreciation of what was the 'current state of the art' in 'conventional' telecommunications. The wideband telecommunications cable systems of the time used coaxial cables carrying traffic in an (analogue) frequency division multiplexed form although there was intensive development aimed at utilizing some of this cable stock for digital transmission. Microwave line-of-sight systems also used analogue FDM transmission whilst for the inter-exchange networks, digital PCM systems were being considered at 2 Mbit/s rates. Firmly in research was work on the TE_{01} millimetric waveguide system which was to operate over a hollow pipe waveguide about 5 cm in diameter, to be made and installed with great precision along super-straight duct routes but capable of carrying huge numbers of telephone channels.

Noting that light does not travel too well over long distances in the atmosphere on foggy or rainy days, it was perhaps not surprising to find that much early work was aimed at using beams of light, probably from HeNe lasers, propagating inside tubes running in similar ducts to the proposed millimetric waveguide system. To recollimate the beam at suitable intervals, it was proposed to use very simple lenses formed from heated gas, located at intervals of about 100 m. As with the millimetric waveguide, such a system potentially offered huge data capacity over very long distances. We know from common experience that an unpolluted dry atmosphere, such as one enjoys in desert regions, allows very clear transmission over huge distances, say 100 km, whilst the dispersion of the gaseous medium is very small and should allow distortion-free transmissions at high data rate. The only problem lies in the fact that the ducting, being of large mechanical dimensions and super straight, did not lend itself to low cost installation on normal telecommunications routes.

1.1.4   Glass materials and fibers

The next interesting development occurred in 1966, although it was not widely recognized as such until something like 5 years later. Charles Kao and George Hockham published a paper proposing that an optical fiber be used as a dielectric optical waveguide that could be enclosed in a telecommunications type cable and carry light signals over useful distances. Initially, the proposal looked rather stupid, since the attenuation of most known optical glasses was of order 1 dB/m so that a block or rod glass a few metres long obviously attenuated light severely. The idea that a glass fiber might carry light over a reasonable distance, say 1 km, seemed preposterous. However, rather soon after this publication, we find it followed by several others reporting measurements by Kao and others on the attenuation of pure fused-silica glass and indicating that in that one glass system, attenuation levels of less than 10 dB/km already existed in bulk glasses. In parallel with this, intensive studies of the attenuation mechanisms in glasses highlighted various elements, Rayleigh scattering and (UV) band edge absorption, water and metal ion impurity absorption. These studies suggested that the fundamental loss mechanism in the 900 mm region was likely to be Rayleigh scattering with a value in the region of 1 dB/km at a wavelength 1 μm and scaling as λ^{-4} with wavelength. However, it was to be many years before such values could be realized in fibers.

Most early work on fiber fabrication was influenced, unduly in retrospect, by the accepted wisdom of the fiberglass manufacturing industry which already made light conducting fibers for a variety of short distance applications, such as probing into inaccessible spaces. Such fibers only needed to be a few metres long and were formed from compound glasses, often the rod-in-tube method whereby a rod of 'core' glass was surrounded by a 'tube' of cladding glass and the resulting assembly drawn down by conventional glass blowing techniques.

The Corning Glass Company, having extensive experience of working with and manufacturing silica which has to be worked at very high temperatures (1600 to 1800°C) compared with 600–900°C for compound glasses, adopted a different approach which in 1970 led to the next major advance. They reported measurements on a fiber with an attenuation of about 20 dB/km at about 900 nm wavelength. Once it was realized approximately how this result had been achieved (and it should be noted that the details were closely guarded by the company) there followed a period of intense interest in techniques for the fabrication of silica glass fibers. Today, this work has led to all the standard production processes, with the Corning Outside Vapour Deposition (OVD) and the ATT-Bell Laboratories. Modified Chemical Vapour Deposition (MCVD), both described during the mid 1970s and now the basis of much of the world's production. Later, the Japanese variant of Vapour Axial Deposition (VAD) was reported and together, the trio reign supreme barring minor variations.

In each, the key factor is that the glass is formed directly by oxidation of vapour that has been produced from multiply distilled liquid starting material that, as a result, is extremely pure and in particular has very low levels of water and transition metal ions.

1.1.5   System objectives and the first real fibers

In any aimed R&D programme, it is always useful to have some idea of the target to be aimed for. In the case of optical fiber communications, this was identified soon after Charles Kao's first publication by F. F. Roberts working at the British Post Office Research Laboratories, now better known as British Telecom Research Laboratories. At the time, work was underway on the development of digital transmission techniques to exploit the existing analogue coaxial cable media already installed and since much was already known about their transmission characteristics, it was possible to identify what repeater spacing was likely to be achievable. Aiming at the CEPT standard digital multiplexed transmission rate of 140 Mbit/s, it was expected that repeaters would be required at intervals of 2 km or so, and since they would contribute a major fraction of the overall system cost and the same was expected to be true of an optical system, the target of at least 2 km repeater spacing was chosen. Given a plausible output power for a semiconductor laser of about 1 mW and the sensitivities then available with silicon avalanche photodiode detectors, it was relatively simple to establish that an insertion loss per section of about 40 dB was likely to be possible, leading to a requirement of 20 dB/km fiber-in-cable attenuation. Today, this sounds almost trivial but with commercially available fiber bundles having attenuations of 1–10 dB/m (i.e. 1000–10 000 dB/km), it did not look so at the time. However, the existence of this target makes all the more understandable the great surge of interest caused by the Corning Glass Works announcement in 1970 at a conference at the IEE in London, UK of the experimental achievement of that figure, results which were rapidly verified by a number of laboratories who were asked to measure the test fiber.

It might be thought that signified the arrival of fibers for serious systems experiments. However, the Corning fiber of that time was extremely fragile as a result of the particular processing they had adopted and it was some years before fibers appeared that could readily be handled. Then they had to be packaged in cables that could be installed in standard cable ducts and the general confidence and capability established to launch system studies 'in the field' and it was about 1977 before this stage had been reached. However, by then, a number of other more mundane battles had been fought and won.

1.1.6   Graded index versus single mode

The foremost of these battles concerned the structure of the core to be used within the fibers. The initial proposals by Charles Kao had concentrated on 'single mode' fiber with a small, uniform core structure guiding with a V value of less than 2.405. However, this raised major concerns about the problems of launching, field splicing, repair and connectorization since it led to a requirement for alignments tolerances measured in fractions of a micron. The thought of doing this in the normal field splicing environment (i.e. muddy manhole) was not enticing and alternative designs were sought. A larger core fiber would almost inevitably be multimode and since different modes normally travel with different group velocities, this would be expected to lead to massive pulse spreading as light travelled by all the different physical ray paths through the fiber. The proposed solution was the graded-index fiber in which a parabolic refractive-index core profile was substituted for a uniform core. Theoretically, it could be shown that this would lead to massive reduction in pulse spreading, say from 50 ns/km down to less than 1 ns/km. However, to achieve such values required very high precision control of the core refractive-index profile and this problem concentrated many minds during the 1970–75 period and was still concentrating production engineers' minds 5 years later. Suffice it to say that by about 1976, graded-index fibers with nominal multi-mode pulse spreadings of about 1 ns/km were becoming commercially available in limited quantities.

In parallel with this again, much work had been directed at the problem of splicing and repair in the field and a variety of reasonably well engineered solutions were available at least for field testing. GaAs laser sources were also becoming available with plausible lifetime specifications so that the potential for impressive demonstrations was emerging, especially when one notes that, by then, fiber attenuations not of 20 dB/km but of 3–5 dB/km were available.

1.1.7 The first systems and gearing up for production

In the UK, the first experimental systems were installed in 1977, at about the same time as a number of other trial installations in the USA, Japan and mainland Europe. Two experimental links operating over graded-index fiber ran from the British Telecom Research Laboratories, one carrying data at a rate of 8 Mbit/s over an unrepeatered distance of 13 km and the other at a rate of 140 Mbit/s over an unrepeatered section of 8 km. In each case, the very small cables involved were installed in existing and crowded cable ducts running along a route from the laboratories into the nearby town of Ipswich. The trials were run very much as laboratory experiments with intensely detailed assessment and monitoring to ensure that the maximum information was obtained on the behaviour of fibers, splices and systems in the field environment. Some unexpected results were obtained but the overall conclusions were that the fiber cable had considerably exceeded the designers' expectations and very attractive systems performance was possible.

Later in 1977, a much more fully engineered system was installed along a 9 km route from Stevenage to Hitchin. This 140 Mbit/s system included two remotely powered optical regenerators at the 3 and 6 km points and once again was fully successful. The net result was a rapid rise in confidence that optical technology had arrived and serious system planning should begin. British Telecom issued specifications for fiber systems to be delivered in 1980/81 and to be known as Proprietary Optical Line Systems, since each manufacturer was invited to propose his own proprietary designs. The routes offered included a very wide variety of terrains and bit rates of 8, 34 and 140 Mbit/s so that they offered a challenging range of new business opportunities which were seized upon with enthusiasm. Other countries throughout the world started broadly similar programmes at about the same time.

A necessary adjunct to the field deployment of working systems was the establishment of new manufacturing facilities for both fibers and fiber cables as well as all the associated components and test gear, together with training of the field staff in the new technologies. Were we able to return to the period 1978–82, we would find ample evidence of such a build up and training, so that by the first years of the new decade, fiber transmission was well established as a working medium for telecommunications transmission both intercity and between major switching nodes within the cities: in UK parlance, in the Trunk and Junction network and in the US parlance, in the Long Lines and Inter-Office Trunk networks.

1.1.8   The first single mode systems

Just as these major developments were occurring and it looked as if the decade were set for the massive deployment of graded-index fiber cable systems, another development occurred that caused a major rethink. Many years earlier, it had been realized that operation of fiber at a longer wavelength, typically close to 1300 nm rather than the 830–900 nm region used by the first systems would bring benefits in dispersion since the material dispersion in silica glass was identically zero in this region. Moreover, as the fiber production processes improved, it began to emerge that with very dry (OH free) glasses, attenuation figures well below the 3–5 dB/km at 850 nm could be achieved at the longer wavelength of 1300 nm, with values of 0.5–1 dB/km being widely reported. Later, it became apparent that even lower figures could be achieved at 1550 nm of around 0.2 dB/km. However, neither semiconductor laser sources nor detector diodes were initially available, and as a result progress towards systems at these wavelengths was initially slow. Nevertheless, by about 1980, the key components were becoming available and the realization grew that neither material dispersion nor attenuation would limit the system but rather it would be mode dispersion in even the best graded-index fiber. The conclusion was clear; single mode fibre would now offer further massive gains in performance provided that the engineering problems of splicing and handling it in the field could be overcome. There followed during the early years of the 1980s a worldwide rush to develop single mode fibre systems, coupled with almost monthly reports of the latest 'hero' experiment as repeater spacing were stretched from 30 km to 200 km and bit rates from 140 Mbit/s to many Gbit/s. The results are now history and have led to truly dramatic simplifications in the design and implementation of inter-city systems as well as new families of applications such as unrepeatered undersea systems spanning sea-channels of up to 200 km length, trans-oceanic fiber systems as well as massive reductions in the cost of transmitting the regular inter-city traffic. This was the real fiber revolution.

1.1.9   Pulling the plug on transmission!

Closely following on the heel of these advances have been others that promise equally dramatic advances in engineered transmission systems in the future, and with those, changes in the very nature of the whole telecommunications network. The demonstrations of coherent communications systems at BTRL in 1981/82 showed that for the first time, one could exploit the optical spectrum in an analogous manner to that of the radio spectrum. But since the low attenuation window in optical fiber centred on 1550 nm, spans a spectral width of about 20 000 GHz, there is almost infinite space available for further exploitation.

Since those early experiments, it has been realized that it is not necessary to suffer the complexity of fully coherent transmission and detection to take advantage of this spectrum space, and today there is much discussion of Dense Wavelength Division Multiplexing (Dense WDM), whereby many wavelength channels can be carried by a single fiber, perhaps as many as 100 in a practical system context leading to a 100 fold increase in capacity. This might not be of great commercial significance, since with other things being equal, it would imply 100 times the number of regenerators. However, with recent developments in optical amplifiers, it begins to look as if a complex repeater can be replaced by a simple fiber amplifier at all points in the network except at the terminating node. Thus the stage is set for a massive growth in the capability of existing installed cable at very low marginal cost. Once this happens, a major rethink on the nature and operation of the network as a whole, including the switching nodes, will become obligatory.

Related developments seem set to shake the transoceanic transmission market. The discovery that non-linear effects in the optical fiber can be used to completely cancel bandwidth limiting dispersive effects through the use of soliton transmission, coupled with the development of optical fiber amplifiers, promises transoceanic communication systems of 9000–10 000 km in length without regenerators, transparent to data from one end to the other, a truly stunning development. Such engineered systems seem likely to emerge within 5 years.

1.1.10 Fiber to the home

So far we have referred only to the use of fiber within the main lines of the telecommunications network. However, it has long been clear that the long term cost trends will favour the use of fibers into the home in place of copper wires provided that the customer seeks broader bandwidth services than telephony. Videophone might be such a service but so also might new forms of entertainment and information television, including the new High Definition TV operating at about 1200 line resolution and gives film quality pictures. The arguments for and against such a deployment centre primarily upon the political and commercial rather than the scientific and technical since there are a number of very attractive technical proposals for implementing such schemes together with many working trial systems.

The technical challenges in such applications are nevertheless considerable, since cost is of prime importance and it seems likely that before they can become widespread, extensive use of optoelectronic integrated circuits (OEICs) will be necessary to greatly reduce the component counts in the terminals and simplify their assembly and testing. Moreover, innovative ways of using fibers may also be necessary, so that some of the ideas for passively branching optical networks whereby one fiber feeds signals to tens, hundreds or even thousands of homes may also be necessary before fully cost effective deployment is possible. Answers to these questions remain in the future at present.

1.1.11   The optical 'inner space', communications within computers and switches

Finally, we should note that optical technology is now mounting a major assault on another application previously the exclusive domain of electrical technology. This is the field of communications within electronic processors or switching machines. Already fibers are used to connect racks of equipment together and to connect computer peripherals, like disc stores, to mainframes etc. However, there remains much greater scope for penetration by optics. For example, optics is already tackling the backplane wiring problem and optical 'printed' circuit boards exist, with optical waveguides in place of metal tracks. Optics may also provide a means for superior communications within large processor chips as well as for a much enhanced pin-out from the chip.

Such applications remain firmly in the sector of Research and/or Development but such rapid progress is being made that this may not be true much longer. Optics will then be seen to be the truly supreme communications medium.

1.2   SOME MAJOR FOUNDATIONS FOR DEVELOPING TECHNOLOGY

Data rate, volume of optical information channels, and repeater spacing are some important measures of performance for an advanced fiber communication system. In the present state-of-the-art technology, some experimental systems with 10–20 Gbit/s data rate [1.1], or with several tens of optical channel multiplexing [1.2], or with 200–300 km repeater spacing [1.3] have been demonstrated that excited the large emphasis to develop these advanced techniques to application. All these new systems will be determined by some major foundations such as the materials (including fibers, compound semiconductor and MQW structure materials, as well as integrated optics materials), the devices (narrow linewidth, frequency stable and frequency tunable lasers, high speed devices, fiber amplifiers, other fiber components and integrated waveguide optics devices) as well as some basic techniques.

The convenient basic techniques to offer such highly interesting characteristics are probably the frequency-division multiplexing technique, the high speed-time-division multiplexing technique, the coherent optics transmission techniques, the broadband local area network, and ISDN, as well as photonic switching techniques.

1.2.1   Optical fiber and fiber components

One of the most fundamental factors supporting or affecting the potential systems and technology is optical fiber. The longer repeater spacing, the broader bandwidth (which means higher information data rate), even the more functions that fibers could play are all determined by improving the fiber materials, construction and characteristics, such as further decreasing the loss and dispersion, making the zero dispersion matching with the minimum loss in spectra, as well as developing the multi-function of fibers, not only the transmission components (as a passive medium), but also some active function devices, doped-fiber amplifiers and lasers (as an active medium). All of these will have a contribution to the developing technology.

LOSS AND DISPERSION REDUCTION, MATCHING EACH OTHER IN SPECTRA

Losses and bandwidth are two of the most important characteristics of fibers. So the loss spectra, the dispersion spectra, minimizing both of them and matching the spectra are all hot topics in fiber research and development. In addition, the ease of light coupling and polarization maintaining are also highly valuable for opto-electronic systems especially coupling fibers with various kinds of semiconductor devices and integrated optics waveguide devices used in modern techniques. All of these characteristics depend on the fiber design and production.

Absorption and scattering of light travelling through a fiber are the main sources of the losses. As expressed in Fig. 1.1, for both multimode and single mode fiber, the loss depends strongly on wavelength. On the short wavelength side ($< 1.5\,\mu$m), losses are mainly determined by the intrinsic (Rayleigh) non-elastic scattering contribution which is the reciprocal of the fourth power of the wavelength. The minor contribution is caused by some imperfect fiber waveguide structure in the production [1.4]. The theoretical limit of these two terms is not clear but trying to improve the minimum loss of 0.15 dB/km at 1.55 μm is practical [1.5]. On the long wavelength side, loss is caused by the intrinsic absorption. Hydrogen affects the fiber and is caused by different kinds of processes. One of these is the vibrational resonant

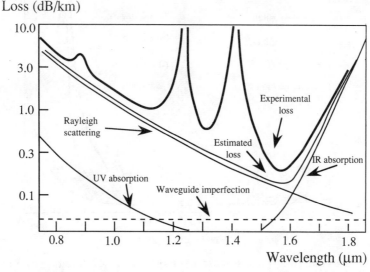

Fig. 1.1 Loss spectrum of SiO$_2$ fibers.

absorption at certain frequencies of molecular hydrogen H_2, such as 1.24, 1.69, as well as 1.89 μm, as shown in Fig. 1.2 [1.5]. The hydrogen is more or less unavoidably diffused into the silica fiber in production processes. The diffusion process is reversible and the induced loss saturates in about as little as 10 hours. So, the saturation value of the optical loss is needed to predict the long term loss behaviour. Another loss is caused by OH absorption which is formed between the diffused hydrogen and the constituent oxygen atoms. This process is irreversible and the resultant loss mainly depends on the types of dopant and their concentrations. The OH absorption peak appears at 1.40 μm as also shown in Fig. 1.2.

The long term loss increase caused by hydrogen gas diffusion was estimated by extrapolating data from elevated temperature experiments [1.6]. The sample was a germanium-doped single mode fiber. The dominant factor influencing the transmission bandwidth of fiber systems is chromatic dispersion, which depends on the fiber geometry, refractive-index profile and material properties. Pulse-broadening in single mode fibers occurs because injection laser sources radiate a spectrum of wavelengths. Chromatic dispersion effects make simultaneously produced wavelengths propagate with different group velocities along the fiber and therefore broadens the output pulse beyond its initial time slot. The bit rate capacity transmitted in the fiber is then determined by how close successive input pulses can be spaced without causing significant overlap or intersymbol interference between pulses in adjacent time slots reaching the receiver. Single mode fibers can have bandwidths of the order of 1000 GHz/km × nm, which correspond to several ps/km broadening of semiconductor pulses [1.7]. To achieve the goal of the lowest possible loss simultaneously wth the highest possible bandwidth, more sophisticated structures must be considered. Two classes of fiber fall into this category: 'dispersion-shifted' fibers, where the wavelength of zero dispersion is moved from 1.3 μm to 1.55 μm to coincide with the wavelength of minimum loss in silica dioxide fibers,

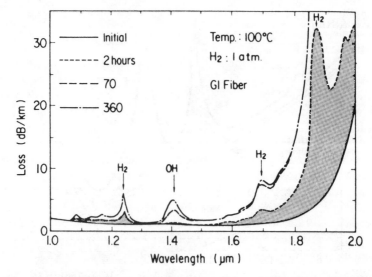

Fig. 1.2 Fiber loss caused by intrinsic absorption of H_2 and OH. (From [1.5] Kimura (1985) *J. Lightwave Technol.* **LT-6**(5), 611–619. Reproduced by permission. ©1985 IEEE.)

and 'dispersion-flattened' fibers in which low dispersion is made to be over an extended wavelength range close to 1.3 and 1.55 μm. The former enable an increased (repeater spacing)$\times$(bandwidth) product to be realized in direct-detection systems when sources with significant linewidths are used, and may be useful in very long coherent systems operating at multigigahertz line rates. The latter have potential applications in wavelength multiplexed systems operating at high data rates, where a uniform performance over a wide range of wavelengths is desirable.

ACTIVE FIBER AND FUNCTIONAL FIBER DEVICES

The subject of rare earth doped fiber amplifiers and lasers has recently received growing attention in lightwave technology. Optical fibers with a number of rare earth dopants are now available in silica. They will provide low cost, easily produced sources, and amplifiers at a variety of wavelengths, including the 1.3, 1.55 and 2–3 μm regions in the infra-red wavelengths that are important to both current and projected future requirements in the field of optical fiber telecommunication systems.

ER-DOPED OPTICAL FIBER AMPLIFIERS

Er^{3+}-doped fiber amplifiers are promising, particularly in 1.5 μm optical transmission systems. They have several attractive features: high optical gain, and low insertion loss because of their compatibility with optical fiber systems. There, efforts have been directed towards exploring the potential of Er-doped optical fiber amplifiers [1.8–1.11]. The possibility of 30 dB optical amplification with several mW of pumping power has already been reported. Also, optical amplification of high-speed optical signals has been demonstrated [1.12], and noise characteristics and signal excited state absorption characteristics have been investigated [1.13, 1.14]. Pumping wavelengths for most of the amplifiers over a wavelength of more than 1.4 μm has been demonstrated [1.15]. Such wavelengths are expected to provide more efficient optical amplifiers.

An optimized Er^{3+}-doped fiber optically pumped wavelength is an interesting topic for fiber amplifiers as well as fiber lasers; there has been some progress in 1989. 810 nm has been the preferred pump wavelength owing to the ready availability of the laser diode. However, the 810 nm pump band of Er^{3+} suffers from the serious disadvantage of pump excited-state absorption (ESA). As a result, much of the pump light is wasted in exciting ions from the metastable level to a still higher energy level, rather than producing stimulated emission. This is reflected in the reported slope efficiency of 17% [1.16], obtained for Er^{3+} sensitized with Yb^{3+}.

A recent report has demonstrated that the most efficient wavelength to pump Er^{3+}-doped fiber amplifiers is 980 nm, since this pump band is entirely free from ESA. Further, laser diodes operating at 980 nm are a realistic proposition and their development will, as is shown here, enable the production of a particularly useful class of fiber lasers with very high pump conversion efficiencies operating in the third telecommunications window. Data show a slope efficiency of 51% which can be compared with only 17% for Er^{3+} pumped with a diode array operating at 806 nm.

In addition to the above fundamental work, it is also important to study the performance of Er-doped optical fiber amplifiers in a transmission experiment. To date, a long span system of over 300 km using an Er-doped silica fiber optical amplifier without a conventional repeater had been reported. 1.8 Gbit/s transmission over 210 km using an Er-doped fiber amplifier with 20 dB gain [1.17] and 267 km with 1.2 Gbit/s optical transmission using two in-line LD-pumped Er-doped optical fiber amplifiers [1.18] were all demonstrated.

One of these transmission experiments using the fiber amplifier as a linear repeater is reported in [1.19]. 1.8 Gbit/s pseudorandom (PN, $2^{15} - 1$), return to zero (RZ) directly modulated optical pulses were emitted from a 1.552 μm distributed feedback laser diode (DFB LD). The average fiber input power was -0.2 dBm. After propagating through 118.9 km of dispersion shifted fiber (DSF, zero dispersion wavelength $\lambda_0 = 1.55$ μm, total loss $L = 26.0$ dB), the optical pulses were amplified by 90 m Er-doped fiber amplifier. Pump power was adjusted to 76 mW. The average input power to the amplifier signal gain was -26.2 dBm and 25 dB, respectively. The amplified signal was detected by an InGaAs-APD after propagating DSF 92 km in length with a loss of 19.1 dB. Average received power at a BER of 10^{-9} was -31.0 dBm. An optical fiber laser has a more easier coupling with fiber systems, so it is much more attractive for the application. Lasing output spectra of a fiber laser can be controlled by varying the glass host in which the rare earth ions lie. Lasing from rare-earth-doped fluorozirconate glass fibers is now possible at 1050, 1350, 1380 and 1550 nm in the near infra-red. The output at about 1350 nm is of particular value, as silica fiber lasers can be made to operate only with considerable difficulty in this region. Fluorescence exists which, in addition, make lasing possible in the important 2.08, 2.3 and 2.7 μm mid-infra-red bands. Such sources would have applications in telecommunications.

Both continuous wave and pulsed outputs are possible; fiber lasers have been operated to give Q-switched and mode locked operation and picosecond pulses are now achievable. Recently, picosecond optical pulse generation from a 1.48 μm laser diode pumped Er-doped fiber ring laser was successfully demonstrated by harmonic mode-locking, transform-limited pulses with a duration of 7.6 ps at repetition rates of 30 GHz being achieved [1.19]. An efficient operation of an Nd doped phosphate glass single mode fiber laser at 1.364 μm was used. The lasing threshold of 5 mW and slope efficiency of 10.8% is for a diode-pumped operation [1.20].

CO-DOPED OPTICAL FIBERS

One method of enhancing the fiber lasing is co-doping the Er^{3+} with Yb^{3+}. The Yb^{3+} absorbs much of the pump light and cross relaxation between adjacent ions of Er^{3+} and Yb^{3+} allows the absorbed energy to be transferred to the Er^{3+} system (Fig. 1.3) [1.21], thus providing another route for pumping the laser. The effect of co-doping the fiber with Yb^{3+} is clearly shown in Fig. 1.4 [1.21] (the dopant ratio of Yb^{3+} : Er^{3+} is 30 : 1). With Yb^{3+} present, light can be absorbed both directly into the Er^{3+} band (at 0.8 μm) and to the long wavelength side of it.

Fiber lasers can be made to tune over a substantial portion of their fluorescence spectra and, most importantly, they are available with extremely narrow bandwidth,

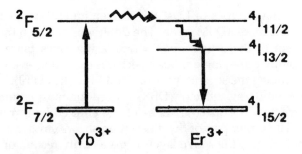

Fig. 1.3 Relevant energy levels of co-doped fiber. (From [1.21] Barnes *et al.* (1989) *J. Lightwave Technol.* **LT-7**(10), 1641–1645. Reproduced by permission. ©1989 IEEE.)

single longitudinal mode, outputs. They are therefore available for application to coherent communication and other fields requiring a monochromatic output.

1.2.2 Semiconductor lasers

Some of the most important advanced lightwave systems, such as dense frequency division or coherent transmission systems, are expected to use lasers. The candidates for such lasers in the low-loss wavelength region of optical fibers are $1.52\,\mu$m He–Ne lasers, $1.33\,\mu$m Nd : YAG lasers, $1.32\,\mu$m lithium neodymium tetraphosphate (LNP) lasers, and 1.3–$1.6\,\mu$m InGaAsP semiconductor lasers. But semiconductor laser diodes (LDs) are the most promising oscillators and the most interesting objects of research because of their long life and high efficiency. Semiconductor lasers have been the target of intensive study by many research organizations ever since the first successful CW room-temperature operation in 1970. But for the above potential applications, semiconductor lasers have still some characteristics to be improved, such as modulation speed, linewidth minimizing, frequency stabilization and frequency tuning, and phase noise reduction.

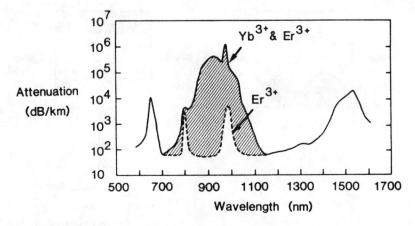

Fig. 1.4 Attenuation spectrum of co-doped fiber. (From [1.21] Barnes *et al.* (1989) *J. Lightwave Technol.* **LT-7**(10), 1641–1645. Reproduced by permission. ©1989 IEEE.)

There are two major types of LDs, the conventional Fabry–Perot (FP) type and the dynamic single mode (DSM) type. The difference between these two structures lies in the design of the active area, where lasing takes place. The DSM LD is characterized by its pure single-mode-spectrum operation even when modulated at high speed. Among the several types of DSM LDs, the distributed feedback (DFB) LD and distributed-Bragg reflector (DBR) LD are the most common lasers [1.22].

Fig. 1.5 shows how the spectral shape of the emitted light from the two diodes is determined. The upper part of the illustration shows cross sections of the FP and DFB LDs. In the FP LD, the active layer forms a cavity resonator having the shape of a thin narrow stripe with cleaved mirror facets at both ends, enabling standing waves of light to form in it. Therefore, the spectral shape of the emitted light is determined by the frequencies of standing waves that exist within the gain bandwidth of the active layer. As shown in the lower portion of this figure, the spectrum of the FP LD is composed of a number of spectral lines, resulting in an effective spectral width of about several nm.

DFB AND QUANTUM WELL DFB LASERS

On the other hand, in the DFB LD, there is a corrugation structure close to the active layer. In addition, the light-emitting facet is processed with antireflective coating to reduce mirror reflection so that the FP mode of oscillation is suppressed. Because the corrugation provides distributed feedback for light in a very narrow spectral range, the lasing occurs with a pure single frequency. The lasing spectrum is thus a single line having a very narrow spectral width even when the diode is modulated with a high-speed signal, as shown at the bottom of Fig. 1.5. The intensive research efforts to develop DFB LDs over the past several years have pushed forward the technical level to the point where some DFB LDs are beginning to be introduced in practical systems. Spectral linewidth, required for the various heterodyne and homodyne

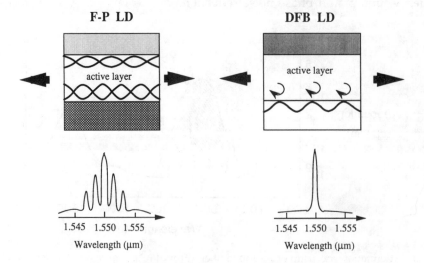

Fig. 1.5 Spectral shape for the FP-LD and DFB-LD.

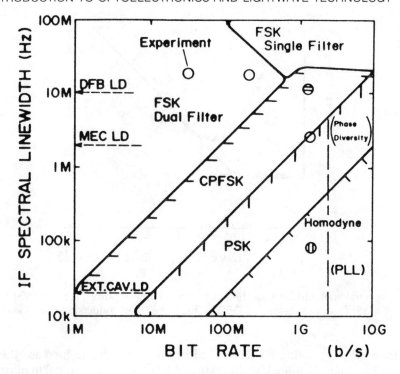

Fig. 1.6 Spectral linewidth required for coherent detection systems. (From [1.25] Kobayashi and Ito (1987) *J. Lightwave Technol.* **LT-6**(11), 1623–1633. Reproduced by permission. ©1987 IEEE.)

systems, is summarized in Fig. 1.6 [1.25]. Less than a few MHz seems to be a target when applications for differential phase shift keying (DPSK) or continuous phase frequency shift keying (CPFSK) systems at up to several gigabits per second are considered [1.24].

Spectral linewidths of 1.3 and 1.5 μm DFB LDs, measured with delayed self heterodyne or homodyne systems, are shown in Fig. 1.7 [1.25] as a function of the inverse output power. In the figure, results are described for 1.3 μm DFB LDs with asymmetric facet reflectivity and for 1.5 μm DFB LDs with a phase shifted grating structure with both facets antireflection coated.

Results on DFB LD cavity length dependence were reported experimentally [1.26]. A long cavity length, 1.2 mm, gives a linewidth as small as 1.7 MHz at 1.55 μm wavelength. A long cavity is effective in decreasing the spectral linewidth. However, because of the narrowed mode separation, such a long cavity DFB LD is likely to suffer from mode jumping, compared to DFB LDs with the usual 300–400 μm cavity length.

Recently, some progress in narrow spectral linewidth DBR lasers and multi-quantum well (MQW) lasers was successfully demonstrated. The linewidth of 3.2 MHz was obtained at $P = 1.5$ mW in a GaInAsP/InP distributed Bragg reflector (DBR) laser [1.27]. It was found that narrow spectral linewidths can be obtained constantly for the DBR lasers with long and low optical loss corrugation regions,

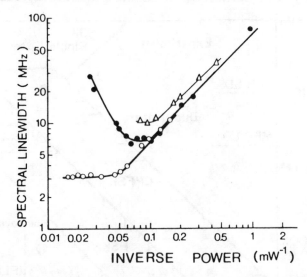

Fig. 1.7   Spectral linewidth as a function of reciprocal output power. (From [1.25] Kobayashi and Ito (1987) *J. Lightwave Technol.* **LT-6**(11), 1623–1633. Reproduced by permission. ©1987 IEEE.)

and with a small coupling coefficient. Moreover, in order to increase the output power of these narrow linewidth DBR laser, AR coating was applied to output facet. The degradation of the linewidths by the AR coating was insignificant. A linewidth of 2.5 MHz was obtained in a relatively new material GaInAs/AlInAs MQW laser diode [1.28]. Room-temperature CW operation ridge-type stripe lasers was confirmed in the 1.55 μm wavelength range. Spectral behaviour of the GaInAs/AlInAs MQW lasers was evaluated from the viewpoint of spectral linewidth and chirping characteristics. An obtained minimum value of the linewidth was as small as 2.5 MHz at an output power of 10 mW in a diode with a cavity length of 750 μm. Small chirping characteristics were also confirmed in which, for example, 1.5 Å chirp width was observed at 1 GHz direct modulation with a modulation depth of 67%. A proposal and fundamental properties of a distributed-reflector (DR) laser, consisting of an active section and a passive distributed reflector section, are introduced. This laser, in principle, has the properties of high efficiency and high-power performance in dynamic-single-mode operation, thus eliminating the problems of low device efficiency in DFB lasers and low output power in DBR lasers [1.29]. A simple model of coupled phase-shift DFB structures are analyzed and its feasibility of narrow-linewidth emission is shown [1.30]. It was found that the linewidth can be made as narrow as 0.2 MHz by coupling two DFB laser units. This linewidth is about five times narrower than the value expected theoretically in the conventional structure. A GaInAsP/InP distributed feedback buried ridge structure laser diodes emitting at 1.52 μm have been fabricated on material grown by two step low-pressure metal-organic chemical vapour deposition, with a second-order corrugation on the GaInAsP guiding layer. The minimum CW threshold current of 5 mA and single longitudinal mode operation with a side mode suppression ratio of 40 dB from 20 to 100°C with a temperature evolution of 0.64 A/°C have been obtained. At only 20 mA above threshold, a bandwidth of 9.6 GHz and a relative intensity noise (RIN) lower than -150 dB/Hz at 4 GHz have been measured [1.31].

Frequency stabilization is carried out by extracting an error signal and feeding it back to countermodulate the laser temperature or injection current. A frequency stability of 10^{-12}–10^{-11} at an average time of 100 s has thus far been achieved [1.32], where this characteristic actually means the frequency traceability of a semiconductor laser to the frequency reference. The oscillation frequency of a semiconductor laser can be locked to such a frequency reference as a Fabry–Perot interferometer of an absorption spectral line of gaseous atoms or molecules.

An absorption line exhibits a highly stable frequency over a long term, permitting the development of a semiconductor laser having an absolute standard frequency. An absorption line of NH_3, which is found in the 1.5 μm wavelength region, for example, exhibits a centre frequency shift toward a temperature change $(\Delta f/f)\Delta T$ of 10^{-18} deg^{-1} through the second-order Doppler effect [1.33], while a Fabry–Perot cavity whose housing is made of a low-expansion material, such as super-invar or quartz glass, shows a frequency shift of 10^{-7}–10^{-6} deg^{-1}. An InGaAsP DFB laser stabilized to an NH_3 absorption line at 1.52 μm has been reported [1.34]. Besides NH_3, such gaseous molecules as CO_2, H_2O, and CH_3Cl possess absorption spectra in the 1.3–1.7 μm wavelength region. Important for future systems, then, are the detailed assignment of absorption lines and the search for new absorption media.

An external cavity structure employing a long cavity structure and having a mirror or a diffraction grating also increases the effective cavity Q value, and thus serves to reduce the spectral linewidth [1.35]. An external-cavity 1.5 μm InGaAsP laser, in which the laser facet facing an external grating was anti-reflection-coated, achieved a spectral linewidth below 1 kHz [1.36]. Furthermore, the linewidth reduction factor and operation stability [1.37] of external cavity lasers have been extensively discussed as promising solutions.

Frequency tunable laser diodes are key devices for coherent optical communication systems or FDM systems. Very fast signal capture as well as large frequency range signal tracking have been achieved by applying them to local oscillator light sources [1.38]. Frequency division multiplexing (FDM) signals are also easily attained by using them as light sources [1.39]. Several types of frequency tunable laser diodes, such as tunable DBR or DFB LDs, and tunable MQW-DBR LDs have been reported so far. Their tunable range was improved to be almost sufficient for multi-channel FDM system applications. However, a problem had remained in the spectral linewidth. Some results of tunable DBR LDs showed less than 16 MHz linewidth in the entire 1000 GHz frequency tuning range [1.40]. A tunable quantum well (QW) DFB LD was reported recently [1.41]. The noise characteristics of semiconductor lasers get more and more attention as they obviously have a great effect on advanced techniques. Laser FM noise basically stems from the spontaneous emission coupled to a lasing mode. In semiconductor lasers, AM noise, i.e. photon number fluctuation, generated through the same mechanism as FM noise, competes with carrier number fluctuation and then enhances FM noise through the dispersion characteristics of the laser medium [1.42]. This FM noise enhancement factor is the so-called α parameter [1.43, 1.44]. Although FM noise possessing the f^{-1} power spectrum has been observed in a low-frequency region [1.45], its cause is still being discussed. The theory of noise in semiconductor lasers has been well established by using the quantum mechanical Langevin equation [1.46]. This method has additionally been studied to extend it to a treatment of the density-matrix master equation and the quantum mechanical Fokker–Planck equation [1.42]. Moreover, it is essential to take a close look at the electron system present in semiconductor lasers by means of a quantum statistical treatment.

Negative frequency feedback control of semiconductor lasers reduces FM noise and spectral linewidth [1.47, 1.48]. Theoretically, negative frequency feedback control offers the possibility of spectral linewidth decreasing below the modified Schawlow–Townes limit [1.48]. Reported experiments confirm, however, that a time delay around a feedback loop dominates the FM noise reduction factor [1.49]. Reduction of the loop time delay, which can be achieved by utilizing Optoelectronic Integration Circuit (OEIC) technology [1.50], is indispensable for ensuring the success of this method.

High speed single mode laser diode and picosecond optical pulses from semiconductor lasers are of great importance for optical communications with a huge capacity of bit rate. A self-aligned process was applied to lower the stray InP junction capacitance to as low as 1.6 pF. A record bandwidth of 13 GHz in the 1.5 μm wavelength region was demonstrated [1.51].

The advantages of using laser diodes as the picosecond light source for an electro-optic sampling system was also realized [1.52]. For these purposes, various generation methods, such as active and passive mode locking, colliding pulse mode locking, and gain switching, have been investigated. By active mode locking, short pulses (0.58 ps) have been generated [1.53]. Subpicosecond pulses are also successfully produced by other means of mode locking [1.54–1.56], and by gain switching, a short pulse of 6.7 ps generated from a conventional 1.3 μm InGaAsP laser diode was demonstrated [1.57].

A high-power single mode LD for extending the repeater spacing of intensity modulation/direct detection transmission system is still in great demand. Several kinds of structures have been proposed to increase the output power; one of these is to increase the power of the DFB laser, as it is a single longitudinal mode laser. For example, a stable operation of 1.55 μm distributed feedback lasers with a narrow linewidth and high power is demonstrated for coherent transmission systems [1.58]. Improving the geometrical uniformity of the active region and thinning the active laser are attempted by using MOCVD and a dry-etching technique to realize both a narrow linewidth and high power. A narrow linewidth of 1 MHz was achieved at a high power of 20 mW for a distributed feedback laser with a thin active layer of 0.07 μm and a long cavity of 1.2 mm. Even though the cavity was quite long, the low threshold current of 20 mA was observed. The current versus output power characteristics and the measured linewidth as a function of the inverse power for this laser are reported in [1.58].

Another kind, the so-called buried crescent LDs and VIPS LDs, has been developed to offer more than 100 mW or even 200 mW output power. Of course, it is a single transverse mode laser. The curved active layer and long cavity length increased the power, improved the transverse mode and the ease of fiber coupling. The optimized front facet is favored to reduce the high power intensity damage. The ultimate limiting factors affecting the output power from single mode fiber pigtail are LD to SM fiber coupling efficiency. The actual output power from the SM fiber pigtail is about 10 mW for produced LDs. But the lifetimes, and also the long term reliability at such high power, will be the subject of further development. For example, CW highpower output was obtained from a GaInAsP 1.3 μm V-grooved inner stripe on P-substrate (VIPS) lasers [1.59]. An output over 200 mW was obtained at room temperature using a 700 μm long cavity laser with 98% reflectivity of the front and rear facets. The fundamental transverse mode operation was confirmed up to 170 mW. Since the beam was isotropic and narrow, high coupling efficiency to a single mode fiber was obtained. The fiber coupled power was measured using a 9 μm core single mode fiber with two spherical collimation lenses [1.59]. A coupled power over 110 mW into a single mode fiber was achieved with a coupling efficiency of 58%.

1.2.3 Integrated optical waveguide device

Progress in optoelectronics and lightwave technology (such as optical fiber, single-mode semiconductor lasers and optical fiber communications) have also stimulated the research and development of guided wave optical devices. Integrated optics which basically employ single mode guided wave techniques are expected to contribute to rugged, cost-effective, and batch-fabricated components such as modulators/switches, multiplexers, tunable filters, frequency shifters, polarization controllers, signal routing, and timing devices.

Such functions can be realized by electro-optic waveguide components. At present the performance of LiNbO$_3$ devices is superior to other comparable waveguide components. For about a decade, research efforts in electro-optic LiNbO$_3$ waveguide components have resulted in a high status of development with regard to technology, device performance, and fiber coupling. This may be characterized by intensity modulators with a beyond-10-GHz modulation bandwidth and about 100 mW drive power [1.60, 1.61], high-speed low-insertion loss PSK modulators for coherent systems [1.62], 8×8 non-blocking switch arrays [1.63], multichannel wavelength multiplexers [1.64], and 0.5 dB insertion loss of fiber coupled 1.3 μm LiNbO$_3$ channel waveguides [1.65].

The reproducible fabrication of stable waveguide devices on LiNbO$_3$ by standardized technologies, and efficient waveguide to fiber coupling are essential for system applications. The common in-diffusion of titanium stripes for channel waveguides on LiNbO$_3$ (Ti:LiNbO$_3$) is the first thing with respect to technological tolerances and reproducible fabrication conditions. The attachment of fibers to LiNbO$_3$ devices requires special means if polarization preserving fibers are demanded. Generally, the electrodes of electro-optic devices must be decoupled from the evanescent field of waveguide modes. The electrode fabrication crucially influences the device stability.

Electro-optic guided wave devices such as intensity modulators, PSK modulators, and multiplexers are utilized for terminal functions, whereas switched branching structures and switch arrays are examples of network functions. Polarization control or polarization independent properties are required for network components and receiver terminal components if standard communication fibers are utilized. Application areas and state-of-the-art performance characteristics of electro-optic LiNbO$_3$ devices are illustrated by several selected examples.

Terminal and network functions of electro-optic $LiNbO_3$ waveguide devices are demonstrated for an assessment of their present abilities to provide high-speed operation, low insertion loss and high-performance characteristics of complex electro-optic waveguide configurations.

HIGH-SPEED WAVEGUIDE MODULATORS

External $LiNbO_3$ intensity modulators for laser transmitters of direct detection fiber-optic systems offer the advantages of large signal GHz-modulation-bandwidth with low modulation voltage and high on–off ratio, elimination of dynamic laser chirping, and phase modulation in coherent systems. In general, devices for transmitter terminal functions require—above all—low insertion loss and a high modulation bandwidth to compete with electronic and direct laser modulation techniques.

Phase-shift-keying (PSK) yields the highest signal-to-noise ratio in coherent systems. Here, external $LiNbO_3$ phase modulators are particularly attractive since they provide a clean phase modulation. Efficient and low loss $LiNbO_3$ PSK modulators are presently used in coherent system experiments [1.58]. The electro-optic modulators with a modulation bandwidth in excess of 1.5–2 GHz and low voltages require travelling-wave electrodes. Coplanar transmission line electrodes of 2–3 μm thickness show the best RF performance. The coplanar line are tapered for a connection to 50 Ω microstrip or coaxial lines. Their wave impedance is commonly chosen lower than 50 Ω to allow for broader electrode stripes with less attenuation. Alternatively, capacitively loaded 50 Ω coplanar lines are used to improve, for example, the electrical–optical overlap.

High-speed intensity modulators employing Mach–Zehnder interferometers [1.60, 1.61, 1.66], directional couplers (DCs) [1.67], X-switches [1.68], and cutoff modulation have been demonstrated on $LiNbO_3$. Integrated Mach–Zehnder interferometers yield the largest modulation bandwidth. One recognizes that push–pull operation is achieved with a simpler electrode configuration on Z-cut material. DC modulators/switches have the advantage of being four-port devices. The extra port can be used for monitoring the laser power. High-speed low insertion loss switches for $\lambda = 1.3 \mu$m and $\lambda = 1.56 \mu$m have been reported and used in system experiments [1.69].

DC switches with a low crosstalk are achieved by applying the $\Delta\beta$ reversal principle [1.70], where electrode segments are driven by voltages of opposite polarity. The concepts of $\Delta\beta$ reversal and of travelling wave electrodes are combined by connecting the electrode segments to a coplanar transmission line (for 1.3 μm wavelength, ± 8 V switching voltage, >6 Gbit/s switching speed, 2 dB insertion loss) [1.71].

WAVEGUIDE SWITCH ARRAYS

Future wide band communication systems may require optical switching networks, especially for routing high speed digital signals and for frequency multiplexed channels in coherent communication systems. Direct switching of the optical signals without electro-optic conversion by using optical waveguide switches may be preferable because of the high switching speed, lack of limitation in signal bit rate, bidirectional switching, and conservation of optical wavelength. Optical switching networks have been proposed and demonstrated for switched branching applications [1.72], time division switching [1.73], and $N \times N$ space division switching. As an example of the integration capability of LiNbO$_3$ devices, we discuss $N \times N$ switch arrays.

Among others [1.74], there are mainly two types of $N \times N$ matrix architectures: the blocking or statically non-blocking matrix of the rearrangeable type [1.75], and the non-blocking full crosspoint array, the so-called busbar structure [1.63]. Both of them are shown in Fig. 1.8.

The rearrangeable matrix has the advantage that the number of switch points $N(N-1)/2$ is considerably reduced in comparison to the busbar structure with N^2 switches, where N is the number of input and output ports. However, since several switches are involved in establishing an interconnection, some information is lost during the time required for the rearrangement of the signal path, in the case of changing an interconnection. In the busbar structure, the activation of only one switch is sufficient to build up an interconnection from any of the inputs to any of the outputs. Therefore, this structure is most suitable for application in high data rate transmission systems. Most of the reported switch arrays have been realized by using directional couplers [1.64, 1.76] or X-switches. As an alternative to the well known DC, the X-switch has been developed especially as a simple switch of compact size for the integration in large switching networks. Its operation principle is based on the electro-optically controlled two-mode interference in the intersection region [1.77, 1.78].

Recently, excellent performance characteristics of realized switch arrays have been reported: one 8×8 matrix of the busbar type with 6.8 dB maximum insertion loss and -30.50 dB average single-switch crosstalk, several 4×4 switch arrays of the busbar structure, and of the rearrangeable type with single-switch crosstalks lower than -35 dB, and total insertion losses of the order of 5 dB.

1.2.4   Some advanced lightwave techniques

Some advanced lightwave techniques considered as major foundations affecting future applications here are coherent optical modulation and detection, dense frequency-division multiplexing/demultiplexing, high speed time-division multiplexing/demultiplexing, network switching for spatial/time/frequency-multiplexing in very wide band networks (local area or ISDN).

HETERODYNE DETECTION TECHNOLOGY

The present research targets for coherent lightwave transmissions are longer repeater span, high speed and dense optical multiplexing. Optical heterodyne and homodyne detection is a key technology for long repeater span and high-speed transmission, and optical frequency-division multiplexing is a key technology for dense optical multiplexing.

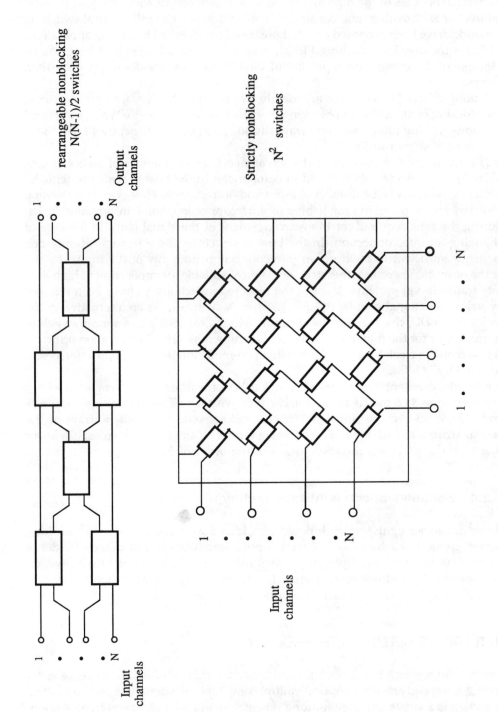

Fig. 1.8 Architectures of $N \times N$ switch array.

Heterodyne detection offers various advantages over direct detection. These include receiver sensitivity improvement and electrical signal processing of a received signal at an IF electrical band. The first advantage is well known and will significantly affect future long haul transmission systems. The second advantage will play a significant role in chromatic dispersion compensation in very high-speed transmissions, by eliminating amplified spontaneous emission noise caused in optical amplifers and suppression of interfering signals in optical FDM. These advantages will be discussed in Chapter 2. The most important item is laser diode FM/PM noise. Generally speaking, the laser diode spectrum width requirement is 10^{-1}–10^{-4} times transmission speed. However, the precise spectrum requirement depends on modulation/demodulation schemes. The conventional DFB/DBR LD can be applied to ASK or FSK modulation systems. However, higher receiver sensitivities require narrower LD spectrum widths. The phase noise influences error rate performance [1.79]. Some narrow linewidth DFB/DBR LDs, such as monolithic integrated external cavities, have been developed. Since the required linewidth becomes tolerable as data rate increases, the differential detection of DPSK has been recently achieved by using these advanced DFB LDs [1.80]. One advantage of optical FSK is that it does not require an external modulator, thus allowing a higher launch power and a more compact transmitter. However, frequency response uniformity in optical FM modulation is a requirement for LD direct modulation. Use of a multielectrode DFB LD is one way to obtain flat frequency response, which avoids waveform degradation [1.81, 1.82]. FM frequency response curves of a conventional DFB LD and a three electrode DFB LD are reported in [1.79]. The flat FM response region of the multi-electrode DFB LD ranges from dc to 600 MHz with FM deviation of 1.3 GHz/mA. This improved FM response DFB LD is a promising light source for FSK modulation systems [1.83]. In heterodyne detection systems, polarization compensation is required unless polarization holding fibers are used. In conventional fibers the polarization state changes slowly, less than 1 kHz, but with endless rotation, and its deviation distribution spreads with cable length and time [1.84]. From a practical viewpoint, there are two major ways to compensate for the polarization state. Optically, the combination of a quarter wave and a half wave plate in fiber retarders or electro-optic devices potentially realizes compensation [1.85]. Electrically, polarization diversity achieves polarization compensation [1.86].

Optical heterodyne detection requires electrical circuits and optodetectors with broader bandwidth than is required for direct detection. Therefore, it is necessary to develop low noise wide-band electrical circuits, especially preamplifiers, to realize coherent high-speed transmission. The required bandwidth for PSK or ASK is about twice as large as the baseband signal bandwidth. Required bandwidth for FSK depends on the IF band demodulation scheme and FM modulation index.

In the present direct intensity modulation-direct detection systems, transmission distances of very high-speed systems are limited not by S/N but by transmitted waveform degradation caused by LD frequency chirping. Even if chirping is eliminated, possibly by an external modulator, direct detection repeater span is restricted by the propagation time difference between modulation sidebands. The calculated repeater span of PSK modulation-heterodyne detection and intensity modulation-direct detection is shown as a function of transmission speed [1.79].

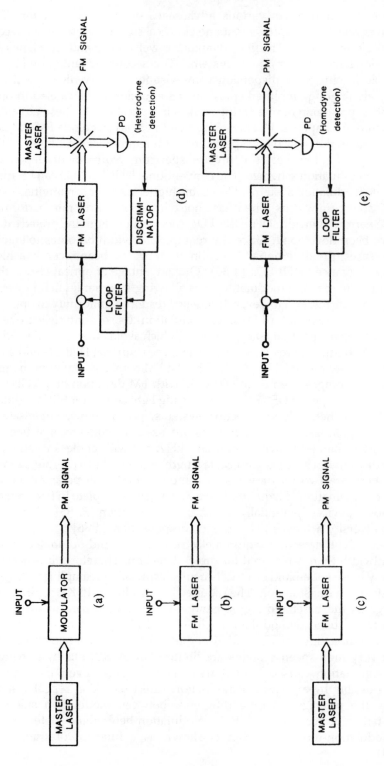

Fig. 1.9 Transmitter configurations for some modulation schemes. (From [1.91] Kimura (1987) *J. Lightwave Technol.* **LT-5**(4), 414–428. Reproduced by permission. ©1987 IEEE.)

Coherent transmission systems are more sensitive to chromatic dispersion than direct detection systems [1.87]. Since heterodyne detection systems can compensate for waveform distortion caused by fiber chromatic dispersion in the IF band, they can transmit up to the loss limit [1.88].

COHERENT MODULATION AND DEMODULATION

Angle modulation is a promising modulation scheme in coherent optical transmission because of the various advantages. Of great interest along this line is MFSK (multiple-frequency-shift-keying), which improves receiver sensitivity through multilevel signalling. For PSK signal generation a conventional guided-wave electro-optic phase modulator can be used, which is installed just behind a frequency stabilized CW laser as shown in Fig. 1.9(a) [1.91]. Besides external phase modulators, direct modulation of semiconductor lasers can be applied to FSK and PSK signal generation. The basic transmitter configurations for these modulation schemes are shown in Fig. 1.9. There are:

(a) phase modulation using an external electro-optic modulator;
(b) frequency modulation by direct modulation of FM semiconductor laser injection current;
(c) phase modulation using an injection locking technique;
(d) frequency modulation in optical FMFB configuration;
(e) phase modulation in an optical PLL configuration.

An optical FM signal is obtained by directly modulating the injection current of conventional single-longitudinal mode semiconductor lasers [1.90]. Direct frequency modulation characteristics are determined by the carrier modulation effect in the high-modulation frequency region, and by the temperature modulation effect in the low-frequency region. Amplitude and phase characteristics of direct frequency modulation have previously been measured in detail [1.90]. Although the FM response of semiconductor lasers is not uniform, a frequency shift of 100 MHz to 1 GHz is easily obtained without serious intensity modulation. Direct frequency modulation of semiconductor lasers has the advantage of a simple transmitter configuration as shown in Fig. 1.9(b). The inherent spectral linewidth of a semiconductor laser, which may result in one drawback for this method, can be suppressed, for example, by an external cavity configuration at the expense of FM efficiency degradation [1.35]. Non-uniform FM characteristics can also be compensated for by electrically equalizing the modulation input current [1.92].

Optical phase modulation is achieved by direct current modulation of a semiconductor laser into which external coherent laser light is injected [1.93], as shown in Fig. 1.9(c). The output signal phase relative to the input light phase is zero when the injected laser frequency is exactly tuned to the input signal frequency. It changes by $\pi/2$ rad when the injected laser frequency is detuned away from the input light frequency to the locking bandwidth limit. The cutoff modulation frequency in this method is determined by the injection locking bandwidth. Even if the inherent spectral linewidth of the injection locked laser is rather broad, it is reduced to the level of the injected signal linewidth [1.94].

In conjunction with electrical negative feedback techniques, direct frequency modulation of semiconductor lasers generates optical FM/PM signals having uniform modulation frequency characteristics and narrow spectral linewidths [1.95, 1.96]. As shown in Fig. 1.9(d), a fraction of the output signal from a frequency modulated semiconductor laser is demodulated by heterodyne frequency discrimination detection with master laser light having a stable frequency and narrow linewidth. After being phase-reversed, the demodulated signal is fed back to counter-modulate the injection current of the modulated laser. This negative feedback loop structure, which is normally called frequency modulation feedback (FMFB), suppresses both the FM noise and non-uniform FM response to the semiconductor laser [1.95].

A feedback loop structure using the homodyne detection scheme shown in Fig. 1.9(e), that is, the phase-locked loop (PLL), is effective for PM signal generation [1.96]. For this phase modulation method, the loop bandwidth of the PLL determines the cutoff modulation frequency. Since the loop delay time of these schemes in practice limits the effective modulation bandwidth, its reduction, for example, by integrating the feedback loop components, is indispensable.

Optical signals are demodulated by heterodyne or homodyne detection using a local oscillator. FM noise of the local oscillator, as well as that of this transmitter laser, causes S/N degradation in receivers through FM–AM or PM–AM conversion, and determines a lower limit of error-rate performance [1.92]. Excess AM noise of the local oscillator, which is due to the resonance characteristics of a semiconductor laser, also deteriorates the S/N, and then, in this case, degrades the receiving signal level [1.92]. In order to remove the effect of local oscillator FM noise, a semiconductor laser must be used whose spectral linewidth is suppressed, for example, by an external cavity configuration.

FREQUENCY-DIVISION MULTIPLEXING (FDM)

Optical frequency division multiplexing (FDM) increases transmission capacity and network flexibility by utilizing the very large bandwidth potential of optical fibers [1.97]. In this arrangement, a number of frequency channels can be placed adjacent to each other to provide a large capacity of transmitted signals.

The noise of electrical amplifiers increases significantly at higher frequencies, and optical transmitted waveform at bit rates over 10–30 Gbit/s is degraded by fiber chromatic dispersion [1.98]. Therefore, optical FDM, which simultaneously transmits a number of optical signals by different carrier frequencies, will have future applications in large-capacity telecommunication and local broad-band information distribution networks.

The concept of optical 'wavelength' division multiplexing is the same as that of optical 'frequency' division multiplexing in the sense that the frequency of an electromagnetic wave uniquely corresponds to as specific wavelength. For large optical carrier spacings, for example, 20 000 GHz frequency spacing (0.1 μm 'wavelength' spacing), it seems convenient to use the term 'wavelength'. On the other hand, for narrow channel spacings, for example, 5 GHz frequency spacing (0.000 04 μm or 0.04 nm 'wavelength' spacing), it might be convenient to use the term 'frequency'.

Optical FDM has the following two advantages: (1) increased multiplexed optical channels; (2) decreased transmission characteristic variations of optical component characteristics, such as optical fiber dispersion and loss, optical amplifier gain, and mirror reflectivity among optical channels.

The first thing we would be concerned with in FDM is the frequency stable and narrow linewidth laser sources as just mentioned above. The next is the technique of multiplexing and demultiplexing. Star couplers and directional couplers are used to combine optical carrier waves having different optical frequencies. However, coupling losses in these devices are fairly large, which makes them unfavorable choices for $N:1$ optical multiplexers except for $N:N$ local distribution. Use of optical filters results in high efficient $N:1$ coupling for optical FDM. Since the channel spacing of FDM is three or four orders of magnitude smaller than WDM channel spacing, WDM filters cannot be utilized for FDM systems. WDM mainly utilizes dielectric thin film filters or gratings as optical filters. However, these filters cannot combine and separate narrowly spaced optical waves. The filter configuration develop for microwaves or millimeter waves is applied to FDM systems [1.99, 1.100].

Demultiplexing is a fundamental technique for FDM. There are some ways to separate optical signals. One is to use an optical filter, the other is to use heterodyne detection. Typical configurations: (a) optical filtering type, (b) separate optical heterodyne type, and (c) all-in-one optical heterodyne detection type are shown in Fig. 1.10 [1.98].

Type (a) can increase the transmission capacity per fiber without decreasing repeater span. Therefore, optical filtering is useful for long-haul large-capacity transmission as an alternative for high-speed transmission. In addition, optical filtering will play a significant role in future all-optical telecommunication network systems, such as optical frequency self-routing, optical FDM switching, and other optical signal processing systems. It should be noted that optical filtering is independent of optical modulation and demodulation schemes. Thus, it allows both direct detection as well as heterodyne detection.

Type (b), the separate optical heterodyne-detection type, filters more packed optical signals, making use of frequency selectively of an intermediate frequency (IF) band circuit in heterodyne detection [1.101]. However, in this configuration, the received optical power per channel decreases as the number of optical channels increases.

Type (c), an all-in-one heterodyne-detection scheme, also separates high-density multiplexed optical carrier waves.

In this configuration, extremely broad-band preamplifiers and IF amplifiers are necessary for large channel systems. Thus, both type (b) and type (c) optical heterodyne detection systems can be used for local distribution, metropolitan area network (MAN), or multichannel short links.

An example of the transmission experiment is performed in the configuration shown in Fig. 1.11 [1.98]. Frequency-stabilized LD light waves are individually modulated by external modulators and coupled to a single mode fiber by a coupler. At present, since the silica waveguide used has a large birefringence, a polarization controller must be installed in front of the optical demultiplexer.

Another similar experiment of the channel selection and stabilization technique for a waveguide-type 16-channel FS–SW has been demonstrated [1.103]. A 16-channel FS–SW consists of four serially connected periodic filters integrated on

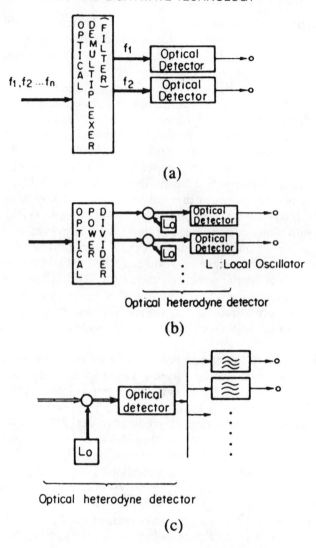

Fig. 1.10 Demultiplexing configurations. (From [1.98] Nosu *et al.* (1987) *J. Lightwave Technol.* **LT-5**(9), 1301–1308. Reproduced by permission. ©1987 IEEE.)

one chip. The configuration of periodic filters is the same as that of an asymmetrical Mach–Zehnder interferometer. The frequency spacings of the periodic filters are 5, 10, 20, and 40 GHz, respectively. The temperature of the FS–SW chip is stabilized to within +0.01°C. The desired channel could be selected from 16 optical channels with a 5 GHz channel separation. The frequency fluctuation width for the observe short period was 65–250 MHz and the frequency drift during 30 min was 20–100 MHz. An average crosstalk level of less than − 20 dB was successfully achieved in the 'ON' state of the FS–SW stabilization.

A very attractive example for FDM is a densely spaced FDM coherent multi-channel broadcast system which has a high potential for use in distribution

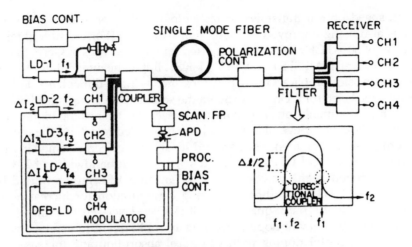

Fig. 1.11 A FDM transmission experiment. (From [1.98] Nosu *et al.* (1987) *J. Lightwave Technol.* **LT-5**(9), 1301–1308. Reproduced by permission. ©1987 IEEE.)

networks. One such demonstration system consists of six optical channels spaced by 2.2 GHz, the minimum frequency interval possible without adjacent channel interference. The channels are generated by multiple quantum well distributed Bragg reflector (MQWDBR) lasers operating at 1.53 μm wavelength. These lasers provide a stable single-frequency signal, tunable continuously over 1000 GHz. They also provide a narrow linewidth (2–4 MHz) small enough for coherent detection application. Modulation is carried out by FSK at 200 Mbit/s with a modulation index of two using a pseudorandom NRZ bit stream with a $2^{15}-1$ pattern length. The six optical channels multiplexed by a 16×16 optical fibre coupler are reported [1.104].

Another experimental 155.52 Mbit/s coherent broadcast network with 16 channels using conventional DFB laser operation at 1540 nm was reported in [1.105] and each laser module was mounted on a circuit board. It includes absolute laser frequency stabilization by locking to the 1509.554 nm absorption line of ammonia gas, and fiber span of 74.4 km as well as a balance polarization-diversity single-filter FSK heterodyne receiver. At a channel spacing of 8.5 GHz a power penalty of 0.3 dB is incurred due to adjacent channel crosstalk.

HIGH SPEED TRANSMISSION AND TIME DIVISION MULTIPLEXING

The high data rate of an optical fiber system is, in principle, affected by the laser modulation bandwidth, fiber dispersion, photodiode bandwidth and the speed of electronic devices. Systems for 1.6 Gbit/s systems are ready for service, and 8 Gbit/s systems are currently being studied. Several pioneering works have been reported which show that source and detector bandwidths are being improved to adapt to higher data rates.

To exploit transmission media for the 10–100 Gbit/s data rates, we should consider dispersion flattened optical fibers [1.106] or incorporated non-linear optical propagation in fibers [1.107], that is fiber solitons.

In high-speed optical transmission technologies, the speed of GaAs electronic devices is also being improved at these high data rates. For example, high cutoff frequencies of 50–80 GHz have been reported for various types of GaAs transistors, such as HEMT, HBT, MESFET, and PBT [1.108]. But at the moment, for ten or more Gbit/s transmission application, it becomes increasingly difficult to develop the necessary digital electronic circuits to avoid the so-called electronic speed bottleneck.

The resonant frequency of a semiconductor laser determines the highest internal modulation frequency. Several methods have been tried to extend the modulation bandwidth. Resonant frequencies of around 10 GHz have been obtained in InGaAsP DFB lasers in the 1.3 and $1.5\,\mu m$ bands [1.109, 1.110]. A bandwidth of 22 GHz has been demonstrated at the wavelength of $1.3\,\mu m$ in a VPE grown buried hetero-structure laser [1.111]. Another interesting configuration for increasing optical differential gain is the multi-quantum-well (MQW) laser. Further increasing the differential gain by heavily doping accepter in multi-quantum-well active region [1.112]. High acceptor doping reduces optical absorption and increases the net stimulated emission in the multi-quantum-well structure active layer. The principle was confirmed by an experiment using Be doped AlGaAs/GaAS MQW lasers [1.112] with a resonance frequency of 30 GHz with an output power of 160 mW. The idea underlying this device could be applied to $1.5\,\mu m$ devices. Several optical receiver schemes are compared in [1.113] for receiving signal levels over a wide range of data rates. The receiving level of APDs degrades when the data rate exceeds a few Gbit/s because of their junction capacitances, carrier transit times, and avalanche build-up times. Optical heterodyne detection is sensitive up to about 10 Gbit/s. When the data rate is higher than 10 Gbit/s, however, an extremely high level of local oscillator power is needed to give a low receiving level.

One method to relieve this electronic speed bottleneck is to extend the well known techniques of electrical multiplexing into the optical domain.

In optical time division multiplexing (OTDM), a high bit-rate data stream is constructed directly by time-multiplexing several lower bit-rate optical streams. Similarly, at the receiver end of the system, the very high bit-rate optical signal is demultiplexed to several lower bit-rate optical signals before detection and conversion to the electrical domain. This approach to optical time division multiplexing and demultiplexing moves the demand for high-speed performance away from electronic devices such as transistors, and places it on optical and optoelectronic devices such as pulsed semiconductor lasers and optical switches. The time division multiplexing approach is a purely digital technique and is therefore compatible with the concept to an all-digital network that combines switching and transmission. In addition, optical time division multiplexing offers system design flexibility, including the possibility of adjustable bandwidth allocation in different baseband channels and the possibility of simple system hardware in which only a single transmitter laser is required for all channels.

The potential of optical time division multiplexing and demultiplexing for very high bit-rate PCM systems has been recognized for more than two decades [1.114, 1.115], but until recently [1.116–1.118], there have been few system level demonstrations of the technique at multigigabit-per-second bit rates. The implementation of very high bit-rate OTDM systems has been slow because electronic multiplexing has usually served adequately and because the necessary hardware, such as high-speed optical switched and compact pulsed semiconductor lasers, has only recently reached a sufficient state of refinement. These experiments with time division multiplexing and demultiplexing have been made possible by improvements in lasers and Ti:LiNbO$_3$ switch/modulators [1.119]. A block diagram of the four-channel system is shown in Fig. 1.12 [1.120]. In this system, the baseband data rate is 4 Gbit/s and the electrical time delays τ are 62.5 ps, corresponding to one bit period at 16 Gbit/s. A common 4 GHz clock drives the four transmitters via a series of microwave delay lines that are adjusted to provide correct timing of the optical pulses.

Very high bit-rate point-to-point transmission systems, multiuser systems and time-multiplexed photonic switching networks are some of the most important systems in this field. System architectures describe the requirements on individual system components, and transmission system experiments of these systems will be discussed further in Chapter 4.

PHOTONIC SWITCHING IN COMMUNICATION

In modern communication systems narrow-band telephone services and high-speed computer data and wide-band video information services are expected to be integrated in one communications network. So, many studies on ISDN are being made in an attempt to provide enhanced services that integrate voice, data and video. In order to provide high-speed transmission lines, high-speed broad-band switching systems must be constructed. As for the optical transmission systems, high-bit-rate

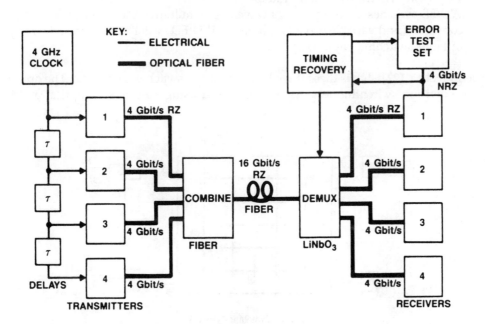

Fig. 1.12 Block schematic of a 4-channel OTDM system. (From [1.120] Tucker *et al.* (1988) *J. Lightwave Technol.* **LT-6**(11), 1744. Reproduced by permission. ©1988 IEEE.)

optical transmission systems are widely used, and optical subscriber systems are also beginning to be used in commercial applications. On the other hand, optical switching systems are in the research stage in many laboratories worldwide. However, optical switching systems are expected to play a key role in realizing future communication systems in which broad-band, bit-rate independent switching will be needed.

High-speed broad-band switching systems require high-speed operation in which electronic circuits have many problems, such as high power consumption and low isolation. Optical switching technology is a highly promising candidate for solving these problems. It may be said that switching systems, using optical switching technologies, will most likely provide the future high-speed, broad-band services.

Fig. 1.13 shows the configuration of an optical switching system. Electrical signals are transformed into optical signals by laser diodes (LDs) or light emitting diodes (LEDs). The optical signals from the sending terminals are transmitted through optical fibers to the receiving terminals via the optical switching network. The received optical signals are transformed into electrical signals by photodiodes (PDs) or avalanche photodiodes (APDs) [1.121].

Optical switching networks can be categorized into three types: optical space division (SD) switches, optical time division (TD) switch, and optical wavelength (frequency) division (WD/FD) switches. Each network has different elements and features.

(1) SD switches are composed of optical matrix switches.
(2) TD switches are composed of multi/demultiplex switches, highway switches, and optical memories for time slot exchanger. TD switches are well suited to work with TD transmission systems.
(3) WD/FD switches are composed of wavelength multi/demultiplexers, wavelength converters and variable wavelength filters. WD/FD switches have the advantage of large switching capacity.

Fig. 1.14 [1.122] shows these three kinds of optical switching networks. The optical space division switching network is constructed using a number of optical matrix

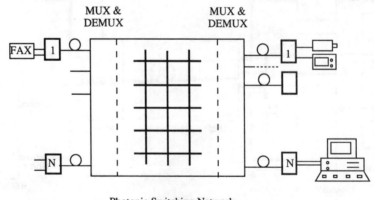

Photonic Switching Network

Fig. 1.13 Photonic switching.

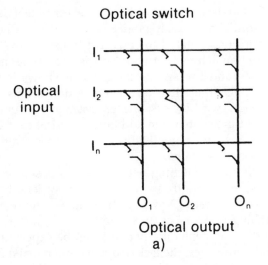

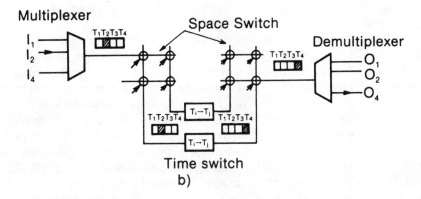

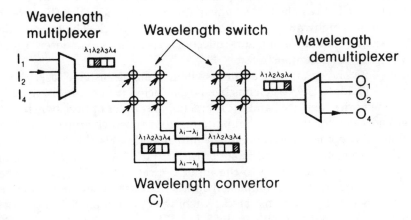

Fig. 1.14 Three kinds of photonic switching networks. (From [1.122] Yasui and Goto (1987) *IEEE Commun. Mag.* **25**(5), 10–15. Reproduced by permission. ©1987 IEEE.)

switches as shown in Fig. 1.14(a). It requires no other optical functional device, except the optical matrix switch, so it can be easily fabricated. However, a large number of matrix optical switches are necessitated to construct large-size systems.

The optical time division switching network consists of time multiplexers/demultiplexers, space switches, and time switches. Fig. 1.14(b) shows an STS (Space–Time–Space) construction optical time division switching network as an example.

To realize optical time division switching systems, optical memories or optical delay lines are required for optical time switches in addition to optical switches. Optical time division switching systems have a good affinity for the existing time division multiplexed optical transmission systems.

The optical wavelength division switching network is composed of wavelength multiplexers/demultiplexers, wavelength switches, and wavelength converters. Fig. 1.14(c) shows an SWS (Space–Wavelength–Space) construction optical wavelength division switching network as an example.

The optical wavelength division switching network can be expected to result in a flexible switching network. However, the lack of optical functional devices makes it difficult to construct the network at present. In recent years, there are substantial progresses in switching components and other devices, such as memories for TDM-switching, and frequency components for FDM-switching, as well as the experimental systems and architectures.

A 32-line optical space division switching system has also been developed [1.123]. It uses 8×8 lithium niobate (LiNbO$_3$) directional coupler optical matrix switches [1.124]. Wavelength multiplexing is adopted for a two-way switching network consisting of folded three-stage and one-stage switching networks. The former offers video telephone services. The latter offers TV multicasting services.

An optical time division switch using optical fiber delay lines is presented [1.125]. The optical time switch consists of a delay-line block, laser diode (LD) switches, and a 4×1 optical coupler block. Time-multiplexed signals on an input highway might be delayed by passing every delay line. The LD switches select one of these delayed signal. They are combined onto an output highway by the 4×1 optical coupler block. Time-multiplexed signals on an input highway might be delayed by passing every delay line. The LD switches select one of these delayed signals. They are combined onto an output highway by the 5×2 optical coupler block. The laser diode used in an LD switch module is an anti-reflection-coated InGaAsP laser. Operation at 256 Mb/s has been demonstrated.

Figure 1.15 shows an optical time switch, utilizing bistable laser diode optical memories [1.125]. The switch has been constructed using four bistable laser diode optical memories, and 1×4 and 4×1 directional coupler optical switches (OSW) as write/read gates for them, accommodating four video channels. The 1×4 optical switch injects optical signals sequentially into individual bistable diodes. Each bistable laser diode remains in the 'off' state or turns to the 'on' state, and memorizes the binary optical data according to the input optical signal. The 4×1 optical switch constructs an output time-multiplexed optical highway, where time slot interchange is accomplished. The demonstrated switching system operation speed is 256 Mb/s in 1986 [1.126], and 512 Mb/s in 1988 [1.127].

Optical frequency division switching systems with phase-shift-controllable DFB LD filters have been developed in most recent years. This system consists of a

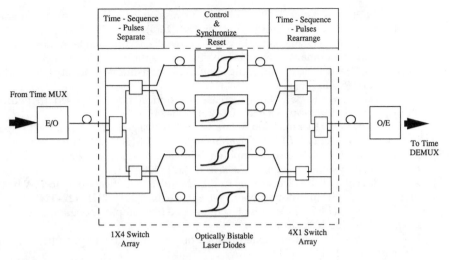

Fig. 1.15 Photonic switch using bistable LDs.

wavelength multiplexer, a wavelength switch and a wavelength demultiplexer. The wavelength multiplexer consists of modulators, which intensity-modulate light carriers supplied from wavelength reference light sources according to input CH1-n signals, and an optical combiner. In the wavelength switch, an input WDM signal is split and parts are led to individual tunable wavelength filters, each of which extracts a specific wavelength signal. The output signal from the tunable wavelength filter is then converted to an electronic signal by an optical–electronic converter. A pre-assigned-wavelength light carrier is intensity-modulated by a modulator according to the electronic signal from the optical–electronic converter. Therefore, a specified wavelength signal, selected by the tunable wavelength filter, can be converted to a different wavelength signal. The wavelength demultiplexer consists of an optical splitter, fixed wavelength filters and optical-electronic converters.

Components for switching in the optical domain offer substantially different characteristics to electronic switches. However, except in special cases these characteristics do not immediately map well onto the network requirements as currently perceived so that there remains great challenge in establishing a viable technology [1.128].

REFERENCES

[1.1] M. Shirasaki *et al.* (1990) "20 Gbit/s no-chirp intensity modulation by DPSH-IM method and its fiber transmission through 330 ps/nm dispersion" *Electron. Lett.* **26**(1), 33–35.

[1.2] W. I. Way *et al.* (1990) "160 channel FM-video transmission using optical FM/FDM and subcarrier multiplexing and an Erbium doped optical fiber amplifier" *Electron. Lett.* **26**(2), 139–142.

R. Welter *et al.* (1989) "Sixteen-channel coherent broadcast network at 155 Mbit/s" *J. Lightwave Technol.* **LT-7**(10), 1438–1444.

C. Lin *et al.* (1988) "Wavelength-tunable 16 optical channel transmission experiment at 2 Gbit/s and 600 Mbit/s for broadband subscriber distribution" *Electron. Lett.* **24**, 1215–1216.

[1.3] T. Imai *et al.* (1990) "Over 300 km CPFSK transmission experiment using 67 photon/bit sensitivity receiver at 2.5 Gbit/s" *Electron Lett.* **26**(6), 357–358.

[1.4] H. Yokota *et al.* (1986) "Ultra-low-loss pure silica core single mode fiber and transmission experiment" *OFC'86*, Feb. 1986, Atlanta, GA, paper PD3.

[1.5] T. Kimura (1988) "Factors affecting fiber-optic transmission quality" *J. Lightwave Technol.* **LT-6**(5), 611–619.

[1.6] K. Noguchi *et al.* (1985) "Loss increase for optical fibers exposed to hydrogen atmosphere" *J. Lightwave Technol.* **LT-3**(2), 236–243

[1.7] L. G. Cohen, W. L. Mammel and S. Lumish (1982) "Dispersion and bandwidth spectra in single-mode fibers" *IEEE J. Quantum Electron.* **QE-18**(1), 49–53.

[1.8] R. Noe *et al.* (1990) "Optical amplifier with 27 dB dynamic range in a coherent transmission system" *Photon. Technol. Lett.* **2**(2), 120–121.

[1.9] A. Takada *et al.* (1990) "Picosecond laser diode pulse amplification up to 12W by laser diode pumped Erbium-doped fiber" *Photon. Technol. Lett.* **2**(2), 122–124.

[1.10] R. J. Mears *et al.* (1987) "Low-noise Erbium-doped fiber amplifier operating at 1.54 μm" *Electron. Lett.* **23**, 1026–1028.

[1.11] J. T. Whitley (1988) "Laser diode pumped operation of Er^{3+}-doped fiber amplifier" *Electron. Lett.* **24**, 1537–1539.

[1.12] C. R. Giles *et al.* (1988) *Post-deadline paper CLEO* PD9.

[1.13] R. I. Laming *et al.* (1988) "Pump excited-state absorption in Erbium-doped fibers" *Opt. Lett.* **13**, 1084–1086.

[1.14] R. Olshansky (1988) "Noise figure for Erbium-doped optical fiber amplifiers" *Electron. Lett.* **24**, 1363–1365.

[1.15] E. Snitzer (1988) *Post-deadline paper OFC* PD2.

[1.16] R. Wyatt *et al.* (1988) "Efficient operation of array pumped Er^{3+} doped silica fiber laser at 1.5 μm" *Electron. Lett.* **24**, 1362–1363.

[1.17] A. Takada *et al.* (1989) "1.8 Gb/s transmission over 210 km using an erbium-doped fiber laser amplifier with 20 dB repeater gain in a direct detection system" *IOOC'89*, Kobe, Japan, July 18–21, TD 21B3-3.

[1.18] N. Edagawa *et al.* (1989) "267 km, 1.2 Gbit/s optical transmission experiment using two in-line LD-pumped Er-doped optical fiber amplifiers and an electroabsorption modulator" *IOOC'89*, Kobe, Japan, July 18–21, TD 21B4-1.

[1.19] A. Takada and H. Miyazawa (1990) "30 GHz picosecond pulse generation from actively mode-locked Erbium-doped Fiber laser" *Electron. Lett.* **26**, 216–217.

[1.20] S. G. Grubb (1990) "Diode-pumped 1.36 μm Nd-doped fiber laser" *Electron. Lett.* **26**, 121–122.

[1.21] W. L. Barnes *et al.* (1989) "Er^{3+}–Yb^{3+} and Er^{3+} doped fiber lasers" *J. Lightwave Technol.* **LT-7**(10), 1641–1645.

[1.22] Y. Suematsu *et al.* (1985) "Dynamic-single-mode lasers" *Optica Acta* **32**(9/10), 1157–1173.

[1.23] T. Nakagami and T. Sakurai (1988) "Optical and optoelectronic devices for optical fiber transmission systems" (1987) *IEEE Communications Magazine* **26**(1), 28–33.

[1.24] M. Shikada *et al.* (1987) "Optical devices for coherent optical fiber transmission systems" *GLOBCOM'87*, Tokyo, Japan **1**, 694–698.

[1.25] K. Kobayashi and I. M. Ito (1988) "Single frequency and tunable laser diodes" *J. Lightwave Technol.* **LT-6**(11), 1623–1633.

[1.26] K. Y. Liou *et al.* (1987) "Linewidth-narrowed distributed feedback injection lasers with long cavity length and detuned Bragg wavelength" *Appl. Phys. Lett.* **50**(9), 489–490.

[1.27] M. Takahashi *et al.* (1989) "Narrow spectral linewidth 1.5 μm GaInAsP/InP distributed Bragg reflector (DBR) lasers" *IEEE J. Quantum Electron.* **QE-25**(6), 1280–1287.

[1.28] Y. Matsushima *et al.* (1989) "Narrow spectral linewidth of MBE-grown GaInAs/AlInAs MQW lasers in the 1.55 μm range" *IEEE J. Quantum Electron.* **QE-25**(6), 1376–1380.

[1.29] K. Komori *et al.* (1989) "Single-mode properties of distributed-reflector lasers" *IEEE J. Quantum Electron.* **QE-25**(6), 1235–1244.

[1.30] T. Kimura and A. Sugimura (1989) "Coupled phase-shift distributed-feedback semiconductor lasers for narrow linewidth operation" *IEEE J. Quantum Electron.* **QE-25**(4), 678–683.

[1.31] M. Krakowski *et al.* (1989) "Ultra-low-threshold, high-bandwidth, very-low-noise operation of 1.52 µm GaInAsP/InP DFB buried ridge structure laser diodes entirely grown by MOCVD" *IEEE J. Quantum Electron.* **QE-25**(6), 1346–1352.

[1.32] T. Tako *et al.* (1983) "Frequency control of semiconductor lasers" *CLEO'83, Baltimore, MD, May 17–20* paper WB5.

[1.33] K. Shimoda (1973) "Frequency shifts in methane-stabilized lasers" *Japan J. Appl. Phys.* **12**(9), 1393–1402.

[1.34] T. Yanagawa *et al.* (1984) "Frequency stabilization of 1.5 µm InGaAsP distributed feedback laser to NH_3 absorption lines" *Appl. Phys. Lett.* **45**(8), 826–828.

[1.35] S. Saito *et al.* (1982) "Oscillation center frequency tuning, quantum FM noise and direct frequency modulation characteristics in external grating loaded semiconductor lasers" *IEEE J. Quantum Electron.* **QE-18**(6), 961–970.

[1.36] R. Wyatt (1985) "Spectral linewidth of external cavity semiconductor lasers with strong, frequency-selective feedback" *Electron. Lett.* **21**(14), 658–659.

[1.37] B. Tromborg *et al.* (1984) "Stability analysis for a semiconductor laser in an external cavity" *IEEE J. Quantum Electron.* **QE-20**(9), 1023–1032.

[1.38] S. Yamazaki *et al.* (1988) *Technical Digest of 14th ECOC'88, Brighton* pp. 86.

[1.39] M. Shibutani *et al.* (1989) *Technical Digest of OFC'89, Houston, ThC2.*

[1.40] T. L. Koch *et al.* (1988) "Continuously tunable 1.5 µm multiple-quantum-well GaInAs/GaInAsP distributed-Bragg-Reflector lasers" *Electron. Lett.* **24**, 1431.

[1.41] M. Kitamura *et al.* (1989) "Narrow spectral linewidth operation in frequency tunable quantum well distributed feedback laser diodes" *IOOC'89, Kobe, Japan, July 18–21, TD 19A-2.*

[1.42] Y. Yamamoto (1983) "AM and FM quantum noise in semiconductor lasers—Parts 1 and 2" *IEEE J. Quantum Electron.* **QE-19**(1), 34–58.

[1.43] C. H. Henry (1982) "Theory of the linewidth of semiconductor lasers" *IEEE J. Quantum Electron.* **QE-18**(2), 259–264.

[1.44] K. Vahara and A. Yariv (1983) "Semiclassical theory of noise in semiconductor lasers— Parts 1 and 2" *IEEE J. Quantum Electron.* **QE-19**(6), 1096–1109.

[1.45] F. G. Walther and J. E. Kaufman (1983) "Characterization of GaAlAs laser diode frequency noise" *OFC, New Orleans, LA, Feb. 28–Mar. 2,* paper TuJ5.

[1.46] H. Haug (1969) "Quantum-mechanical rate equations for semiconductor lasers" *Phys. Rev.* **184**(2), 338–348.

[1.47] S. Saito *et al.* (1985) "Frequency modulation noise and linewidth reduction in a semiconductor laser by means of negative frequency feedback technique" *Appl. Phys. Lett.* **46**(1), 3–5.

[1.48] Y. Yamamoto *et al.* (1985) "Theory of a negative frequency feedback semiconductor laser" *IEEE J. Quantum Electron.* **QE-21**(12), 1919–1928.

[1.49] M. Ohtsu and S. Kotajima (1985) "Linewidth reduction of a semiconductor laser by electrical feedback" *IEEE J. Quantum Electron.* **QE-21**(12), 1905–1912.

[1.50] S. R. Forrest (1985) "Monolithic optoelectronic integration: A new component technology for lightwave communications" *J. Lightwave Technol.* **LT-3**(6), 1248–1263.

[1.51] Y. Hirayama *et al.* (1989) "High-speed 1.5 µm-aligned constricted mesa DFB lasers grown entirely by MOCVD" *IEEE J. Quantum Electron.* **QE-25**(6), 1320–1324.

[1.52] J. M. Wiesenfeld *et al.* (1987) "Electro-optic sampling using injection lasers" *SPIE Proc.* **795**, 339–344.

[1.53] S. W. Corzine *et al.* (1988) "Actively mode-locked GaInAsP laser with subpicosecond output" *Appl. Phys. Lett.* **52**, 348–350.

[1.54] J. P. van der Ziel *et al.* (1981) "Subpicosecond pulses from passively mode-locked GaAS buried optical guide semiconductor lasers" *Appl. Phys. Lett.* **39**, 525–527.

[1.55] H. Yokoyama *et al.* (1982) "Generation of subpicosecond coherent optical pulses by passive mode locking of an AlGaAs diode laser" *Appl. Phys. Lett.* **40**, 105–107.

[1.56] P. P. Vasil'ev *et al.* (1986) "Subpicosecond pulse generation by a tandem-type AlGaAs DH laser with colliding pulse mode locking" *IEEE J. Quantum Electron.* **QE-22**, 149–151.

[1.57] H. F. Liu *et al.* (1989) "Gain-switched picosecond pulse (< 10 ps) generation from 1.3 μm InGaAsP laser diodes" *IEEE J. Quantum Electron.* **25**(6), 1417–1425.

[1.58] K. Sato *et al.* (1989) "1.55 μm narrow-linewidth and high-power distributed feedback lasers for coherent transmission systems" *J. Lightwave Technol.* **7**(10), 1515–1519.

[1.59] S. Oshiba *et al.* (1987) "High-power output over 200 mW of 1.3 μm GaInAsP VIPS Lasers" *IEEE J. Quantum Electron.* **QE-23**(6), 738–742.

[1.60] T. Sueta and M. Izutsu (1982) "High speed guided-wave optical modulators" *J. Opt. Commun.* **3**, 52–58.

[1.61] R. A. Becker "Traveling-wave electrooptic modulator with maximum bandwidth-length product" *Appl. Phys. Lett.* **45**, 1168–1170.

[1.62] R. C. Alferness *et al.* (1986) "Low-loss broad band Ti:LiNbO$_3$ waveguide phase modulators for coherent systems" *Electron. Lett.* **22**, 309–310.

[1.63] P. Granestrand *et al.* (1986) "Strictly nonblocking 8×8 integrated optical switch matrix" *Electron. Lett.* **22**, 816–817.

[1.64] R. C. Alferness and R. V. Schmidt "Tunable optical waveguide directional coupler filter" *Appl. Phys. Lett.* **33**, 161–163.

[1.65] K. Komatsu *et al.* (1986) "Titanium/magnesium double diffusion method for efficient fiber-LiNbO$_3$ waveguide coupling" *Topical Meeting on Integrated and Guided Wave Optics, Atlanta, GA,* paper PDP 2-1.

[1.66] C. M. Gee *et al.* (1983) "17 GHz bandwidth electrooptic modulator" *Appl. Phys. Lett.* **43**, 998–1000.

[1.67] F. Auracher *et al.* (1984) "High-speed $\Delta\beta$-reversal directional coupler modulator with low insertion loss for 1.3 μm in LiNbO$_3$" *J. Opt. Commun.* **5**, 7–9.

[1.68] C. S. Tsai *et al.* (1984) "A 8.5 GHz bandwidth single mode crossed channel TIR modulator and switch in LiNbO$_3$" *Topical Meeting on Integrated and Guided-Wave Optics, Kissimee, FL* paper PD5.

[1.69] S. K. Korotky *et al.* (1985) "4 Gbit/s transmission experiment over 117 km of optical fiber using a Ti:LiNbO$_3$ external modulator" *J. Lightwave Technol.* **LT-3**, 1027–1030.

[1.70] H. Kogelnik and R. V. Schmidt (1976) "Electrooptically switched couplers with alternating $\Delta\beta$" *IEEE J. Quantum Electron.* **QE-12**, 396–401.

[1.71] E. Voges and A. Neyer (1987) "Integrated-optic device on LiNbO$_3$ for optical communication" *J. Lightwave Technol.* **5**(9), 1229–1245.

[1.72] J. E. Watson *et al.* (1986) "A polarization-independent 1×16 guide-wave optical switch integrated on lithium niobate" *J. Lightwave Technol.* **LT-4**, 1717–1721.

[1.73] H. Haga *et al.* (1985) "An integrated 1×4 high-speed optical switch and its applications to a time demultiplexer" *J. Lightwave Technol.* **LT-3**, 116–120.

[1.74] R. A. Spanke (1986) "Architectures for large nonblocking optical space switches" *IEEE J. Quantum Electron.* **QE-22**, 964–967.

[1.75] H. F. Tayor (1974) "Optical-waveguide connection networks" *Electron. Lett.* **10**, 41–43.

[1.76] G. A. Bogert *et al.* (1986) "Low crosstalk 4×4 Ti:LiNbO$_3$ optical switch with permanently attached polarization maintaining fiber array" *J. Lightwave Technol.* **LT-4**, 1542–1545.

[1.77] A. Neyer (1983) "Electrooptic X-switch using single-mode Ti:LiNbO$_3$ channel waveguides" *Electron. Lett.* **19**, 553–554.

[1.78] A. Neyer *et al.* (1985) "A beam propagation method analysis of active and passive waveguide crossing" *J. Lightwave Technol.* **LT-3**, 635–642.

[1.79] K. Nosu and K. Iwashita (1988) "A consideration of factors affecting future coherent lightwave communication systems" *J. Lightwave Technol.* **6**(5), 686–694.

[1.80] S. Yamazaki *et al.* "A 1.2 Gbit/s optical DPSK heterodyne detection transmission system using monolithic external cavity DFB LDs" *OFC'87, Reno, NV, Proc.* 48–51.

[1.81]   S. Yamazaki et al. (1985) "Realization of flat FM response by directly modulating a phase tunable laser diode" Electron. Lett. **21**, 283–285.

[1.82]   Y. Yoshikumi and G. Motosugi (1987) "Multielectrode feedback laser for pure frequency modulation and chirping suppressed amplitude modulation" J. Lightwave Technol. **LT-5**, 516–522.

[1.83]   K. Iwashita et al. (1987) "Optical CPFSK 2 Gbit/s 202 km transmission experiment using a narrow-linewidth multielectrode DFB LD" Electron. Lett. **23**, 1022–1023.

[1.84]   T. Matsumoto and T. Imai (1987) "Polarization fluctuation in installed optical fiber cables and its compensation" CLEO'87, Baltimore, MD paper WL2.

[1.85]   T. Imai et al. (1985) "Optical polarization control utilizing an optical heterodyne detection scheme" Electron. Lett. **21**(2), 52–53.

[1.86]   T. Imai et al. (1986) "Polarization diversity technique for optical coherent detectors" OFC'86, Tokyo, Japan, Proc. 283–286.

[1.87]   A. F. Elrefaie et al. (1987) "Chromatic dispersion limitations in coherent optical fiber transmission systems" Electron. Lett. **23**(14), 756–758.

[1.88]   N. Takachio and K. Iwashita (1988) "Compensation of fiber chromatic dispersion in optical heterodyne detection" Electron. Lett. **24**(2), 108–109.

[1.89]   Y. Ohishi et al. (1986) "Transmission loss characteristics of fluoride glass single-mode fiber" Electron. Lett. **22**(20), 1034–1035.

[1.90]   S. Kobayashi et al. (1981) "Direct frequency modulation in AlGaAs semiconductor lasers" IEEE J. Quantum Electron. **QE-17**(6), 946–959.

[1.91]   T. Kimura (1987) "Coherent optical fiber communication" J. Lightwave Technol. **LT-5**(4), 414–428.

[1.92]   S. Saito et al. (1987) "S/N and error rate evaluation for an optical FSK heterodyne detection system using semiconductor lasers" IEEE J. Quantum Electron. **QE-19**(2), 180–193.

[1.93]   S. Kobayashi and T. Kimura (1982) "Optical phase modulation in an injection locked AlGaAs semiconductor laser" IEEE J. Quantum Electron. **QE-18**(10), 1662–1669.

[1.94]   S. Kobayshi et al. (1981) "Optical FM signal amplification and FM noise reduction in an injection locked AlGaAs semiconductor laser" Electron. Lett. **17**(22), 849–851.

[1.95]   M. Okai et al. (1990) "Corrugation-pitch-modulated MQW-DFB laser with narrow spectral linewidth (170 kHz)" IEEE Photon. Technol. Lett. **2**, 529–530.

[1.96]   G. Wenke and S. Saito (1985) "Stabilized PSK transmitter with negative electrical feedback to a semiconductor laser" Electron. Lett. **21**(15), 653–655.

[1.97]   O. E. Delange (1970) "Wide-band optical communication systems: Part 2, frequency-division-multiplexing" Proc. IEEE **58**(10), 1683.

[1.98]   K. Nosu et al. (1987) "Optical FDM transmission technique" J. Lightwave Technol. **LT-5**(9), 1301–1308.

[1.99]   H. Kumazawa and I. Ohtomo (1977) "30 GHz band periodic branching filter using traveling wave resonator for satellite applications" IEEE Trans. Microwave Theory Tech. **MTT-25**(8), 683.

[1.100] I. Ohtomo et al. (1971) "Two cavity type channel dropping filters for millimeter wave guided wave communication system" IEEE Trans. Microwave Theory Tech. **MTT-19**(5), 481.

[1.101] Y. K. Park et al. (1987) "Crosstalk and prefiltering in a two channel ASK heterodyne detection system without the effect of laser phase noise" OFC/IOOC'87, Reno, NV paper PDP13.

[1.102] H. Toba et al. (1985) "450 Mbit/s optical frequency-division multiplexing transmission with an 11 GHz channel spacing" Electron. Lett. **21**, 656.

[1.103] K. Oda et al. (1989) "Channel selection and stabilization technique for a waveguide-type 16-channel frequency selection switch for optical FDM distribution systems" Photon. Technol. Lett. **1**(6), 137–139.

[1.104] B. Glance *et al.* (1989) "Densely spaced FDM optical coherent system with near quantum-limited sensitivity and computer-controlled random access channel selection" *Electron. Lett.* **25**(14), 883–885.

[1.105] R. Welter *et al.* (1989) "Sixteen-channel coherent broadcast network at 155 Mbit/s" *J. Lightwave Technol.* **7**(10), 1438–1444.

[1.106] B. J. Ainslie and C. R. Day (1986) "A review of single-mode fibers with modified dispersion characteristics" *J. Lightwave Technol.* **LT-4**(8), 967–979.

[1.107] A. Hasegawa and F. D. Tappert (1973) "Transmission of stationary nonlinear optical pulses in dispersive dielectric fibers: 1, Anomalous dispersion" *Appl. Phys. Lett.* **23**(3), 142–144.

[1.108] Y. Yamauchi and T. Ishibashi (1987) "Application of AlGaAs/GaAs HBT's for wide-band direct-coupled amplifers" *Electron. Lett.* **23**(4), 156–157.

[1.109] K. Kamite *et al.* (1986) "DFB laser with bandwidth larger than 9 GHz" *IEEE Int. Semicond. Laser Conf.*, *Kanazawa, Japan*, *Oct. 14–17* paper M-4.

[1.110] I. Mito *et al.* (1986) "High frequency response characteristics of $\lambda/4$ shifted DFB-DC-PBH-LD" *Conf. Optical and Radio Wave Electron. IECE Japan* paper 231.

[1.111] R. Olshansky *et al.* (1987) "InGaAsP buried heterostructure laser with 22GHz bandwidth and high modulation efficiency" *Electron. Lett.* **23**(16), 839–841.

[1.112] K. Uomi *et al.* (1987) "High relaxation oscillation frequency of GaAlAs multiquantum well lasers" *Japan J. Appl. Phys.* **23**(16), 839–841.

[1.113] T. Kimura (1988) "Factors affecting fiber-optic transmission quality" *J. Lightwave Technol.* **6**(5), 611–619.

[1.114] T. S. Kinsel and F. S. Chen (1972) "Experimental evaluation of an optical time division demultiplexer for twenty-four channels" *Appl. Opt.* **11**, 1411–1418.

[1.115] M. Thewalt (1981) "Time domain multiplexing of signals on an optical fiber using mode-locked laser pulses" *IBM Tech. Disclosure Bull.* **24**, 2473–2475.

[1.116] P. R. Prucnal *et al.* (1986) "TDMA fiber-optic network with optical processing" *Electron Lett.* **22**, 1218–1219.

[1.117] P. R. Prucnal *et al.* (1987) "12.5 Gbit/s fiber-optic network using all-optical processing" *Electron. Lett.* **23**, 629–630.

[1.118] R. S. Tucker *et al.* "16 Gbit/s fiber transmission experiment using optical time-division multiplexing" *Electron. Lett.* **23**, 1270–1271.

[1.119] H. Haga *et al.* (1985) "An Integrated 1×4 high-speed optical switch and its applications to a time demultiplexer" *J. Lightwave Technol.* **LT-3**, 116–120.

[1.120] R. S. Tucker *et al.* (1988) "Optical time-division multiplexing for very high bit-rate transmission" *J. Lightwave Technol.* **LT-6**(11), 1744.

[1.121] M. Sakaguchi and K. Kaede (1987) "Optical switching device technologies" *IEEE Commun. Mag.* **25**(5), 27–32.

[1.122] T. Yasui and H. Goto (1987) "Overview of optical switching technologies in Japan" *IEEE Commun. Mag.* **25**(5), 10–15.

[1.123] S. Suzuki *et al.* (1987) "A 32-line optical space-division switching system using 8×8 optical matrix switches" *NEC Research and Development, no. 87*, 44–50.

[1.124] T. Matsunaga and M. Ikeda (1985) "Experimental application of LD switch modules to 256 Mbit/s optical time switching" *Electron. Lett.* **21**(20), 945–946.

[1.125] S. Suzuki *et al.* (1986) "An experiment of high-speed optical time-division switching" *IEEE J Lightwave Technol.* **LT-4**(7).

[1.126] S. Suzuki *et al.* (1986) "An experiment on high-speed optical time-division switching" *J. Lightwave Technol.* **LT-4**(7), 894–899.

[1.127] T. Shimoe *et al.* (1988) "Experimental optical switching system based on 512 Mbit/s time-division technique" *OEC'88, Tokyo, Japan* paper PDP 3A2-3.

[1.128] J. E. Midwinter (1988) "Photonic switching technology: component characteristics versus network requirements" *J. Lightwave Technol.* **6**(10), 1512–1519.

2

COHERENT LIGHTWAVE
TRANSMISSION TECHNOLOGY

2.1 PRINCIPLES OF COHERENT LIGHTWAVE TRANSMISSION

Despite the rapid advance of lightwave technology over the past two decades, the currently used optical fiber communications are in a sense as primitive as electrical communications in their early stages, because neither of these communications makes use of the phase information of the carrier. Direct modulation of the source (on–off keying) and direct detection at the receiver using a pin diode or Avalanche Photodiode (APD) have been the mainstays of lightwave systems since their infancy; in other words, both of these are non-coherent communications.

The modulation/demodulation scheme employed in the currently used fiber communications is called the intensity modulation/direct detection (IM/DD) scheme. In the IM, the light intensity is modulated linearly with respect to the input signal. The term DD stems from the fact that the signal is detected directly at the optical stage of the receiver. On the other hand, in the history of electrical communications, the heterodyne scheme has become common since 1930, and is now widely used even in pocket radio sets. Sophisticated coherent modulations such as FM, PM, frequency-shift keying (FSK), and phase-shifting keying (PSK) are also widely used in radio broadcasting and electrical communications.

It became clear that coherent detection offers many important advantages with respect to conventional combination intensity modulation/direct detection (IM/DD): greatly enhanced frequency selectivity, conveniently tunable optical receivers, the possibility of using alternative modulation formats (such as FSK and PSK), and improved receiver sensitivity.

A similar situation concerns IM/DD fiber communication and when it will be replaced by more sophisticated optical coherent systems according to the current status.

The term *optical coherent* is used here to refer to any technique employing non-linear mixing between two optical waves. (In the radio communications literature, the term 'coherent' is used to refer to detection techniques in which the absolute phase of the incoming signal is tracked by the receiver. Many of the techniques which have come to be labelled coherent in optical communications, such as envelope detection of ASK, are explicitly incoherent in the radio literature) [2.1].

Typically, one of these is an information-bearing signal, the other is a locally generated wave (the local oscillator or LO), and the mixing is done using a photodetector. The result of this *heterodyne* process is a modulation of the detector photocurrent at a frequency equal to the difference between the frequencies of the signal $A_s\cos\omega_1 t$ and the LO $A_{LO}\cos\omega_2 t$:

$$I_d = A_s^2 + A_{LO}^2 + 2A_s A_{LO}\cos(\omega_1 - \omega_2)t. \tag{2.1}$$

The new signal is called the intermediate frequency or IF. A quantum mechanical treatment of optical heterodyne detection for $h\nu \gg kT$ has shown that, unlike the case of heterodyne detection at radio frequencies, the detector current, in this case the photon count rate, does not contain the double frequency and sum frequency components [2.2]. This IF signal contains any information, in the form of amplitude, frequency, or phase modulation, which had been present on the original optical signal. Detection of this information is accomplished using standard radio techniques. When the signal and LO frequencies are identical, the process is called *homodyne* detection and the information appears directly at baseband frequencies (i.e. near zero frequency). In this regard, homodyne detection is similar to conventional direct detection which uses no LO and in which optical signals are converted directly to electrical currents by the photodetector. Since the amplitude of the IF signal is proportional to the product of the signal and LO amplitudes, the IF power can be made arbitrarily large compared with noise from subsequent electronic amplifiers. Therefore coherent detection provides a mechanism for overcoming thermal noise, as well as detector noise, which limits direct detection sensitivities. The limiting noise, for sufficiently high LO levels, is the quantum noise present in the detected photocurrent. This noise also increases linearly with LO power so that the electronic signal-to-noise ratio approaches a constant shot-noise limited value for large LO [2.3]. For binary digital transmission, the exact value of this limiting sensitivity varies from 9 to greater than 80 photons/bit of signal required for a 10^{-9} bit error rate (BER), depending on the details of the modulation and detection techniques used. These sensitivity values can be compared with the quantum limited direct detection sensitivity of 10 photons/bit [2.4]. While this theoretical sensitivity is as good as, or better than, those of the coherent schemes, actual sensitivities obtained in the long-wave length range of interest (1.3–1.5 μm) are limited by detector and electronic amplifier noise to values at least 40 times larger than theory. On the other hand, coherent detection has been demonstrated with sensitivities within a factor of two of the theoretical value.

From the time of the invention of lasers in 1960, efforts to utilize coherent properties of laser light for optical communications existed. Signal transmission using the optical frequency or phase modulation scheme, instead of intensity modulation, was one of the initial focuses of research interest. Optical heterodyne or homodyne detection was shown in 1962 to improve the signal-to-noise ratio (S/N) above that of direct detection. The information capacity of various communications systems including heterodyne and homodyne receiver systems was discussed [2.5]. A heterodyne-detection transmission experiment using 3.39 μm He–Ne lasers demonstrated a significant S/N improvement in 1967 [2.6]. The concept of frequency division multiplexing using coherent detection schemes, that really is coherent optical communication, was first proposed in 1970 [2.7] just two decades ago.

Improvement in the 1970s in the lifetime of semiconductor lasers as well as in the loss characteristics of optical fibers led to a large optical communications success. Simple and reliable modulation–demodulation technologies, such as direct intensity modulation of semiconductor lasers and direct power detection using avalanche photodiodes (APDs) or p–i–n photodiodes, are used in the present fiber systems. These systems already surpass conventional coaxial cable and radio relay systems in terms of transmission performance. In these systems, however, the coherent laser light properties are not fully utilized, but rather, the optical energy of the noisy carrier wave is used instead to convey information. The transmitted signal spectrum is not transform-limited, but typically spreads to 1 THz [2.8]. These features signify that the high level of success gained by present fiber-optic systems results mainly from the low-loss and broad-band characteristics of optical fibers. Furthermore, the features indicate that present optical fiber transmission systems still have room for further development from the viewpoint of modulation–demodulation technology. Of course, in this decade, there have not been so many advances in this field.

In the 1980s, the research and development of coherent optical fiber communications techniques have expanded rapidly. There is currently a resurgence in coherent optical transmission research in communications laboratories throughout the world. This recent activity is a result of the astonishing progress of optoelectronic components and techniques, such as the single mode fibers with very low loss and dispersion, semiconductor lasers with narrow spectral linewidth of several MHz or even kHz and various kinds of coherent optical modulating and demodulating techniques for fiber transmission, as well as the urgent need of very long repeaterless distance and very large data volume of transmission. Coherent communication seems really nearer the applied stage than ever before.

In the last two years (1989–1991), because of the doped fiber amplifier coming into application (see Chapter 5), it is not so imperative for improving sensitivity by coherent transmission. At the same time as the development of DWDM/FDM transmission and direct detection (see Chapter 3), it is also not so urgent for dense channels of transmission by coherent systems. So, in the next decade, there is a new tendency to combine DWDM/FDM with doped fiber amplifiers. But, in any case, it is intrinsic to use the phase and polarization characteristics of coherent transmission to enhance the capability of information carrying. From the viewpoint of long term development, we should not relax in improving the technology of coherent transmission.

2.1.1   Operation of coherent transmission

PRINCIPLES OF OPERATION [2.4]

Unlike the conventional intensity modulated/direct detection (IM/DD) transmission, where the amplitude-modulated optical signal is converted directly into an amplitude-demodulated electrical output, the coherent transmission, in which amplitude, frequency, and phase of a coherent optical carrier wave are utilized to modulate and demodulate high capacity information efficiently, are expected to improve system performance toward long repeater spacing and large information capacity. It first adds to the signal a locally generated optical wave and then detects the combination [2.9]. The resulting photocurrent carries all the information of the original signal, but is at a frequency low enough (GHz) so that further signal processing can be performed using conventional electronic circuitry as shown in Fig. 2.1. This method offers significant improvements in receiver sensitivity. In the 1.3–1.6 μm lightwave band, for example, an ideal coherent receiver requires a signal energy of only 10–20 photons per bit to achieve a BER of 10^{-9}, far fewer than the roughly 1000 photons required by today's APDs. Even though the doped-fiber optical amplifiers could offer compensation over the line losses, the intrinsic high sensitivity still has much benefit for long distance communication. A further advantage of coherent reception, because of its improved wavelength selectivity compared to direct detection, a coherent receiver might permit FDM or DFDM systems with channel spacings of only, say, 100 MHz, instead of the 100 GHz required with conventional optical multiplexing technology.

Conceptually, the simplest type of coherent reception, although not necessarily the most practical, is achieved with a homodyne receiver. For homodyne detection to work, the local semiconductor laser in a receiver must lock on to the phase of the incoming carrier. Homodyne systems match the frequency of the local oscillator to that of the incoming beam, extracting the modulating signal directly. Such systems need only moderate bandwidths in the electronic receiving circuitry but make high demands of the local oscillator. But while homodyne is likely to be the more sensitive form of coherent detection, heterodyne detection, which deliberately introduces a difference between the incoming and local frequencies, is easier to implement at lower bit rates and is more likely for the first commercial systems.

Fig. 2.2 [2.4] shows the principle of a homodyne detection, in which the local oscillator is phase locked to the incoming optical carrier. The optical signal is first combined with the much stronger local oscillator wave in the same polarization using a partially reflecting plate called a beam splitter of a fiber directional coupler. Usually, signal power is more precious than local oscillator power, so the beam splitter is made almost completely transparent, and consequently it reflects only weakly. The

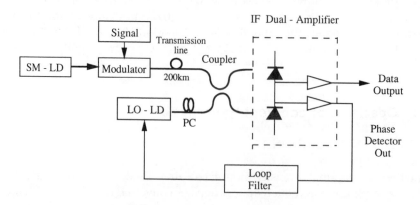

Fig. 2.1 Schematic of coherent transmission.

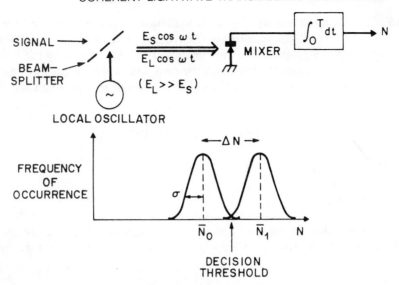

Fig. 2.2 Homodyne receiver. (From [2.4] Henry (1985) *IEEE J. Quantum Electron.* **QE-21**(12), 1862–1879. Reproduced by permission. ©1985 IEEE.)

combined signal and local oscillator waves illuminate a photodetector, called the mixer, whose average output current is proportional to the total optical power averaged over many optical cycles (P_{opt}). For a p–i–n mixer,

$$R = \frac{\eta P_{\mathrm{opt}}}{h\nu} \overset{\Delta}{=} (E_L + E_S)^2 \tag{2.2}$$

where R is the average generation rate of photoelectrons, η is the quantum efficiency of the detector, and $h\nu$ is the photon energy. E_L and E_S, which are proportional to the envelopes of the local oscillator and signal fields incident on the photodiode, are defined in normalized units such that the simple square-law relation of (2.2) holds. The output of the photodetector is integrated for the duration of a data bit T, and the result is compared to a threshold to determine if a binary '0' or '1' was transmitted. Firstly, the sensitivity of the homodyne receiver for OOK, in which a binary '1' is transmitted as an optical pulse ($E_S > 0$), and a binary '0' is represented by the absence of optical energy ($E_S = 0$). The expected number of counts at the integrator output for the '0' and '1' signal state is

$$\bar{N}_0 = E_L^2 T$$

$$\bar{N}_1 = (E_L + E_S)^2 T \sim (E_L^2 + 2E_L E_S)T \tag{2.3}$$

where the approximate equality is based on the assumption $E_L \gg E_S$.

When an individual bit is detected, the integrator output, of course, will not be precisely $\bar{N}_0$ or $\bar{N}_1$ because the photoelectrons are generated at random intervals. This randomness is the basis of shot noise, which is seen whenever current flows through a diode. The integrator outputs will therefore be distributed around

$\bar{N}_0$ and $\bar{N}_1$, in accordance with Poisson statistics, as shown by the histogram in Fig. 2.2. The error probability (BER) of the receiver is related to the fraction of each distribution on the 'wrong side' of the threshold level. The narrower the distributions are relative to their separation, the lower the error rate will be. Under our assumption of a strong local oscillator ($E_L^2 T > 100$), the distributions are approximately Gaussian with peak-to-peak separation

$$\Delta N = \bar{N}_1 - \bar{N}_0 = 2E_L E_S T \qquad (2.4a)$$

and width

$$\sigma \sim \sqrt{\bar{N}_1} \sim \sqrt{\bar{N}_0}. \qquad (2.4b)$$

Thus, tabulated values of the error function can be used to calculate error rates [2.10]. To achieve 10^{-9} BER requires $\Delta N/\sigma \approx 12$, so from (2.4a) and (2.4b) we find

$$E_S^2 T = 36. \qquad (2.5)$$

But from (2.2) $E_S^2 T$ is simply the expected number of photoelectrons per bit when the mixer is used as a simple direct detector (i.e. $E_L = 0$). That is, to achieve 10^{-9} BER with OOK homodyne, the average energy of each optical pulse must be sufficient to produce 36 direct-detection photoelectrons. We can take this result a step further by noting that, for anti-reflection-coated InGaAsP p–i–n diodes, the quantum efficiency approaches unit, so 10^{-9} BER corresponds to an average received optical energy of 36 photons per pulse. For the usual case where '1's and '0's are equally probable, an OOK data stream is on only half the time. Thus, the required number of photons per bit of information is half the required number per pulse, or 18.

If phase shift keying (PSK) is used instead of OOK, even greater receiver sensitivity can be achieved. In this case information is impressed on the phase of the transmitted wave and the two states of the received wave are denoted as $E_S \cos \omega t$ and $E_S \cos(\omega t + \pi)$ respectively. Using arguments similar to those in the OOK case

$$\Delta N = (E_L + E_S)^2 T - (E_L - E_S)^2 T = 4E_L E_S T \qquad (2.6a)$$

and

$$\sigma = \sqrt{(E_L^2 T)} \qquad (2.6b)$$

the condition $\Delta N/\sigma = 12$ then implies

$$E_S^2 T = 9. \qquad (2.7)$$

Thus, for PSK homodyne detection with an ideal photodiode ($\eta = 1$) an average signal energy of 9 photons per bit is required to achieve 10^{-9} BER. (For PSK the optical signal is on all the time so we need not distinguish between photons per pulse and photons per bit, as we did in the OOK case.) The improved sensitivity of PSK compared to OOK stems from the fact that for a given signal field E_S the value of ΔN using PSK is twice as large as with OOK. PSK with homodyne detection provides the best sensitivity that can be achieved with the simple coherent receiver structure of Fig. 2.2. The transmission distance at 1 Gbit/s is about 50% greater than with conventional intensity modulation and APD detection.

Homodyne receivers, although they are the most sensitive, are also the most difficult to build because the local oscillator must be controlled by an optical phase-locked loop. The heterodyne receiver, in which the local oscillator frequency is deliberately offset from the signal frequency, is considerably easier to implement. In this case the useful portion of the mixer output appears at a frequency called the intermediate frequency (IF), which is the difference between the optical signal and local oscillator frequencies. The desired data stream is extracted from the IF signal with standard radio-frequency demodulation techniques. The price paid for elmination of the optical phase-locked loop in the heterodyne receiver is a 3 dB degradation in sensitivity compared to homodyne. A semiclassical way to view this impairment is to recognize that in a heterodyne receiver the signal and local oscillator are constantly slipping in phase. The mixer output (2.2) is most sensitive to the incoming signal when the signal and local oscillator are aligned in phase, either parallel or antiparallel. When they are in quadrature, mixer sensitivity is negligible. The IF signal, which carries the desired data, is an average over these 'good' and 'bad' conditions. With homodyne, on the other hand, the signal is always aligned with the local oscillator, so mixer response to the signal is maximized. Since the homodyne receiver takes full advantage of the incoming signal, it is not surprising that its sensitivity is greater. More precisely, for given signal and local oscillator powers, the power available from the mixer output in a heterodyne receiver is just half that available in the homodyne case. Since the shot-noise power for both is the same (because the local oscillator power are equal), the sensitivity of the heterodyne receiver is 3 dB poorer.

For coherent homodyne on–off keying, a bit-error rate of 1 in 10^9 requires 18 photons per bit of information. Phase-shift keying, rather than on–off keying, produces even greater sensitivity in a homodyne receiver. Because the information is impressed on the phase of the transmitted wave rather than on its amplitude, the transmitter is never turned off and the effective sensitivity of the receiver doubles. Phase-shift keying thus needs only 9 photons per bit for a maximum error rate of 1 bit in 10^9. The trouble with homodyne receivers is that the local oscillator must stay in phase with the incoming carrier.

Coherent heterodyne receivers, with local oscillator frequency offset from the carrier frequency, are much easier to build. The useful portion of the mixed waves is at an intermediate frequency (IF), the difference between the carrier and local oscillator frequencies. The data steam is extracted from the IF signal by standard radio demodulation techniques.

Heterodyne reception entails a penalty, however, because the carrier and local oscillator constantly slip out of phase with each other. The receiver is most sensitive at the instant when the signal and local oscillator are in phase. When they are out of phase by 90°, sensitivity approaches zero. The IF signal averages those good and bad conditions. Indeed, with on–off keying, heterodyne detection is down by 6 dB, and an error rate of 1 bit in 10^9 demands 36 photons per bit. Because it is the easiest to implement, however, this form of coherent communications may be the first to go into service, although heterodyne frequency-shift keying is a close competitor and will likely be not far behind. Oddly enough, simple direct detection of on–off keyed signals in theory needs only 10 photons per bit and so should be competitive with the best coherent schemes. But the best direct detection receivers run 17 to 20 dB below that ideal. Noise levels limiting sensivity, both from the photodetector and from the electronic amplifier, may soon be cut by a few decibels, but direct-detection systems will still be 10 dB or more below the ideal or 10 photons per bit. On the other hand, the best coherent receivers, while clearly impractical right now, are already only about 3 dB away from the ideal.

ADVANTAGES OF COHERENT TRANSMISSION

The advantages of coherent optical transmission have been much more thoroughly discussed in many papers [2.3, 2.8, 2.11, 2.17]. Coherent optical fiber transmission systems have several preferred features compared to the conventional IM/DD system. They will have a signal receiving level limited only by the signal shot noise provided that the local oscillator had sufficient optical power at the receiver. Receiving signal levels to achieve a prescribed error rate is markedly improved by the coherent optical fiber transmission systems in the place of the IM/DD system. Optical heterodyne detection converts the carrier frequency from about 200 THz in the optical region to around several GHz in the RF region. Advanced signal processing techniques such as FSK or PSK signal modulation and demodulation, frequency multiplexing and demultiplexing, and local oscillator frequency stabilization have been established and utilized in the microwave and millimeter wave circuit. Optical heterodyne or homodyne detection is sensitive to the mismatch in wavefront, frequency, phase, and polarization states between signal and local oscillator waves. The mode selective character is advantageous to suppress background light noise and also opens the possibility of a spatial mode multiplexing system, while stringent requirements are imposed on spatial alignment and frequency stability.

THE RECEIVING SENSITIVITY IN COHERENT TRANSMISSION

A comment is added here on the relation of heterodyne and coherent schemes. Heterodyne detection is practically a premise for coherent communication. However, coherent modulation is never a premise for the heterodyne system; an IM signal could also be detected by a heterodyne scheme.

The greatest advantage of a heterodyne/coherent system is the improvement of the equivalent receiver sensitivity, more exactly, the reduction of the minimum receiving signal level for achieving a prescribed bit-error rate, for example 10^{-9}. This improvement is attributed to two effects [2.12]. One is the improvement of the S/N (signal-to-noise ratio) at the output end of the receiver preamplifier by the use of the heterodyne scheme. The other is the improvement brought about by the use of a coherent modulation/demodulation scheme [2.13].

Various detection schemes for achieving high S/N are first compared [2.14]. In optical detection, i.e., in the optoelectronic (OE) signal conversion, an absolute

S/N limitation exists due to the fact that light is not continuous but consists of photons, which produce a flow of discrete electrons at the electrical terminals of the photodetector. Therefore, an ideal photodetection is not a noise-free detection but a quantum-noise-limited detection, giving, as shown in

$$(S/N)_{ideal} = P_S/2hf\Delta f \qquad (2.8)$$

where P_S denotes the received signal power, h is Planck's constant ($=1.38 \times 10^{-34}$ J s), f the optical signal frequency, and Δf the bandwidth of the receiver.

The main noise source of coherent detection is thus the quantum noise of the transmitted signal light. This is significantly different from the fact that APD multiplication noise and load-resistance thermal noise dominate the receiver sensitivity in conventional direct detection. The APD receiver in the long wavelength region requires about 700 photon/bit for binary on–off signals to achieve an error rate of 10^{-9} [2.15]. Coherent detection can be accomplished with a smaller photon number than direct detection. The required receiving signal level, for example, is 18 photon/bit for binary FSK homodyne detection and about 1.4 photon/bit for coded 32-level FSK coherent detection [2.16]. The theoretical channel capacity of optical communications systems derived by quantum mechanical treatment corresponds to a receiving level of 0.02 photon/bit [2.5].

FDM PLAY BEST BENEFIT IN COHERENT TRANSMISSION

The second advantage of a heterodyne (or homodyne) system, in addition to the sensitivity improvement, is the possibility of sharp receiver frequency selectivity which is equal to the frequency selectivity in the IF amplifier. So, a system's channel density (channels per unit bandwidth) could go to 1000 times that for direct detection. Thus, frequency division multiplexing (FDM) with fine carrier separation becomes possible [2.17, 2.18].

Just as direct optical detection is analogous to the way a primitive crystal radio detects broadcast signals, so coherent optical detection is much like the modern radio's superheterodyne. And just as such radios vastly improved broadcasting, so coherent optical techniques open up new possibilities for telecommunications: the techniques combine the powerful signal processing of radio with the virtually limitless bandwidth of optical wavelengths. Each optical fiber can carry at least as much information from point to point as the entire range of the electromagnetic broadcast spectrum in use today.

MODULATION BANDWIDTH IMPROVED IN COHERENT TRANSMISSION

Modulation speed in the direct intensity modulation of semiconductor lasers is limited to the resonance frequency, above which modulation efficiency is inversely proportional to the square of the modulation frequency. The direct frequency modulation of semiconductor lasers [2.19] relaxes the limitation caused by the resonance property. This is because the frequency shift normalized by the unit modulation current, namely the FM efficiency, is inversely proportional to the modulation frequency which is higher than the resonance. Electro-optic waveguide modulators [2.20] also achieve a phase modulation whose bandwidth is broader than the direct intensity modulation limit.

DISSOLUTION OF BANDWIDTH LIMITATION IN OPTICAL FIBERS

Frequency dispersion of single mode fibers imposes one of the important bandwidth limitation factors on transmission systems, since an optical signal generated by the direct intensity modulation of semiconductor lasers is accompanied by an unintended spectral spread due to the carrier modulation effect. In coherent systems, however, the limitation is alleviated because the transform-limited spectrum is obtained by modulation the single-frequency carrier wave [2.8].

Coherent modulation–demodulation schemes demonstrate the possibility of achieving an approximate 20 dB improvement in receiver sensitivity over conventional intensity modulation direct detection schemes. At a data rate of 1 GHz, coherent transmission theoretically admit a transmission loss totalling up to about 90 dB, which corresponds to a fiber length of more than 400 km. Long-haul terrestrial or undersea systems are promising applications of coherent optical fiber transmission. Coherent modulation–demodulation schemes can be applied to subscriber systems, especially distribution networks where many terminals simultaneously receive signals. This is because the large signal gain obtained by coherent techniques can compensate for the bridging loss which constitutes the dominant loss factor in passive distribution systems.

2.1.2 Modulation and demodulation techniques [2.21]

The scheme of modulation used to impose data on the carrier strongly affects sensitivity. Frequency-shift keying (FSK), on–off keying (OOK), phase-shift keying (PSK), and other coding methods all have their proponents. Just as with heterodyne and homodyne detection, the different schemes offer different tradeoffs between ultimate capability and ease of implementation. FSK, for example, is straightforward and tolerant of wide laser lines. But other considerations enter as well: PSK promises higher bit rates, and both PSK and ASK offer higher ultimate sensitivity.

Basic formats of different modulation–demodulation techniques, local oscillator in coherent systems as well as some practically experimental configurations are discussed in this section. Configurations of various coherent optical fiber transmission systems are shown in Fig. 2.3 [2.21]. Modulation schemes such as optical amplitude shift keying (ASK) heterodyne-envelop and coherent detection in Fig. 2.3(a), frequency shift keying (FSK) heterodyne-discrimination and dual filter detection in Fig. 2.3(b), phase shift keying (PSK) heterodyne-differential and coherent detection in Fig. 2.3(c), local oscillator configuration for ASK and PSK homodyne detection in Fig. 2.3(d) and coherent optical amplifier repeater systems in Fig. 2.3(e) are illustrated. Here, Pol.C. represents a polarization control device.

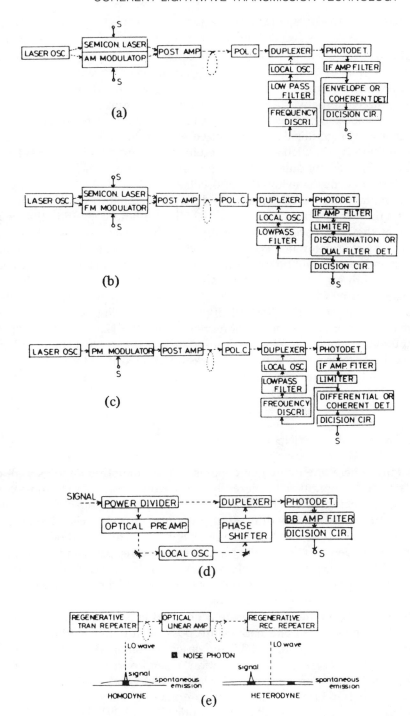

Fig. 2.3 Basic configurations of coherent fiber systems. (From [2.21] Yamamoto and Kimura (1981) *IEEE J. Quantum Electron.* **QE-17**(6), 919–935. Reproduced by permission. ©1981 IEEE.)

AMPLITUDE SHIFT KEYING (ASK) FORMAT

With binary ASK, '0's and '1's are transmitted by complete amplitude modulation. Modulation must be achieved without significant carrier frequency shift.

In the ASK heterodyne envelope detection and ASK heterodyne-coherent detection schemes, three alternatives exist for transmitter configuration: (1) combination of a frequency stabilized laser oscillator and an external amplitude modulator, (2) direct modulation of a longitudinal mode stabilized semiconductor laser under a large signal drive condition [2.22] such as distributed feedback and distributed Bragg reflector lasers, and (3) direct modulation of a semiconductor laser which is injection locked by another frequency stabilized laser oscillator [2.23].

The advantage of method (1) is to modulate the amplitude of the optical carrier wave at a modulation frequency up to 1–2 GHz without degrading the frequency stability of the laser oscillator. However, a high power laser oscillator and low insertion loss modulator are necessary, since a compact optical post amplifier with high gain and large output power for the ASK signal seems to be difficult to construct at present. Both travelling-wave type and Fabry–Perot type semiconductor laser amplifiers are used for the ASK signal. A Fabry–Perot type suffers from its small saturation output power of −10 to −5 dBm, while a travelling-wave type is expected to have a saturation output power of more than 10 dBm [2.24]. Method (2) suffers from spurious frequency modulation caused by injection current change [2.25] and limited coherence property. Method (3) uses injection locking by another frequency stabilized and coherent laser signal to suppress frequency noise of the directly modulated semiconductor laser.

FREQUENCY SHIFT KEYING (FSK) FORMAT

Binary FSK, where one frequency denotes a '1' and another denotes a '0' can be modulated and demodulated coherently or non-coherently. For the more common case of non-coherent FSK, the '1' and '0' frequencies are spaced by an integral multiple of B Hz. Since it is difficult at present to create precise frequency shifts in semiconductor lasers, an alternative is the FSK single filter method [2.27], where one optical frequency falls entirely outside the receiver passband, and the signal is received as ASK with a 3 dB system penalty. The phase coherent systems such as minimum shift keying (MSK) require less bandwidth but are more demanding and will not be considered in this book.

In the FSK heterodyne-dual filter detection and FSK heterodyne-discrimination detection schemes, three alternatives are possible for transmitter configuration: (1) combination of frequency stabilized laser oscillator and external frequency modulator, (2) direct frequency modulation of semiconductor laser, and (3) direct frequency modulation of semiconductor laser which is injection locked by another frequency stabilized laser oscillator. The external frequency modulation is realized by an electro-optic [2.28] and an acousto-optic effect [2.29]. Two sets of electrodes with a small separation in the orthogonal direction are necessary for a waveguide-type electro-optic frequency modulator. The conventional electro-optic phase modulator driven by a sawtooth input signal is also used to generate an optical FSK signal. Large loss and limited modulation frequency bandwidth are the main shortcomings in a waveguide-type acousto-optic frequency modulator.

Direct frequency modulation of a semiconductor laser is achieved by three methods: (1) modulation injection current in a semiconductor laser [2.25, 2.30, 2.31, 2.34]: (2) monolithic integration of electro-optic phase modulator and laser amplifier [2.32]; and (3) photoelastic modulation by an acoustic wave [2.33]. All these methods utilize a refractive index change in the laser waveguide by the electrical input signal.

The electro-optic modulation may need a complicated device structure and may suffer from a large residual AM by the Franz–Keldysh effect. The residual AM also occurs in the injection current modulation of a conventional laser. The frequency noise reduction by the external coherent signal injection is possible. In the FSK transmitter, however, the injection locking bandwidth must be smaller than the modulation frequency in order not to degrade the frequency modualtion efficiency. Frequency drift and FM quantum noise only at a lower frequency than the modulation signal can be suppressed by the injection. An injection locked semiconductor laser can be used as a post amplifier for FSK and PSK signals. The normalized frequency shift $(\Delta f_{out}/\Delta f_{in}) = (d/dt)(\alpha/\theta)$ versus the normalized modulation frequency of input signal $f_m/\Delta f_L$ is presented in [2.35]. The frequency shift reduction can be suppressed to less than 1 dB by keeping the locking bandwidth Δf_L larger than $1.3 f_m$. Available signal gain in an injection locked laser is related to a locking bandwidth, which is given by

$$\Delta f_L = \frac{f_0}{2Q} \sqrt{\left| \frac{P_{in}}{P_{osc}} \right|} \tag{2.9}$$

where Q, P_{in}, and P_{osc} are a cavity Q value, an injection signal power, and laser power without injection, respectively. The locking bandwidth Δf_L versus the signal gain P_{osc}/P_{in} of the injection locked semiconductor laser amplifier is shown in Fig. 2.4 [2.36]. Here, L, R, and α are cavity length, facet reflectivity and absorption coefficient, respectively. For FSK signal gains of more than 40 and 30 dB are available.

PHASE SHIFT KEYING (PSK) AND DIFFERENTIAL PHASE SHIFT KEYING (DPSK) FORMAT

In binary phase shift keying the phase of the transmitted carrier is shifted from ϕ_1 to ϕ_2 to signal a change from data '0' to '1'. Biphase modulation $(\phi_1 - \phi_2 = \pi)$ gives the highest receiver sensitivity, but a phase locked loop locked to the suppressed carrier is needed for demodulation. Reducing the modulation depth slightly provides a residual carrier, and this eases phase locking. External modulation of the laser diode (e.g., with an integrated optic phase modulator) is necessary, and modulator losses (several decibels) can erode the system advantage of PSK.

In binary DPSK, the signal is differentially encoded before modulation. A '1' is transmitted by changing the waveform phase by 180° between successive bits, while '0' is transmitted by sending a pulse in phase with the previous bit. Demodulation and differential decoding can then be achieved by comparing the phase of one bit

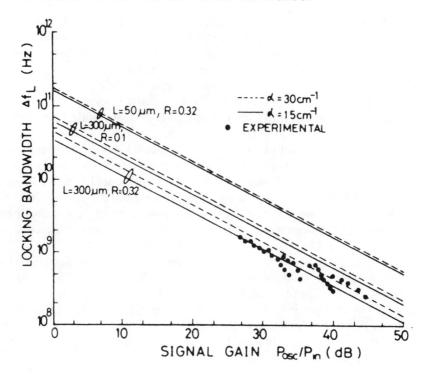

Fig. 2.4 Bandwidth versus signal gain of injection locked LD for FSK. (From [2.36] Kobayashi and Kimura (1980) *IEEE J. Quantum Electron.* **QE-16**, 915–917. Reproduced by permission. ©1980 IEEE.)

with the previous bit, so the system only requires bit to bit phase coherence in the optical sources. There is no need to lock to the suppressed carrier, although there is a sensitivity penalty of the order of 1 dB compared to PSK due to demodulation with the (noisy) previous bit signal, rather than with a clean phase locked carrier. An external phase modulator with its associated losses is still necessary.

As a transmitter configuration for the PSK heterodyne differential detection and PSK heterodyne-coherent detection schemes, combination of frequency stabilized laser oscillator and an external phase modulator is possible. A conventional guided wave electro-optic modulator with, for example, $LiNbO_3$ [2.37] can be used as a phase modulator. Its insertion and coupling loss may be compensated for by the injection locked semiconductor laser post amplifier.

DEMODULATION TECHNIQUE

Demodulation processes of the IF signal obtained by the optical heterodyne detection scheme is principally the same as those in microwave or millimeter wave systems [2.38, 2.39]. Envelope detection and coherent detection has been used to demodulate an ASK signal. Both schemes possess almost the same receiving power level [2.13]. The envelope detection scheme employs a much simpler configuration and is more tolerable for a carrier frequency noise. Frequency discrimination detection and dual filter detection have been used to demodulate and an FSK signal is chosen to be equal to the clock frequency for the discrimination detection, and to be twice the clock frequency for the dual filter detection. Frequency discrimination detection employs a simpler configuration, but imposes a stringent requirement on the carrier frequency noise compared to the dual filter detection. The FSK system with a large frequency deviation and a dual filter detection is tolerable for the carrier frequency noise, but necessitates a large detector bandwidth. Differential detection and coherent detection have been used to demodulate a PSK signal [2.40, 2.41]. The optimum phase shift in a binary system is π for both detection schemes. Differential detection employs a much simpler configuration and is insensitive to carrier frequency noise, but it necessitates a higher receiving level by about 0.5 dB as compared to the coherent detection scheme.

IF demodulation circuitry has been developed at the 1.7 and 4 GHz centre frequencies. A higher IF frequency system will be possible in the near future. The limitation on the data rate seems to be determined from the modulation technique and the photodetector bandwidth.

2.1.3 Experimental configurations of coherent transmission

As much progress was achieved on coherent transmitters and receivers in different laboratories in recent years, the field demonstration of a coherent optical DPSK transmission system in an operational network environment has been reported in Cambridge recently [2.42], with a power budget now exceeding 50 dB at 565 Mbit/s, and error free operation consistent with a BER $< 10^{-13}$. The system configuration used for this experimental demonstration is shown schematically in Fig 2.5 [2.42], and in the photograph in Fig. 2.6 [2.42]. Miniaturized long external cavity (LEC) lasers [2.43] were used as the optical sources at the transmitter and local oscillator, each providing a spectral linewidth of < 100 kHz and a launch power of 0 dBm.

Further laboratory experiments have shown the feasibility of coherent transmission through a diode-pumped Er-doped fiber amplifier repeater. This particular system implementation thus shows much promise for eventual deployment in a range of traffic-carrying applications. The laboratory demonstrations called for the development of a new range of devices such as narrow-linewidth single mode optical sources, optical isolators, external modulators, polarization adjusters, and wide-band receivers. The field demonstrations have then required these devices to be redesigned and packaged into stable, robust, hermetic, efficiently-coupled optical assemblies, with the provision of controls for temperature, IF, and state of polarization. The first systems truly capable of field deployment will require continuous control over these parameters for indefinite periods of time; and further iterations of device design are required to overcome performance limitations which are still outstanding. Coherent techniques nevertheless offer a range of possible implementations which are on the verge of practical deployment, offering a potential for increased power budget, much greater access to the spectral capacity of SM fiber, and a natural affinity for use with optical amplifier repeaters.

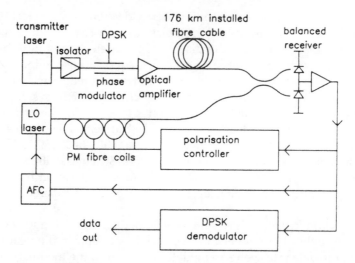

Fig. 2.5 Schematic of the field demonstration system. (From [2.42] Brain *et al.* (1990) *J. Lightwave Technol.* **8**(3), 423–437. Reproduced by permission. ©1990 IEEE.)

A non-repeated fiber transmission experiment over 308 km of conventional optical fiber is reported at 2.488 Gbit/s in a CPFSK heterodyne detection system [2.44]. The system has a sensitivity of -46.7 dBm or 67 photons/bit at 10^{-9} BER and has the maximum allowable attenuation of 55.1 dB without an optical amplifier. The experimental transmission arrangement is shown in Fig. 2.7 [2.44]. The transmitter is a long cavity three-electrode 1.554 μm DFB LD [2.45]. It was directly modulated with a 3.2 mA$_\text{p-p}$, 2.488 Gbit/s NRZ signal (PN 2^7-1) by the central electrode. The modulated signal was transmitted through a 308 km 1.3 μm zero dispersion fiber. The fiber loss, including splicing loss, was 54.3 dB at 1.554 μm. The receiver sensitivity before transmission was -46.7 dBm or 67 photon/bit at 10^{-9} BER. After transmission, the sensitivity degradation was 0.4 dB. The degradation is considered to be caused by fiber chromatic dispersion over 308 km.

A 1 Gbit/s pilor-carrier PSK homodyne transmission system using phase-locked 1.5 μm external cavity semiconductor lasers was reported [2.46]. After 209 km fiber transmission of a $2^{15}-1$ PRBS, the measured receiver sensitivity is -52.2 dBm of 46 photon/bit. The transmitter and LO sources are 1.508 μm InGaAsP ECLs [2.47] which have a beat linewidth of 3 kHz. Each ECL is contained in a Plexiglas box in a $\pm 1°$ C temperature-controlled environment. The fiber-pigtailed, X-cut, travelling-wave LiNbO$_3$ phase modulator requires 9 V to produce a 180° phase shift. Signals are transmitted through either 5 m or 209 km of spooled single mode fiber, the latter having a loss of 50.7 dB and an estimated dispersion of 2.9 ns/nm. The phase detector output is processed by a lag-lead integrator and the correction signal is resistively added with the dc current to the low-biased electrode of the LO laser. Output of the receiver data preamplifier directly yields the baseband data signal, which is filtered using a 700 MHz, three-pole Chebyshev filter. An electronic phase-lock loop is employed for clock recovery.

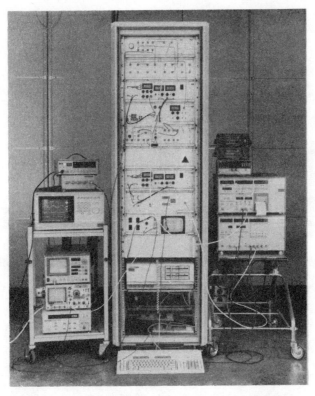

Fig. 2.6 Demonstration system installed at Cambridge. (From [2.42] Brain *et al.* (1990) *J. Lightwave Technol.* **8**(3), 423–437. Reproduced by permission. ©1990 IEEE.)

2.2 QUANTUM LIMITATIONS ON COHERENT TRANSMISSION

2.2.1 Quantum limitations on direct detection [2.11]

Direct detection of light pulses implies a photodetector that converts light energy to electrical signals. The detection mechanism is based upon photon counting, which is subject to statistical fluctuations. More specifically, the photon counting process is a time-varying Poisson process whose intensity function $\lambda(t)$ is directly proportional to the information-bearing data wave. In the case of binary transmission, the choice between a 1 or a 0 is translated into the presence or absence of a burst of optical energy. As an illustration [2.11], consider the passage of a single pulse through an ideal transmission model depicted in Fig. 2.8 [2.11]. In the case of a 1 being transmitted, a square electrical signal turns on the laser or LED and energy is sent into the fiber. In the photodetector, light will be detected due to the electromagnetic energy present. Exactly when in time the photons register on the detector is random. The actual electrical current at the output of this device caused by a photon is a wide-band pulse $gw(t)$ (which is very narrow compared with the signal duration T), where g (gain) is an integer-valued random variable or $g=1$, depending on

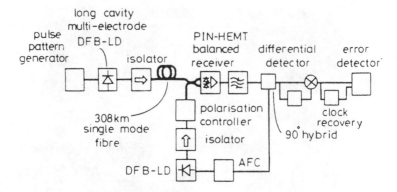

Fig. 2.7 A 300 km 2.5 Gbit/s CPFSK heterodyne system. (From [2.44] Imai *et al.* (1990) *Electron Lett.* **26**, 357–358. Reproduced by permission of IEE.)

whether an (APD) or a p–i–n diode is used. In practical systems where amplification of weak signal s is required, APDs are invariably used.

Assuming that superposition holds for optical fiber transmission, the single-pulse description may be extended to an entire data wave. If 1 transmits a sequence of on or off pulses, then the received signal, defined as the electrical output of the photodetector on which processing is performed, is written as

$$I(t) = \sum_n g_n w(t - t_n) \tag{2.10}$$

where the time points $\{t_n\}$ form a Poisson process having intensity function $\lambda(t)$ with

$$\lambda(t) = \sum_n a_n h(t - nT) \tag{2.11}$$

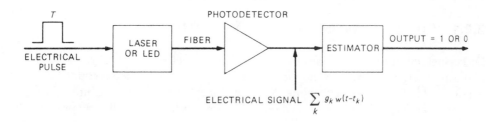

Fig. 2.8 Direct detection. (From [2.11] Salz (1985) *AT & T Tech. J.* **64**(10), 2153–2209. Reproduced with permission. Copyright ©1985 AT & T.)

and $h(t)$ is a square pulse, $\{a_n\}=0$ or 1 are the data levels, $\{g_n\}$ is avalanche gain, $T=$ signalling interval, and $w(t)=$ output pulse of the photodetector.

In this simple model, to detect the jth bit, one integrates the output of the photodetector over the jth T-second interval and compares the random variable with a threshold. If the output is greater than the threshold, a 1 is declared; if it is less, a 0 is declared.

In the ideal situation, when a p–i–n diode is used, $g_n=1$ and when the threshold is set at 0, the average output of the integrator will yield $\int_0^T \lambda(t)dt = \lambda T$ when a 1 is sent and zero output when a 0 is sent. Since the number of counts n with intensity λT is Poisson distributed

$$p(n) = \frac{(\lambda T)^n e^{-\lambda T}}{n!} \tag{2.12}$$

and the chance of making an error is just $\frac{1}{2}p(n=0)$ or,

$$P_e = \frac{1}{2}e^{\lambda T}. \tag{2.13}$$

The average optical energy, photons per bit, is just $P = \frac{1}{2}(\lambda T) + \frac{1}{2}(0)$ and so (2.13) is written

$$P_e = \frac{1}{2}e^{-2P}. \tag{2.14}$$

This is a fundamental limit on the bit error rate and is commonly referred to as the 'quantum limit'.

Equation (2.14) implies that, in order to obtain an error rate of 10^{-9}, about 10 photon/bit are required. This of course is the error rate achieved in the absence of coding. It has been shown that by employing coding the number of required photons per bit is on the order of 2 to 3 provided the information rate is less than a characteristic rate called channel capacity.

When an avalanche detector is used to gain optical amplification, the average value of $\{g_n\}$ may be large but the fluctuations are also large causing amplitude jitter. The penalties incurred by avalanche detectors have been extensively studied. Depending on the type of avalanche detectors used, the loss can be anywhere from 10 to 20 dBs from the quantum limit (see, for example [2.48]). Thus one of the chief motivations for turning to coherent techniques is to minimize this tremendous loss in detector sensitivity.

2.2.2 Quantum limitation on homodyne detection [2.11]

In homodyne detection, the electromagnetic wave at the output of a laser can be represented as

$$s(t) = A\cos\omega_0 t \tag{2.15}$$

where A^2 is proportional to the optical power. Now suppose that this wave is phase modulated so that a 1 results in $A \cos\omega_0 t$ and a 0 results in $-A \cos\omega_0 t$. An ideal homodyne detector adds to the received wave a local carrier wave of amplitude equal to exactly A. So, the sum is

$$s_0(t) = (A \pm A)\cos\omega_0 t. \tag{2.16}$$

When the sum is detected by a photodetector (p–i–n diode) and the output integrated for T seconds, one obtains for the average number of counts λT, either $4A^2T$ or 0. The average transmitted optical energy in this case is $P = A^2T$ and so the probability of a bit error is

$$P_e = \tfrac{1}{2}e^{-4P}.$$

This result indicates a 3 dB improvement over the quantum limit, and it is often referred to as the 'super quantum' limit. Reviewing briefly, to achieve the super quantum limit, the local laser had to know exactly the frequency, phase, and the magnitude of the transmitter laser—a rather ambitious requirement. This detector is depicted in Fig. 2.9 [2.11] with alternative No. 1 used as the input to the photodetector.

   Now suppose that we relax the requirements on the local laser and permit its intensity to be any value B, but still requiring knowledge of the transmitted carrier frequency and phase. Now the combined waves become

$$s_0(t) = (B \pm A)\cos\omega_0 t. \tag{2.17}$$

Again (2.17) is detected by a photodetector and consequently the average number of counts at the output after integration is now $(B \pm A)^2T$, where B is the amplitude of the local laser and it is assumed that $B \gg A$.

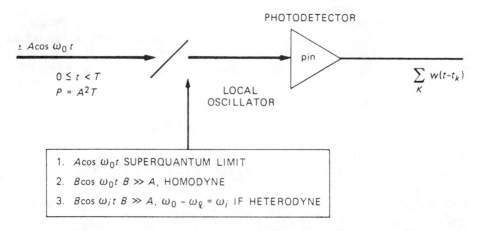

Fig. 2.9 Ideal homodyne and heterodyne techniques. (From [2.11] Salz (1985) *AT & T Tech. J.* **64**(10), 2153–2209. Reproduced with permission. ©1985 AT & T.)

To estimate the resulting bit error rate in this situation we invoke a limit theorem. The theorem has to do with the conditions under which a 'shot noise' process—the output current from the photodetector—is well approximated by a 'white Gaussian' noise process. The main requirement is that the rate of photon arrivals be large. Since B in (2.17) can be made as large as one desires, the average number of photons is proportional to $\lambda T = (B^2 + A^2 \pm 2AB)T$. If the common bias term $(B^2 + A^2)T$ is substracted from λT, there is left an antipodal signal pair $\pm 2ABT$ for the net average counts corresponding to reception of binary 1s and 0s. The variance of the resulting Poisson process also equals λT and, since $B \gg A$ by hypothesis, the variance is essentially TB^2. Now in the limit of large number of counts due to the addition of the local laser to the incoming signal, the output electrical signal can be modelled by

$$s_0(t) = \pm 2AB + n(t) \qquad 0 \leqslant t \leqslant T \tag{2.18}$$

where $n(t)$ is a white Gaussian noise process with double-sided spectral density equal to B^2. Integrating (2.18) from 0 to T results in a Gaussian random variable. The resulting bit error rate is then

$$P_e = \tfrac{1}{2}\mathrm{erfc}(2A^2T)^{1/2} \approx \exp(-2A^2T)$$

$$= e^{-2P} \tag{2.19}$$

which is asymptotically (large P) the quantum limit. We have thus demonstrated that an ideal homodyne detector using a p–i–n photodiode achieves the quantum limit. This is made possible by the availability of large 'local' optical power that provides indirect amplification of the incoming weak optical signal. While providing amplification, the procedure also produces additive noise. This mode of detection is depicted in Fig. 2.9 below [2.11] with alternative No. 2 for the input to the photodetector.

2.2.3 Quantum limitation on heterodyne detection [2.11]

Finally for ideal heterodyne detection, this procedure might be advantageous in some cases to make a frequency translation to an intermediate frequency (IF) and is depicted in Fig. 2.9 below with alternative No. 3 for the input to the photodetector.

To understand the consequences of this approach we proceed as follows. Let the local laser frequency be denoted by ω_1 and the incoming optical frequency by ω_0 such that the IF frequency is $\omega_i = \omega_0 - \omega_1$. The addition of the two waves now results in

$$s(t) = \pm A\cos\omega_0 t + B\cos\omega_1 t \qquad 0 \leqslant t \leqslant T \tag{2.20}$$

where we denote the phase modulation by ± 1 and again require that $B \gg A$.

Expressing $s(t)$ in terms of the envelope and phase about ω_1 results in the representation

$$s(t) = E(t)\cos(\omega_1 t + \beta(t)) \tag{2.21}$$

where

$$E(t) = [(B \pm A\cos\omega_i t)^2 + A^2\sin^2\omega_i t]^{1/2} \tag{2.22}$$

and

$$\beta(t) = \tan^{-1}\frac{\pm A\sin\omega_i t}{B \pm A\cos\omega_i t}. \tag{2.23}$$

The response of the photodiode to the wave (2.21) is again a shot-noise process with intensity function, λ_0, equal to the envelope squared:

$$\lambda_0(t) = B^2 + A^2 \pm 2AB\cos\omega_i t. \tag{2.24}$$

Using the same limit arguments as in the previous section, we first subtract $B^2 + A^2$ from $\lambda_0(t)$, which retains the antipodal signal pair

$$\pm 2AB\cos\omega_i t. \tag{2.25}$$

Because $B \gg A$ the fluctuating noise is white Gaussian with double-sided spectral density $\approx B^2$. Denoting the additive noise by $n(t)$, the equivalent signal-in-noise problem after heterodyning becomes

$$s_0(t) = \pm AB\cos\omega_i t + n(t) \quad 0 \leqslant t \leqslant T. \tag{2.26}$$

This is a standard elementary detection problem and deciding whether a plus or a minus was sent is accomplished by multiplying (2.26) by $\cos\omega_i t$, integrating for T seconds, and comparing the result with a threshold set to zero. The decision statistic is

$$\pm ABT + \int_0^T n(t)\cos(\omega_i t)dt \tag{2.27}$$

where we have neglected the double frequency term. Since the random variable $\int_0^T n(t)\cos(\omega_i t)dt$ has variance equal to $B^2 T/2$, the bit error rate in this case is asymptotically

$$P_e \sim \exp\left(-\frac{A^2 B^2 T^2}{B^2 T}\right) = \exp(-A^2 T) = \exp(-P) \tag{2.28}$$

The exponent is seen to be a factor of 2 smaller than in (2.19) and because of this heterodyne detection is 3 dB inferior to the quantum limit. So, due to the quantum limitation, BERs are e^{-2p} for homodyne, e^{-4p} for super homodyne and e^{-p} for heterodyne respectively.

2.3 HETERODYNE AND HOMODYNE OPERATION [2.49]

2.3.1 Heterodyne operation

In heterodyne operation, the data is converted to an intermediate frequency and the resulting bandpass signals are demodulated. A heterodyne coherent optical receiver is shown in Fig. 2.10 [2.49]. The incoming signal and the optical local oscillator are applied to the input ports of a four-port optical hybrid network.

The signal and LO differ in optical frequency by the desired IF, and the electrical current from the photodetector contains an IF signal containing the modulation, together with direct detected signal and local oscillator components. In practice, the local oscillator will add amplitude and phase noise. With ASK and PSK modulation the IF signal and spectra are as in Fig. 2.11 [2.49]. FSK with moderate deviation will also appear as a conventional IF signal. To keep the IF at the desired value, the local oscillator laser will need to be frequency (but not phase) locked to the incoming signal.

The optical hybrid can be symmetrical and utilize a balanced receiver for optimum performance, or asymmetrical to preserve sensitivity with an unbalanced receiver. In the latter case the available LO power must be larger, and amplitude noise cannot be cancelled. At best, ideal heterodyne operation can achieve a sensitivity of 18 photon/bit as opposed to the 9 photon/bit which should be available in a coherent system. In addition, wide-band-width receivers are needed. It is to overcome these

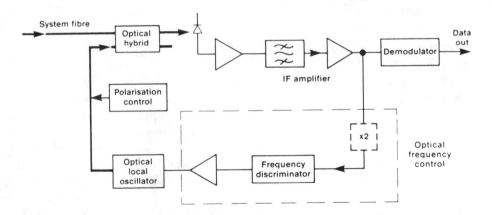

Fig. 2.10 Heterodyne receiver. (From [2.49] Davis *et al.* (1987) *J. Lightwave Technol.* **LT-5**(4), 561–572. Reproduced by permission. ©1987 IEEE.)

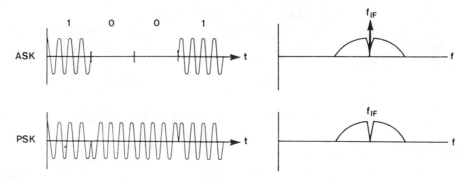

Fig. 2.11 Output signals in a heterodyne receiver. (From [2.49] Davis *et al.* (1987) *J. Lightwave Technol.* **LT-5**(4), 561–572. Reproduced by permission. ©1987 IEEE.)

limitations that the homodyne operation has been considered a desirable goal. However, it will be seen that the necessary conditions to achieve this goal are unrealistically demanding.

2.3.2 Homodyne operation

In a phase locked homodyne receiver, demodulated data emerges directly from one photodiode as a result of synchronous optical mixing. In phase diversity operation, at least two baseband sampled data streams are available from different photodetectors as a result of asynchronous optical mixing. In phase diversity operation, at least two baseband sampled data streams are available from different photodetectors as a result of asynchronous optical mixing.

There are two classes of synchronous homodyne receiver: those deriving phase lock from a full or residual carrier in the incoming signal (carrier phase locked loop) and those using non-linear technique to recover a suppressed carrier (Costas loop or decision driven loop) [2.50]. Consideration of the Costas loop will lead naturally to phase diversity operation.

CARRIER PHASE LOCKED LOOP (CPLL) HOMODYNE RECEIVERS

An outline block diagram of a CPLL receiver is shown in Fig. 2.12 [2.49], with photodiode output signals and spectra in Fig. 2.13 [2.49]. The local oscillator at the frequency of the carrier of the modulated input signal is applied to one arm of a 180° four-port optical hybrid. When phase synchronism with the signal carrier is achieved, the photodiode output is the demodulated data signal. To achieve phase synchronism, a residual pilot carrier must be transmitted: for PSK modulation, incomplete modulation of ±85° gives a carrier component 10 dB below the total signal power. Phase locking to such a pilot carrier has been demonstrated [2.51], and the necessary loop conditions in the presence of signal and LO phase noise has been analysed [2.52, 2.53].

The system has a number of attractive features. It needs only baseband photodiode and amplifier bandwidth and no further demodulator is necessary. A simple symmetrical four-port hybrid can be employed.

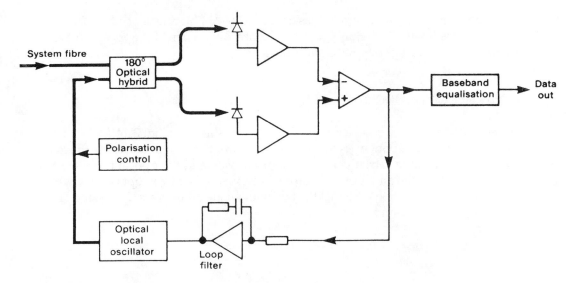

Fig. 2.12 Carrier phase locked loop in a homodyne receiver. (From [2.49] Davis *et al.* (1987) *J. Lightwave Technol.* **LT-5**(4), 561–572. Reproduced by permission. ©1987 IEEE.)

However, there are several practical difficulties. In the first place, direct detection of LO and signal gives a dc output, which masks the desired phase dependent dc signal. Similarly, LO amplitude noise would adversely affect the lock quality. These effects can be overcome by employing a balanced receiver as shown, but drift and noise affecting the balance may well be severe problems, since the receiver must be dc coupled. The data and phase lock signals are both taken from the differential output. Penalties are then incurred because of the need for finite carrier power to phase lock, and due to phase variation on the recovered carrier, caused by phase noise, shot noise, and data to phase lock crosstalk. A total phase variation of $\approx 10^0$ is the maximum permissible to ensure low penalty at a BER of 10^{-9}–10^{-10}. To maintain phase lock to this accuracy, it has been derived [2.53] that a system

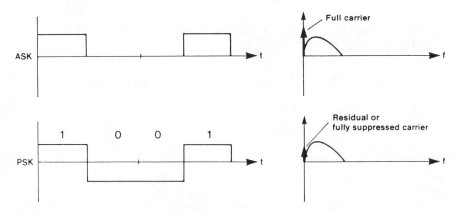

Fig. 2.13 Output signals in a homodyne receiver. (From [2.49] Davis *et al.* (1987) *J. Lightwave Technol.* **LT-5**(4), 561–572. Reproduced by permission. ©1988 IEEE.)

linewidth (due to transmit and LO lasers) of 6×10^{-6} of bit rate is needed. This sets a very severe linewidth requirement and hence the carrier PLL appears to be an unlikely choice for operational systems.

COSTAS LOOP HOMODYNE RECEIVERS

An alternative to CPLL receiver is the Costas loop, shown in Fig. 2.14 [2.49]. Here the carrier (dc) component is recovered by non-linear methods from the data sidebands and so the carrier can be fully suppressed and the optical receivers can be ac coupled. Signal and local oscillator waves are applied to an optical 90° hybrid having a transfer function such that the two photodiode outputs are (neglecting direct detected components):

$$KM(t) \exp(j\omega t + \phi_0)$$

$$KM(t) \exp(j\omega t + \phi_0 + \pi/2)$$

where ω is the angular frequency difference between signal and LO waves and $M(t)$ is the (PSK) modulation. Suitable optical hybrids can take two forms [2.54, 2.55], and we shall discuss them further in connection with phase diversity receivers in the next section.

The Costas loop shares the CPLL receiver advantages of baseband receiver requirements and direct demodulation. Additionally we may note that the receive system may be ac coupled (to the third multiplier output) and there is no restriction on LF data content since, ideally, there is no data/phase lock crosstalk. Mainly for this reason, laser system linewidth requirements derived in [2.55] are reduced to $\approx 3 \times 10^{-4}$ of bit rate for decision directed loops. Linewidth requirements for Costas loops would be similar but this is still a very severe specification.

In view of the necessary conditions, allowing the system to operate synchronously and processing the phase diverse outputs looks an attractive alternative.

2.4 THEORETICAL ANALYSIS OF COHERENT OPTICAL RECEIVERS [2.63]

Theoretical analysis of optical heterodyne receivers has been the subject of interest of many researchers over the past several years [2.56–2.60]. Specifically, the impact of laser phase noise on optical heterodyne communication systems has been the focus of attention of many studies [2.56, 2.57, 2.59]. In particular, [2.56] used the Radon–Nikodym theorem to make an accurate numerical simulation of the probability density function (pdf) of the intermediate frequency (IF) filter output. An analytical approximation of the resulting pdf was then used to evaluate the bit error ratio. The calculations of [2.56] appear to be the most accurate to date, but seem to be rather complicated. A simplified approximate analysis of BER expressed in closed form as a function of main parameters was presented recently [2.61].

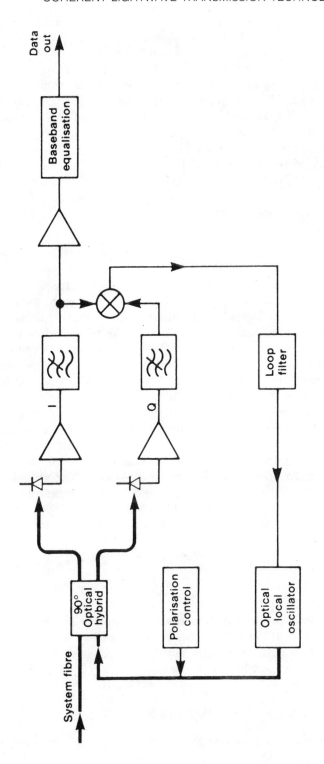

Fig. 2.14 Costas loop in a homodyne receiver. (From [2.49] Davis *et al.* (1987) *J. Lightwave Technol.* **LT-5**(5), 561–572. Reproduced by permission. ©1988 IEEE.)

A simplified approximate theoretical analysis of BER of heterodyne ASK and FSK systems was expressed [2.61] in closed form as a function of main system parameters: laser linewidth, peak IF signal-to-noise ratio (SNR), and IF bandwidth. Two hypotheses (for data=0 and for data=1) by Gaussian probability density function (pdf) were proposed for simplification and we then use the well known approximation [2.62] to compute the BER. The inaccuracy of this theory [2.61] in the case of ASK is 0.4 dB at BER = 10^{-9}; in the case of FSK, it is less than 2.6 dB and is reduced to 1 dB using an empirical correction factor.

In this section, we select the theoretical analysis by Garrett and Jacobsen [2.63] (by permission) as reference material for a coherent optical receiver. Their model, which takes into account the linewidths of the source and local oscillator lasers, includes the receiver noise, filter bandpass characteristics, and in the case of FSK, the modulation index and effects of signal crosstalk and noise correlation between the filters.

MODEL FOR NON-ZERO LINEWIDTH

A coherent optical receiver measures the intermediate frequency (beat frequency between the signal and local oscillator lasers) in a go–no-go sense, corresponding to 0 or 1 outputs. If the lasers have significant linewidth, the measured intermediate frequency will be a random process, reducing the receiver sensitivity to greater or laser extent. To account for the effects of non-zero linewidth for the source and local oscillator lasers, we regard the receiver as measuring a single realization of the intermediate frequency averaged over a measurement interval of duration about a reciprocal IF bandwidth (comparable to or a few times the bit rate). Let Δu be the deviation of the measured intermediate frequency from the nominal centre intermediate frequency, normalized to the bit time T. This measured intermediate frequency deviation is also a random process, with statistics that depend both on laser linewidth and on the measurement bandwidth. Let $p_{IF}(\Delta u)$ be the probability density function for the deviation of the measured IF. The probability $P_c(p_s|\Delta u)$ of the error in detecting the symbol, conditioned on Δu, in terms of the received signal power p_s and receiver parameters could be calculated, using the theory developed in the following section. The overall error probability is then

$$P_T = \int_{-\infty}^{\infty} P_c(p_s|\Delta u) \; p_{IF}(\Delta u) \; d(\Delta u)$$
$$\equiv \int_{-\infty}^{\infty} P_e(p_s, \Delta u) \; d(\Delta u). \qquad (2.29)$$

Here P_e is the error probability density as a function of Δu. In FSK systems, power is transmitted on both 1 and 0 symbols, so that the discussion above applies to each symbol. In ASK systems, no power transmitted on 0 symbols is assumed, and then the discussion above applies only to 1s.

THE PROBABILITY DENSITY FUNCTION (PDF) OF Δu

The pdf for a phase fluctuation $\Delta \phi$ in a measurement time τ for a laser is often expressed as a Gaussian

$$P_{\text{laser}}(\Delta\phi) = \frac{1}{\sqrt{(4\pi^2\Delta\nu_{\text{laser}}\tau)}}\exp(-\Delta\phi^2/4\pi\tau\Delta\nu_{\text{laser}}) \tag{2.30}$$

where $\Delta\nu_{\text{laser}}$ is the full width at half-maximum of the (Lorentzian) power spectral density as measured by an integrating or ensemble-averaging method. This Gaussian form leads to the Lorentzian shape for the power spectral density as observed, and also for example to the expression for the pdf of the output intensity of a Michelson interferometer with delay τ

$$p(I)dI = \frac{1}{\sqrt{(4\pi^2\Delta\nu_{\text{laser}}\tau)}}\sum_0^\infty \left[\exp\left(-\frac{(\Delta\phi'+2\pi k)^2}{4\pi\Delta\nu_{\text{laser}}\tau}\right) \right.$$

$$\left. + \exp\left(-\frac{(\Delta\phi'-2\pi k)^2}{4\pi\Delta\nu_{\text{laser}}\tau}\right) \right] \tag{2.31}$$

where $\Delta\phi'$ is in $\{0,2\pi\}$.

A phase fluctuation $\Delta\phi$ in a measurement interval τ could be interpreted as being equivalent to a frequency deviation between one interval and the next, with the rms frequency deviation being equal to the rms phase deviation divided by the measurement time. So the frequency deviation is defined as a random process related to $\Delta\phi$ by

$$\Delta\omega = \Delta\phi/\tau \tag{2.32}$$

and we envisage the coherent receiver as responding to a $\Delta\omega$ averaged over a reciprocal IF bandwidth around the decision time. So we can write

$$p_{\text{laser}}(\Delta\omega)\ d(\Delta\omega) = p_{\text{laser}}(\Delta\phi)\ d(\Delta\phi)$$

$$= \frac{d(\Delta\omega)}{\sqrt{(4\pi^2\Delta\nu_{\text{laser}}/\pi)}}\exp(-\Delta\omega^2\tau/4\pi\Delta\nu_{\text{laser}}). \tag{2.33}$$

This pdf is thus consistent with the experimentally observable power spectral density of the laser and with interferometric results such as (2.31).

In a heterodyne receiver, we measure single realizations of the difference frequency between the transmitter and local oscillator lasers. If $\Delta\nu_{\text{trans}}$ and $\Delta\nu_{\text{LO}}$ are the linewidths of the transmitter and local oscillator lasers, each of which has a Gaussian pdf for its phase deviation process as in (2.30), then the IF linewidth is

$$\Delta\nu_{\text{IF}} = \Delta\nu_{\text{trans}} + \Delta\nu_{\text{LO}}. \tag{2.34}$$

Finally, normalizing all frequencies and bandwidths with respect to the bit time T, we have

$$\Delta u = \Delta\omega T/2\pi \tag{2.35}$$

$$\Delta\nu = \Delta\nu_{\text{IF}}T. \tag{2.36}$$

If B is the IF bandwidth normalized to the bit time, then the pdf for the normalized intermediate frequency deviation becomes

$$p_{\mathrm{IF}}(\Delta u) = \frac{1}{\sqrt{\Delta \nu B}} \exp \ [-\Delta u^2 \pi / \Delta \nu B]. \tag{2.37}$$

THE CONDITIONAL ERROR PROBABILITY P_c

The ASK receiver is firstly considered here, the model for which is shown in Fig. 2.15 [2.63]. (In Fig. 2.15, DIFF: differentiation to equalize the front-end response; BPF: IF filter in the ASK receiver; BPF_0, BPF_1: filters to pass 0 and 1 symbols in the FSK receiver; ENV DET: ideal envelope detector.) For a received 1 symbol, the total optical power incident on the photodiode can be expressed in terms of the peak signal power p_a and local oscillator power p_L as

$$p(t) = p_a(t) + p_L + 2\sqrt{p_L p_a(t)}\cos \ \Omega_1 t \qquad -T/2 < t < T/2 \tag{2.38}$$

where $\Omega_1 = \omega_L - \omega_s$ is the intermediate frequency, ω_L and ω_s being the optical signal and local oscillator frequencies. For a received 0 symbol, the total incident power is just p_L and there is no signal at the IF. In the case of p_L much larger than p_a, which is desirable and often the case in practice, we can write the modulated part of the resulting photocurrent as

$$i(t) = RM \ 2\sqrt{p_L p_a(t)}\cos\Omega_1 t \tag{2.39}$$

where M is the photodiode gain and $R = nq/\hbar\omega$ is the photodiode responsivity. Here we ignore phase offsets on the IF signal, in line with assuming the use of an ideal envelope detector to demodulate the IF signal. If we put

$$2\sqrt{p_L p_a(t)} = b s_1(t) \tag{2.40}$$

where b is the modulated optical power averaged over the bit time T, and s_1 is the shape factor of the signal pulse envelope for a received 1 symbol, normalized so that

$$\int_{-\infty}^{\infty} s_1(t) \ dt = T. \tag{2.41}$$

Then we can write the signal voltage at the output of the ideal envelope detector as

$$v_1(t) = RMb s_1(t) h_{\mathrm{fe}}(t) h_1(t) \tag{2.42}$$

where h_{fe}, u, and h_1 are the impulse responses of the receiver front end, the differentiator, and the IF filter (see Fig. 2.15(a) [2.63]). Using more convenient frequency domain quantities, here consider an IF signal pulse at a nominal centre frequency Ω with frequency deviation $\Delta\omega$ at the decision time t_d. Then the signal voltage at the output of the envelope detector at the decision time is

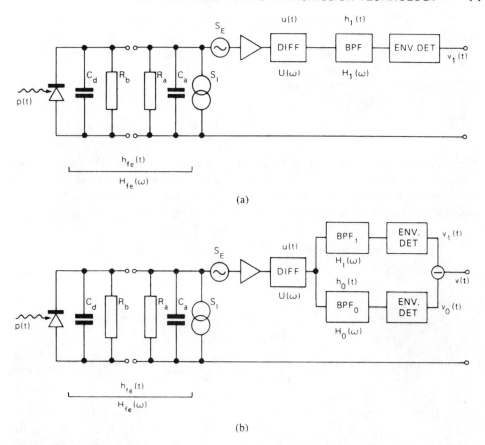

Fig. 2.15 Model of heterodyne ASK receiver (a) and dual-filter FSK receiver (b). (From [2.63] Garrett and Jacobsen (1986). *J. Lightwave Technol.* **LT-4**(3), 323–334. Reproduced by permission. ©1986 IEEE.)

$$v_1(t_d) = \frac{RMb}{2\pi} \int_{-\infty}^{\infty} S_1(\Omega, \Delta\omega) \, H_{fe}(\Omega) \, U(\Omega) \, H_1(\Omega) \, \exp(j\Omega t_d) \, d\Omega \qquad (2.43)$$

where $S_1(\Omega, \Delta\omega)$ is the Fourier spectrum of a pulse with envelope $s_1(t)$. We will assume that the differentiator $U(\Omega)$ exactly compensates for the front-end response $H_{fe}(\Omega)$ and put

$$H_{fe}(\Omega) \, U(\Omega) = Z_0 \qquad (2.44)$$

where Z_0 is an arbitrary gain factor with dimensions of impedance. Let $Y_1(\Omega, \Delta\omega)$ be the Fourier spectrum of the output pulse (corresponding to $H_{out}(\omega)$ in Personick's analysis of direct detection receivers [2.65]), i.e.

$$Y_1(\Omega, \Delta\omega) = RMZ_0 S_1(\Omega, \Delta\omega) \, H_1(\Omega) \, \exp(j\Omega t_d) \qquad (2.45)$$

and let the corresponding time-domain function $y_1(t)$ be normalized as in [2.65] (conceptually by adjusting Z_0)

$$y_1(0) = 1 \qquad \text{for } \Delta\omega = 0$$

$$y_1(kT) = 0 \qquad \text{for } k \neq 0 \text{ and } \Delta\omega = 0 \tag{2.46}$$

so that $H_1(\Omega)$ is a zero-forcing filter for $S_1(\Omega, 0)$. Then

$$v_1(t_d) = \frac{b}{2\pi} \int_{-\infty}^{\infty} Y_1(\Omega, \Delta\omega) \, d\Omega. \tag{2.47}$$

Thus we have the expression for the output signal voltage at the decision time.

The derivation of the mean-square output noise voltage follows [2.66] except that we introduce the stationary shot noise on the photocurrent due to the local oscillator and ignore the signal-dependent shot noise in comparison

$$\langle n^2 \rangle = Z_0^2 \left\{ qRM^{2+x}p_L + \frac{2k\theta}{R_b}S_1 \right\} B_2 + S_E B_3 \tag{2.48}$$

where

$$B_2 = \frac{1}{2\pi Z_0^2} \int_{-\infty}^{\infty} |H_1(\Omega)|^2 \, d\Omega \tag{2.49}$$

$$B_3 = \frac{1}{2\pi} \int_{-\infty}^{\infty} |U(\Omega) \, H_1(\Omega)|^2 \, d\Omega. \tag{2.50}$$

As in [2.65], we normalize the timescale with respect to the bit time, which is particularly appropriate for the coherent receiver since, with the 'strong local oscillator' approximation, the receiver sensitivity expressed in photons per bit time is independent of the date rate. Thus we put

$$\Omega = 2\pi u/T \tag{2.51}$$

and define

$$S_1'(u, \Delta u) = \frac{1}{T}S_1(2\pi u/T, 2\pi\Delta u/T) \tag{2.52}$$

$$Y_1'(u, \Delta u) = \frac{1}{T}Y_1(2\pi uT, 2\pi\Delta u/T) \tag{2.53}$$

$$H_1'(u) = H_1(2\pi u/T). \tag{2.54}$$

Then, from (2.46)

$$\int_{-\infty}^{\infty} Y_1'(u, \Delta u) du = y_1(0)|_{u=0} = 1 \tag{2.55}$$

and

$$v_1(t_d) = b. \tag{2.56}$$

Defining

$$I_{2.1} = \int_{-\infty}^{\infty} \left| \frac{Y_1'(u, 0)}{S_1'(u, 0)} \right|^2 du \tag{2.57}$$

$$I_{3.1} = \int_{-\infty}^{\infty} \left| \frac{Y_1'(u, 0)}{S_1'(u, 0)} \right|^2 u^2 \, du \tag{2.58}$$

and the noise spectral densities

$$n_1 = q p_L M^x / TR \tag{2.59}$$

$$n_2 = \frac{1}{T(RM)^2} \left\{ \frac{2k\theta}{R_b} + S_1 + S_E/R_T \right\} \tag{2.60}$$

$$n_3 = \frac{(2\pi C_T)^2}{(RM)^2} S_E/T^3 \tag{2.61}$$

where $R_T = R_a \parallel R_b$ and $C_T = c_a + C_d$(see Fig. 2.15(a)), we have

$$\langle n_2 \rangle = (n_1 + n_2)I_{2,1} + n_3 I_{3,1}. \tag{2.62}$$

Here n_1 is a white noise spectral density arising from the shot noise on the photocurrent due to the local oscillator, n_2 is the white noise component from the receiver, and n_3 is the frequency-squared receiver noise component. Thus we have expressions for the output signal and noise voltages at the decision time in terms of the optical signal power, filter transfer function, and receiver parameters. These are used to calculate the conditional error probability P_c using Marcum's Q-function [2.66, 2.67]. If the received signal power on 0 symbols is some non-zero fraction ϵ of the power on 1 symbol, i.e. ϵ is the source extinction ratio, the corresponding output voltage is $v_1(t_d)\sqrt{\epsilon}$, and this can be used to evaluate the effects of imperfect source extinction.

In this analysis we ignore $1/f$ noise in the receiver. This noise contribution is most significant around dc while we are considering bandpass filters with one or more zeros at dc and a bandwidth range over which $1/f$ noise is negligible compared with other receiver noise contributions.

The theoretical model for the dual filter FSK receiver is based on Fig. 2.15(b), which is essentially two parallel filters. We define the normalized modulation index as

$$m = (\Omega_1 - \Omega_0) \, T/2\pi = (\omega_1 - \omega_0)T/2\pi. \tag{2.63}$$

Compared with the ASK analysis, we have two extra factors to take into account, namely signal crosstalk and noise correlation. The signal crosstalk between the filters depends on the filter characteristics, the modulation index and the frequency deviation Δu for a particular pulse. It is non-zero even for zero-linewidth lasers and non-overlapping filters since the pulse spectrum will generally extend over the passband of both filters. The noise correlation between the filters is always non-zero for practical, i.e. overlapping, filters.

We can express the output signal voltage at the decision time from envelope detector m when symbol n is received ($m,n,=0,1$) as

$$v_m(\tau_m)|\text{symbol } n \text{ received} = b \int_{-\infty}^{\infty} S'_n(u, \Delta u) \, H'_m(u) e^{j2\pi u \tau_m} du \qquad (2.64)$$

where $\tau_m = t_m/T$ is the normalized decision time. Note that τ_0 and τ_1 can be made different if desired by inserting an appropriate flat delay into one filter arm.

The mean-square noise voltages $\langle n_0^2 \rangle$ and $\langle n_1^2 \rangle$ at the outputs of the two envelope detectors can be calculated as in the ASK system

$$\langle n_{n'}^2 \rangle = (n_1 + n_2) I_{2,n} + n_3 I_{3,n}. \qquad (2.65)$$

The normalized noise cross-correlation coefficient for the two filters is

$$\varrho \equiv \frac{\displaystyle \int_{-\infty}^{\infty} N'(u) \, H_0'^*(u) \, H_1'(u) \, du}{\left[\displaystyle \int_{-\infty}^{\infty} N'(u) |H_0'(u)|^2 \, du \int_{-\infty}^{\infty} N'(u) |H_1'(u)|^2 \, du \right]^{1/2}} \qquad (2.66)$$

where

$$N'(u) = (n_1 + n_2) + n_3 u^2 \qquad (2.67)$$

is the noise spectral density, and $-1 < \varrho < 1$. The conditional probability P_c can be calculated now in terms of the output signal and mean-square noise voltages and the cross-correlation coefficient, using Marcum's Q-function [2.66–2.68].

ASYMPTOTIC ERROR PROBABILITY LIMITS

In the limit of large signal power, we can ignore noise as a source of errors and consider the effects of the frequency deviation alone. In this limit, P_c takes a constant value of unity and the error probability is determined by $p_{IF}(\Delta u)$. Our theory predicts rather different behaviour in ASK and FSK systems.

In the ASK system, we assume that the decision threshold is set adaptively as some fraction a_{th} of the maximum signal power, i.e. for $\Delta u = 0$. Define Δu^- and Δu^+ as the negative and positive frequency deviations for which the output signal voltage for a received 1 is equal to the threshold voltage. Thus

$$\int_{-\infty}^{\infty} S_1'(u, \Delta u^\pm) \, H_1'(u) \, e^{j2\pi u \tau} \, du = a_{th}. \qquad (2.68)$$

For deviations beyond $\Delta u^\pm$ the output voltage is below the decision threshold, and an error sill occur unless the noise takes the voltage above the threshold. Thus in the limit of large signal power the total error probability on received 1 becomes

$$P_{\text{floor}} = \tfrac{1}{4} \, \text{erfc} \left(|\Delta u^-| \sqrt{\frac{\pi}{\Delta \nu B}} \right) + \tfrac{1}{4} \text{erfc} \left(\Delta u^+ \sqrt{\frac{\pi}{\Delta \nu B}} \right). \qquad (2.69)$$

The error probability floor is determined by errors on 1 and may be approached abruptly since the additive error probabilities on 1 and 0 have quite a different dependence on the received power. The error-rate floor depends directly on the IF linewidth $\Delta\nu$ and bandwidth B, and also implicitly on the threshold setting a_{th} and on B through Δu^- and Δu^+.

In the dual-filter FSK system in the limit of large signal power, errors occur when the frequency deviation is so large that more signal passes through the wrong filter than through the correct one, which happens if $\Delta u > m/2$ for a received 0 (assumed to be represented by the lower transmitted frequency) or $\Delta u < -m/2$ for a received 1. Thus in the limit

$$P_{floor} = \frac{1}{2} \text{erfc} \left(\frac{m}{2} \sqrt{\frac{\pi}{\Delta\nu B}} \right) . \tag{2.70}$$

Here, the error-rate floor is dependent on the filter bandwidth and on the modulation index and the IF linewidth.

Note that the condition for the FSK floor, that the frequency deviation is greater than $m/2$, is exactly the condition for an error to occur in a frequency discriminator demodulation system also, since then the output voltage is the wrong side of the decision threshold. Thus (2.70) applies to frequency discriminator demodulation as well.

Equation (2.70) has the same functional dependence as the expression given by [2.69] except for a factor of 2π in the complementary error-function argument. The 'hard floor' observed experimentally by [2.69] is suggested due to some additive error probability caused, for example, by mode-hopping: theory [2.63] shows that the FSK floor due to laser spectral width is approached quite gradually.

SHOT NOISE LIMIT

The 'strong local oscillator' condition is considered here, i.e., all receiver noise sources are negligible compared with the shot noise on the photocurrent due to the local oscillator, with zero linewidth lasers, and bandpass IF filters yielding full raised-cosine output pulses for which $I_2 = 2.2554$ (double-sided). In the ASK system with mean received power $p_s = p_a/2$

$$v(t_d)|1 \text{ transmitted} = b = 2\sqrt{2p_L p_s} \tag{2.71}$$

$$\langle n^2 \rangle = q p_L I_2 / TR. \tag{2.72}$$

For a 10^{-9} error probability, we require [2.66]

$$(S/N)_{voltage} = \frac{v(d_d)}{\sqrt{\langle n^2 \rangle}} = 12.7 \tag{2.73}$$

or

$$p_s TR|q = 20.2\, I_2$$

$$= 45.4 \text{ photon/bit time} \quad \text{on average.}$$

(2.74)

To obtain an expression for the shot-noise limit in the FSK system, we must also assume that the modulation index is high enough for the filters to be considered non-overlapping and uncorrelated, thus implying a large IF for at least the high frequency filter, making the strong local oscillator condition harder to approach in practice.

With this assumption, the output voltage swing between 0 and 1 symbols at the receiver is

$$v_{swing}(t_d) = 2b = 4\sqrt{p_L p_s}$$

(2.75)

$$\langle n^2 \rangle = \langle n_0^2 \rangle + \langle n_1^2 \rangle = 2q p_L I_2 | TR.$$

(2.76)

Thus for a 10^{-9} error probability we require

$$p_s TR|q = 20.2\, I_2$$

$$= 45.4 \text{ photon/bit time} \quad \text{on average.}$$

(2.77)

The difference between these shot-noise limits and the usual value of 40 photon/bit time is partly because we have assumed 'raised-cosine' filters (zero intersymbol interference) instead of taking a normalized bandwidth of unity, and partly because of using ideal envelope detection rather than synchronous IF detection (giving the figure of 12.7 instead of 12 on the right of (2.73)).

IF-SQUARED RECEIVER NOISE

We derive a relation between the I_3 bandwidth integral and the IF. Let $H'_{bb}(u)$ be the filter characteristic translated to baseband, i.e. by U_{IF}. Then, from (2.58)

$$I_3 = 2(Z_0 RM)^2 \int_{-\infty}^{\infty} |H'_{bb}(u)|^2 \, (u + U_{IF})^2 \, du$$

(2.78)

$$= 2I_{3.bb} + U_{IF}^2 I_2$$

(2.79)

where $I_{3.bb}$ is the I_3 value for the baseband filter and the factor of two appears in (2.78) to account for the double sided noise spectrum of the IF bandpass filter. Thus (2.79) is exact and general for symmetric filters.

For $U_{IF} > 1$, $I_2 = 2I_{2.bb}$ $I_{3.bb}$ e.g., for raised-cosine output, rectangular input, $I_{2.bb}$ is 1.1277 and $I_{3.bb}$ is 0.173 64, so that to a very good approximation, $I_3 = U_{IF}^2 I_2$. Hence we refer to the $n_3 I_3$ term in the output noise spectral density as IF-squared noise.

It is generally the dominant receiver noise contribution, emphasizing the importance of keeping U_{IF} low.

THE EFFECTIVE LOCAL OSCILLATOR STRENGTH

The ratio of local oscillator shot noise to the dominant receiver noise term is considered from (2.59)–(2.61) and (2.79)

$$\frac{n_1 I_2}{n_3 I_3} = \left(\frac{q M^{2+x}}{8\pi^2 \Gamma k\theta} \right) \left(\frac{R p_L T^2}{C_T^2 |g_m \, U_{IF}} \right). \tag{2.80}$$

This ratio must be much larger than unity for the strong local oscillator condition to hold. The first parenthesis on the right of (2.80) is $0.7Z \, M^{2+x} V^{-1}$, i.e., close to unity (volt^{-1}). We therefore define the effective local oscillator strength as

$$P_{eff} = \frac{R p_L}{v_{IF}^2 C_T^2 |g_m} \tag{2.81}$$

where $v_{IF} = U_{IF}/T$ is the (unnormalized) IF. We want $p_{eff} \gg 1 \, V$. Note that it is inversely proportional to the IF squared and to the receiver figure of merit C_T^2/g_m.

EXPERIMENTAL EXAMPLE OF A COHERENT OPTICAL RECEIVER

We choose some experimental results of a heterodyne DPSK system [2.70] as an example here to express the dependence of coherent optical receivers with practical factors. The heterodyne DPSK theoretically derives a receiver sensitivity of 21 photon/bit which is about 3.5 dB less than obtained with a PSK homodyne system [2.12]. However, it has the attractive advantage of being much easier to implement in practice because it avoids the need for both optical and electrical phase locked loops at the receiver [2.50, 2.71]. Like PSK, DPSK does impose very stringent requirements on the laser linewidth to bit rate ratio (Δv/BR). Typically this ratio has to be less than 0.3% in order to avoid a bit error rate floor of 1×10^{-10} in a DPSK system [2.72–2.74, 2.77]. In order to meet this requirement on Δv/BR both external cavity grating lasers (ECL) and single frequency He–Ne lasers have been used as sources with an ECL as the local oscillator laser at the receiver. As the IF linewidth of the laser combination was measured to be less than 100 kHz, laser phase noise should not be a problem down to bit rates low as 30 Mbit/s.

An experiment for a 147.75 Mbit/s DPSK heterodyne transmission system is reported [2.70]. Mode selection and, to some extent, power stabilization for the transmitter and local oscillator were achieved by an intracavity etalon of free spectral range of 20 GHz and a finesse of 30. Phase modulation of the transmitter signal was carried out with an X-cut Ti:LiNbO$_3$ travelling-wave phase modulator [2.75], [2.76]. The drive voltage required to achieve 180° phase shift at 147.75 Mbit/s was about 7 V peak-to-peak. After attenuation by fiber and a variable attenuator, the modulated signal was combined with the local oscillator signal in a 1 : 10 monomode fiber directional coupler mixer. One output of the combiner was sent to a p–i–n GaAs FET transimpedance packaged receiver designed for 1.7 Gbit/s direct detection.

The IF output from the receiver module was amplified and filtered by a 5-pole Chebyshev bandpass filter centred around the IF. Part of the IF signal was used to frequency lock the local oscillator laser to the incoming signal. This automatic frequency control (AFC) loop, which had a 1 kHz bandwidth, consisted of a squaring circuit, a frequency discriminator, a high voltage amplifier and the external piezoelectric translator drive of the LO laser. The discriminator input was generated by a squaring circuit which produces a signal at twice the IF value from the suppressed carrier of the phase modulated signal. The advantage of this technique is that IF recovery is achieved with a large signal-to-noise ratio without inflicting a receiver penalty associated with the need for a residual carrier. Long term IF stability of less than ± 1 MHz was achieved with this AFC loop.

Finally, the baseband data was recovered from the IF signal with a delay demodulator. This demodulator consisted of a 3 dB electrical splitter, one arm of which included a delay line equal to one bit period. The delayed and undelayed signals were multiplied by means of a double balanced mixer with 180° input phase difference between the two input ports. This delay demodulator was designed for an integral value of the IF/BR ratio and was used in all subsequent experiments. The IF values were 295 and 591 MHz, which corresponded to IF/BR ratios of 2 and 4, respectively. Detected IF and eye patterns for the $2^7 - 1$ pseudorandom sequence at 147.75 Mbit/s, taken through a short fiber (1 m) at a 1×10^{-9} bit error rate level, are detected.

Bit error rate (BER) versus received power ηP curves for the two IF conditions are reported [2.70]. Also included in this plot are the theoretical curves for shot-noise limited heterodyne detection at 147.75 Mbit/s and experimental measurements obtained for direct detection using the same receiver at 147.754 Mbit/s. For these data, η includes both the quantum efficiency of the photodetector and the fiber–receiver coupling efficiency and has a value of 70%.

2.5 POLARIZATION CONTROL TECHNIQUE

In coherent optical communications systems the optical receiver sensitivity is closely related with the polarization state matching between the signal and local oscillator waves at the receiver end. Even when the transmitted light is linearly polarized, the polarization state at the receiving end is generally elliptically polarized. In an optical heterodyne (or homodyne) receiver, when an optical IC is used in the receiver, a polarization-state control scheme is indispensable [2.3]. The efficiency of coherent reception, without further precautions, will fade unacceptably. Several approaches to this fading problem have been proposed.

One is the use of a polarization-maintaining fiber over the entire length of the optical channel [2.78–2.83]. Polarization-maintaining fibers are of great advantage in developing sophisticated coherent systems using optical integrated circuits; they are often sensitive to the polarization state of input signals.

Second is the use of a State of Control (SOC) device at the receiving end which matches the local oscillator (LO) polarization state with that of the signal by phase compensation. Then, the conventional single mode fibers can be used. In fact, it has been reported that polarization-state fluctuation following 30 km single mode fiber transmission was small and slow enough to be easily suppressed by automatic polarization control devices [2.84]. This method presents the obvious advantage of using well developed low-loss single mode fibers without the need for any structural modification. Phase compensation, basically simulation of the operation of half- and quarter-wave plates, adjusts the polarization state of the transmitted signal to that of the local oscillator light. After extracting a fraction of the transmitted light and detecting its polarization state, phase compensation is accomplished by the feedback control of newly introduced birefringence. The additional birefringence is induced, for example, by applying stress on the optical fiber or by controlling birefringence in a dedicated electro-optic crystal.

Third is the use of the polarization-diversity receiver in which two orthogonal polarization components (modes) of the received optical signal are heterodyne or homodyne detected separately and then added later electrically (in heterodyne receivers, at its intermediate frequency (IF) stage) after an appropriate phase compensation [2.85]. The sum of the two is virtually independent of the SOP of the received signal. In addition, various versions and combinations of these two basic schemes are being investigated. Important parameters in all polarization control schemes are response speed, insertion loss, output stability and device size. Continuous follow up of the input signal polarization state is also an important requirement.

2.5.1   Polarization-state control

All of the polarization-state control schemes proposed so far consist basically of two controlling elements, because the number of degrees of freedom of a polarization state is two, i.e. the ellipticity and deflection angle. The block diagram of the control system using the electromagnetic fiber squeezers was proposed [2.87] as the first proposal of this sort. Suppose that the incident light is elliptically polarized as shown in Fig. 2.16(a) [2.86]. In the scheme, the first electro-optic crystal EOC1 converts, by controlling its birefringence with respect to the x and y directions, the incident polarization state to an 'upright' elliptical polarization (see the centre lower figure in Fig. 2.16(a)). Such a polarization state has the feature that electric field components $E_{x'}, = E_{y'}$, where x' and y' denote coordinates tilted by 45°. Therefore, by controlling the birefringence of the second electro-optic crystal EOC2, which is tilted by 45°, we can convert the polarization state to a linear polarization in the x or y direction. The above explanation applies also to scheme [2.87], using two fiber squeezers; hereafter this polarization-state conversion principle is called Type 1. The device proposed in [2.88], tentatively called here rotatable fiber coils, is based upon a somewhat different principle which is shown in Fig. 2.16(b). The first coil FC1 gives, by bend-induced birefringence, a 90° phase difference to two orthogonal modes, whereas the second coil FC2 gives a 180° phase difference. Hence, as illustrated in Fig. 2.16(b), any elliptical polarization can first be converted, by adjusting the tilt angle of FC1, to a tilted linear polarization. It can then be rotated to another linear polarization with an arbitrary angle by controlling the tilt angle of FC2, because

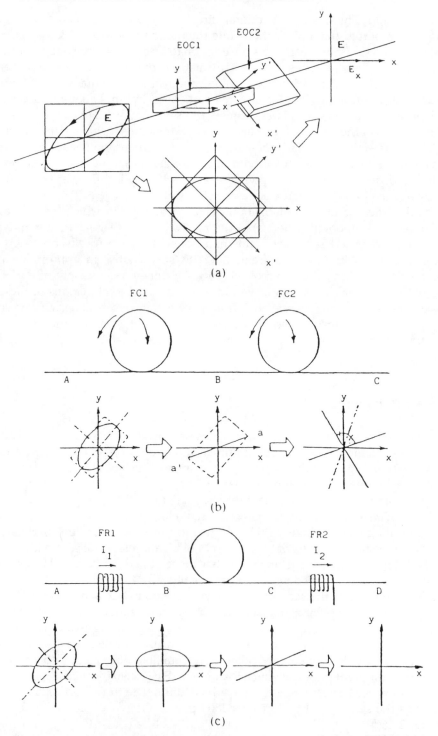

Fig. 2.16 Three types of polarization-state control schemes. (From [2.86] Okoshi (1985) *J. Lightwave Technol.* **LT-3**(6), 1232–1237. Reproduced by permission. ©1985 IEEE.)

a rotatable 180° phase plate functions as a simple deflection-angle rotator for linearly polarized incident light. Such a polarization-state conversion principle is tentatively called Type 2.

A scheme in which quarter-wave and half-wave plates are used in place of the rotatable fiber coils was reported [2.89]. The principle of its operation is entirely identical to that of the device in [2.88], i.e., the Type 2 conversion principle. However, the phase plates have a great advantage in that the rotation is endless in contrast to the limited tilt angle in the device in [2.88].

The third polarization-state conversion scheme shown in Fig. 2.16(c), Type 3 conversion, is used in the Faraday-rotator device.

The requirement for low insertion loss can in practice be satisfied only by all-fiber-type devices. This is because a polarization-state controller is most probably used in a single mode fiber circuit, in which the loss will increase appreciably (at least several decibels, or more) if a non-fiber element is inserted. An all-fiber-type device can be spliced with much lower insertion loss (≈ 0.2 dB). Endlessness in control is an important requirement because the nature of the polarization-state fluctuation in actual fiber channels is not known, and hence 'resetting' might become necessary in those attempts in which the control range is limited. Among those devices only the phase-plate scheme can satisfy this requirement. All the mechanical schemes have more or less the possibility of mechanical fatigue.

The options for providing polarization matching are also restricted by the requirements of field implementation. In a 565 Mbit/s DPSK field deployment experiment [2.42], an automated polarization control system was applied to the output from the LO source. Mechanical stress is again used to vary the birefringence in the transducer; but in this case, strain levels as low as 10^{-4} are sufficient, and are unlikely to damage the fiber. It is applied via piezoelectric cylinders on which the polarization-maintaining (PM) fiber is wound. The implementation is analogous to that of the four squeezers spliced in series, with the principal axes of the fiber cores of adjacent transducers mutually aligned at 45° rotation. A software-driven algorithm regulated the bias allied to each transducer to maximize the IF signal at the output of the receiver, while providing a limitless range of polarization adjustment. Compared with the $LiNbO_3$ technique, the algorithm is slightly slower since four transducers are involved; but the insertion loss is lower (3 dB, and potentially < 1dB), and the 10 W reflectively of the splices allowed the transducers to be placed in the LO arm of the receiver without the use of intermediate optical isolation; also, there is no need for hermetic packaging. The control algorithm used in the field demonstration system was implemented entirely in software; hence its speed of response was slow (≈ 1 s for small excursions in the Poincaré sphere, ≈ 1 min for larger variations involving adjustments to all the transducers). This was adequate to compensate for ambient variations in an otherwise undisturbed fiber, but too slow for staff-induced perturbations. Nevertheless, such changes would be expected to occur only at acoustic rates [2.90, 2.91], and the use of commercial microprocessor-based hardware to drive the dithering of the transducers is expected to increase the speed of response by three orders of magnitude. There is thus every prospect that continuous matching between the states of polarization of signal and LO can indeed be provided for practical field systems using polarization control.

2.5.2   Theoretical analysis of polarization control [2.92]

The state of polarization (SOP) at the end of a long conventional single mode fiber is subject to slow, but potentially large, changes. To ensure reliable communication without interruptions, the tracking range of the polarization control system should be endless. Generally SOP control systems contain one or several retarders, i.e., birefringent elements. They transform the incident SOP by imposing a phase delay between one fundamental polarization mode, which is subsequently referred to as an eigenmode, and the orthogonal eigenmode.

This section describes systems with retarders whose retardations or amounts or birefringence are changed electrically but whose eigenmodes or orientations of birefringence remain fixed [2.92].

DESCRIPTION OF POLARIZATION

The 2×1 Jones vectors and 2×2 Jones matrices [2.93] are widely used for the description of polarization, whereas the Poincaré sphere [2.94] provides easy insight into problems. Both formulations shall be used in this book.

The electrical field of a lightwave at a fixed point can be expressed by the complex vector

$$\mathbf{E}(t) = \begin{bmatrix} E_x(t) \\ E_y(t) \end{bmatrix} = \begin{bmatrix} E_x \\ E_y \end{bmatrix} \exp(j\omega t) = \mathbf{E} \cdot \exp(j\omega t) \tag{2.82}$$

where $\mathbf{E} = [E_y, E_y]^T$ is the (x,y) Jones vector for the orthogonal polarization components in the x- and y-directions. In an optical heterodyne or homodyne receiver the signal and local oscillator fields $\mathbf{E}_S$ and $\mathbf{E}_{LO}$ are superposed in a coupler and squared at the detector. The photocurrent is

$$i(t) = c|\mathbf{E}_S(t) \quad + \mathbf{E}_{LO}(t)|^2$$

$$= c\left\{|\mathbf{E}_S|^2 \quad + |\mathbf{E}_{LO}|^2 + 2|\mathbf{E}_S \cdot \mathbf{E}_{LO}^*|\cos\left(\omega_{IF} t + \varphi_{IF}\right)\right\} \tag{2.83}$$

where

$$\omega_{IF} = \omega_S - \omega_{LO}$$

$$\sin(\varphi_{IF}) = \mathrm{Im}(\mathbf{E}_S \cdot \mathbf{E}_{LO}^*)/|\mathbf{E}_S \cdot \mathbf{E}_{LO}^*|$$

$$\cos(\varphi_{IF}) = \mathrm{Re}(\mathbf{E}_S \cdot \mathbf{E}_{LO}^*)/|\mathbf{E}_S \cdot \mathbf{E}_{LO}^*|.$$

The constant c depends on the field distributions, wavelength, and quantum efficiency of the detector. The normalized electrical power of the IF signal is now defined as the intensity I:

$$I = \frac{|\mathbf{E_S} \cdot \mathbf{E_{LO}^*}|^2}{|\mathbf{E_S}|^2 \cdot |\mathbf{E_{LO}}|^2}. \tag{2.84}$$

Put another way, a light wave $\mathbf{E_S}$ has the normalized power I after passing through an elliptical polarizer which has the transmitted eigenmode $\mathbf{E_{LO}}$.

THE POINCARÉ SPHERE

The Poincaré sphere (Fig. 2.17 [2.92]) uniquely represents each SOP by a point on its surface. The equator carries all linear SOP's like horizontal (H) and vertical (V). P and Q are inclined by $\pm 45°$ to the horizontal. The poles refer to right (R) and left (L) circular polarization. An elliptical SOP has a major axis azimuth α and ellipticity $\beta = \tan(b/a)$ where b and a are the minor and major axes, respectively. On the sphere this is represented by the spherical coordinates 2α and 2β. The equivalent to (2.84) is

$$I = \cos(\widehat{P_S P_{LO}}/2)^2 \tag{2.85}$$

where $\widehat{P_S P_{LO}}$ is the angle between the corresponding points on the sphere. If a light wave passes through a retarder or birefringent medium, its SOP is generally altered. Only two SOPs, the eigenmodes, remain unchanged. A linear retarder with horizontal and vertical linear eigenmodes is represented by the Jones matrix (transition matrix for the Jones vector)

$$\mathbf{L}(0°, d) = \begin{bmatrix} \exp(jd/2) & 0 \\ 0 & \exp(-jd/2) \end{bmatrix} \quad \text{eigenmodes} \begin{array}{c} 1 \\ 0 \end{array}, \quad \text{and} \begin{array}{c} 0 \\ 1 \end{array} \tag{2.86}$$

where d is the phase difference or retardation between the eigenmodes. On the Poincaré sphere this retarder (type A) transforms the incident SOP by an anticlockwise turn of angle d around the HV axis. Examples for this linear retarder are fiber squeezers [2.95, 2.96] with $0°$ azimuth or, in integrated optics, phase shifters [2.97, 2.98].

A linear retarder whose eigenmodes are inclined by $\pm 45°$ with respect to the horizontal (type B) is given by:

$$\mathbf{L}(45°, d) = \begin{bmatrix} \cos(d/2)\, j\, \sin(d/2) \\ j\, \sin(d/2)\cos(d/2) \end{bmatrix} \quad \text{eigenmodes } 1/\sqrt{2}\begin{array}{c} 1 \\ 1 \end{array}, \text{and } 1/\sqrt{2}\begin{array}{c} 1 \\ -1 \end{array} \tag{2.87}$$

Such a retarder turns any SOP around the PQ axis. Fiber squeezers of $45°$ azimuths or integrated optical TE/TM convertors [2.97, 2.98] may be used. Other types of retarders are the circular retarder (type C), e.g., realized by Faraday rotators or a rotation of the coordinate system, or in the general case the elliptical retarder. One retarder type can be realized by another type if it is placed between two retarders of the third type which have $\pm \pi/2$ retardation [2.95]. For example, a fiber squeezer with $45°$ azimuth (type B) may be looked upon as fiber squeezer of $0°$ azimuth (type A) between two fiber sections in which the coordinate system is rotated by $\pm 45°$,

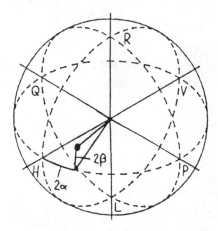

Fig. 2.17 Poincaré sphere. (From [2.92] Noe *et al.* (1988) *J. Lightwave Technol.* **LT-6**(7), 1199–1208. Reproduced by permission. ©1988 IEEE.)

corresponding to $\pm \pi/2$ circular retardation (type C). If a light wave passes through a retarder, the intensity defined by (2.84) or (2.85) can be shown to vary sinusoidally as a function of the retardation d. If the input SOP of the retarder or the input SOP analysed beyond the retarder is an eigenmode, then the intensity will stay constant. If both SOPs are eigenmodes, the intensity will stay constant either at 0 or 1.

OPERATION OF OPTICAL RETARDER

It shall be assumed now that the SOP controllers are entirely situated between the LO and the (polarization insensitive) coupler. The SOP of the LO may be unknown but is certainly constant. The polarization controllers can also be placed between the coupler and the receiver input, if their order in the light path is inverted. The first experimental system [2.99, 2.100, 2.101] had four retarders, only of types A and B. It was tried out both in a heterodyne transmission system and in an arrangement with one laser and a variable SOP analyser. No fundamental differences were observed between the two set-ups. Meanwhile polarization control systems with only three retarders of two types have been proposed [2.102, 2.103]. The experimental system [2.102] shall now be described in detail. It uses one of three possible reset algorithms. Many other retarder configurations and retardation range limits are possible.

The configuration is sketched in Fig. 2.18 [2.92] and a set of SOP transformations is shown in Fig. 2.19 [2.92]. The horizontal input SOP (Jones vector $[1,0]^T$) is transformed by three retarders of retardation d_1, d_2, d_3 and of types B, A, B, respectively. The output SOP is given by matrix multiplications as

$$\mathbf{E}(d_1,\ d_2,\ d_3) = \mathbf{L}(45°,\ d_3) \cdot \mathbf{L}(0°,\ d_2) \cdot \mathbf{L}(45°,\ d_1) \cdot \begin{bmatrix} 1 \\ 0 \end{bmatrix}$$

$$= \begin{bmatrix} \cos(d_3/2)\ \exp(jd_2/2)\ \cos(d_1/2) - \sin(d_3/2)\ \exp(-jd_2/2)\ \sin(d_1/2) \\ j\ \sin(d_3/2)\ \exp(jd_2/2)\ \cos(d_1/2) + j\ \cos(d_3/2)\ \exp(-jd_2/2)\ \sin(d_1/2) \end{bmatrix} . \tag{2.88}$$

If the third retarder is absent or $d_3=0$, (2.91) simplifies to

$$\mathbf{E}(d_1,\ d_2,\ 0) = \left.\begin{array}{l} \exp(jd_2/2)\ \cos(d_1/2) \\ j\ \exp(-jd_2/2)\ \sin(d_1/2) \end{array}\right] . \qquad (2.89)$$

The retardation ranges are now chosen between the limits $k\pi$ and $(k+1)\pi$ for d_1 and between the limits $i\pi$ and $(i+2)\pi$ for d_2, where k,i are integers. The choice allows generation of any desired output SOP. This polarization matching may conveniently be implemented by an automatic control system which modulates ('dithers') d_1 and d_2 around the operation points. The corresponding intensity fluctuations given by (2.89) or (2.88) and (2.84) allow one to determine the intensity gradient, i.e., the partial derivatives $\partial I/\partial d_i$ with respect to the retardation. Integral controllers change the operation points in the direction of increasing intensity to achieve the best matching $I=1$. Alternatively, if the direction of the gradient is inverted, the system will lock onto $I=0$. This minimum is not desired for optical communications but is useful to check for small intensity losses while being insensitive to optical power fluctuations.

The endlessness of the tracking range is established by the following equations which hold at the properly chosen d_1 and d_2 range limits:

$$\mathbf{E}(k\pi,\ d_2+x,\ d_3) = \exp(j(-1)^k\chi/2)\ \mathbf{E}(k\pi,\ d_2,\ d_3) \qquad (2.90)$$

$$\mathbf{E}(k\pi+x,\ d_2\pm\pi,\ d_3) = j(-1)^k\ \mathbf{E}(k\pi-x,\ d_2,\ d_3) \qquad (2.91)$$

$$\mathbf{E}(d_1-(-1)^ix, i\pi,\ d_3+x) = \mathbf{E}(d_1,\ i\pi,\ d_3). \qquad (2.92)$$

These equations mean that the generated SOP theoretically will not change if the retardations are changed by appropriate reset procedures. Only the light-phase may change, which does not deteriorate the function of the optical receiver, as long as the phase changes are slow compared to the bit rate.

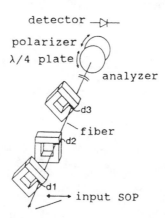

detector

polarizer

$\lambda/4$ plate

analyzer

d3

fiber

d2

d1

input SOP

Fig. 2.18 Schematic of three-retarder system. (From [2.92] Noe *et al.* (1988) *J. Lightwave Technol.* **LT-6**(7), 1199–1208. Reproduced by permission. ©1988 IEEE.)

If d_1 reaches a range limit, e.g., the point $k\pi$, the SOP is not altered by any variations of d_2, because it is an eigenmode. If d_1 has passed the point $k\pi$, d_2 is incremented or decremented by π. The direction is chosen so as not to exceed the d_2 range. As the desired effect this action mirrors the d_1 scaling at the range limit $k\pi$ (2.91). If the tracking system has just moved d_1 beyond the range limit, d_1 will return subsequently within range.

As an alternative d_1 may simply be limited slightly before reaching the nominal range limit. In this case small areas around points H and V of the Poincaré sphere are not accessible by the system. Meanwhile the intensity optimization algorithm will find out how to change d_2 for the best intensity.

If d_2 reaches a range limit, e.g., the point $i\pi$, the reset procedure is more complicated. First d_1 and d_3 simultaneously move in opposite k (if i is even) or equal (if i is odd) directions until d_1 reaches the nearest range limit (2.92). During this operation d_3 takes over the function of d_1. The SOP at the second retarder becomes one of its eigenmodes and d_2 may be changed (2.90) by 2π away from the range limit, i.e. to the other range limit $(i+2)\pi$. At the end d_1 and d_3 are simultaneously led back to their former operation points (2.92). Frequent resets or system blocking are prevented by an appropriate switching hysteresis at the range limits.

AN ERROR-TOLERANT FOUR RETARDER

The system mentioned above is made error tolerant by adding another type A retarder of retardation d_0 at the SOP transformer input [2.89]. The retardation range of d_1 including range limits must be transferred by $\pi/2$. The SOP transformation is directly comparable to Fig. 2.17. The input SOP is no longer horizontal but linear in the 45° direction (point P_0).

Unlike before, this system modulates all four retardations. Integral controllers act on d_1 and d_2 which theoretically can generate any desired output SOP. Non-ideal system behaviour such as imperfect optical elements or input polarization deviations might prevent the system from exactly reaching all required SOPs. Additional proportional first order low-pass controllers acting on d_0 and d_3 overcome this problem. The mean operation point of d_3 is arbitrarily fixed to 0 [2.92].

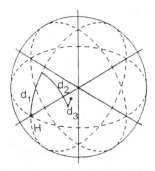

Fig. 2.19 SOP transformations of three-retarder system. (From [2.92] Noe *et al.* (1988) *J. Lightwave Technol.* **LT-6**(7), 1199–1208. Reproduced by permission. ©1988 IEEE.)

2.6 POLARIZATION/PHASE DIVERSITY RECEPTION FOR COHERENT TRANSMISSION

2.6.1 Principles of diversity reception

PHASE-NOISE AND POLARIZATION-FLUCTUATION [2.104]

In a conventional single-detector ASK receiver, a linear state of polarization (SOP) of both the LO and the received signal is assumed. The photodetector current is given by [2.104]

$$i_{\text{Homodyne}} = dA\cos\theta\cos\phi \tag{2.93}$$

$$i_{\text{Heterodyne}} = dA\cos\theta\cos(\omega t + \phi) \tag{2.94}$$

where d is the non-return-to-zero (NRZ) binary signal (it is equal to either zero or one); $A \equiv R\sqrt{P_S P_{LO}}$; R is the detector responsivity; P_S is the signal power on the detector surface; P_{LO} is the local oscillator power on the detector surface; θ is the angle between the SOPs of the signal and local oscillator; ϕ is the combined phase noise of the transmitter and local oscillator; and ω is the intermediate frequency in radians per second.

$$\phi = \int_{-\infty}^{t} \dot{\phi}(t_1) dt_1 \quad \text{(rad)}$$

where $\dot{\phi}(t)$ is the instantaneous angular frequency noise in radians per second. The frequency noise $\dot{\phi}(t)$ can be modelled [2.105–2.107] as a white zero-mean Gaussian random process with the power spectral density (PSD) given by

$$S_{\dot{\phi}}(f) = 2\pi\Delta\nu \quad \text{(rad}^2\text{ Hz)} \quad -\infty < f < \infty \tag{2.95}$$

where $\Delta\nu$ is the full width half maximum (FWHM) linewidth at the 'IF,' i.e.

$$\Delta\nu = \Delta\nu_T + \Delta\nu_{LO} \tag{2.96}$$

where $\Delta\nu_T$ and $\Delta\nu_{LO}$ are the linewidths of the transmitter and local oscillator, respectively. The PSD shape implies the Lorentzian laser lineshape [2.105–2.107].

Inspection of (2.93) reveals that if either θ or ϕ are equal to 90°, then the homodyne current is equal to zero, and all the data are lost. In the case of heterodyning (2.94), the phase noise is not a problem if the intermediate frequency and the bandwidth of the IF filter are sufficiently large [2.106, 2.108, 2.111]: $\cos(\omega t + \phi)$ will go through a maximum at least once per bit, thus ensuring data recovery. However, polarization misalignment remains a problem.

Both phase-noise and polarization-fluctuation problems can be resolved with diversity receivers (Fig. 2.20) [2.104]. In all such receivers, the signal and the LO fields are combined to produce two or more output fields $\{E_k\}$. The fields $\{E_k\}$ are linear combinations of E_S and E_{LO}. We call an optical device used to generate the fields HE_k from the fields E_S and E_{LO} an 'optical hybrid'. The fields $\{E_k\}$ are detected separately.

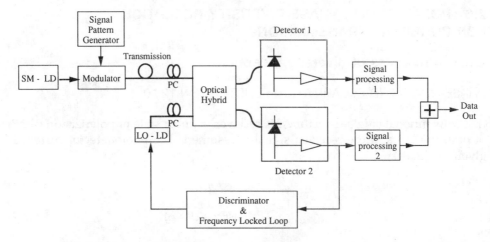

Fig. 2.20 Schematic block of diversity coherent receivers.

OPTICAL HYBRIDS FOR PHASE—DIVERSITY HOMODYNE RECEIVERS

In phase-diversity homodyne receivers, the SOPs of the received signal and of the LO must be the same ($\theta = 0$); this is achieved by means of polarization control devices. Assuming that the SOPs of both input fields are linear, the output fields of symmetric hybrids should be equal to the following.
    In the case of two branches:

$$E_1 = \sqrt{0.5L}\ (E_S + E_{LO}) \tag{2.97a}$$

$$E_2 = \sqrt{0.5L}\ (E_S + jE_{LO}). \tag{2.97b}$$

In the case of three or more branches:

$$E_k = \sqrt{L}\,[E_S + E_{LO}\exp(j360°\ (k-1)/N)/\sqrt{N}] \qquad 1 \leqslant k \leqslant N \tag{2.98}$$

where E_k is the output field of the kth branch, $-10\log L$ is the loss in decibels, and $N > 3$ is the number of branches. An optical device satisfying (2.97) is called the 90° hybrid; an optical device satisfying (2.98) with $N = 3$ is called the 120° hybrid. When the hybrid output fields are detected, the following signal currents appear in the ASK modulation format.
    In the case of two branches:

$$i_1 = LdA\ \cos\ \phi \tag{2.99a}$$

$$i_2 = LdA\ \sin\ \phi. \tag{2.99b}$$

In the case of three or more branches:

$$i_k = (2L/N)dA\ \cos[\phi + 360°\ (k-1)/N] \tag{2.100}$$

where $A \equiv R\sqrt{P_S P_{LO}}$, and the powers are measured at the hybrid inputs. Inspection of (2.102) and (2.103) reveals that the currents $\{i_k\}$ are squared and added together. The result is completely independent of the phase noise and is equal to

$$i_{TOT} = \Sigma i_k^2 = (2L^2/N)R^2 d^2 P_S P_{LO}. \tag{2.101}$$

Expression (2.101) shows that the signal i_{TOT} is inversely proportional to the number of branches N, so that the larger the N, the larger the local oscillator power needed to suppress the receiver noise.

A two-branch receiver is the most advantageous to keep the number of branches N small; the $N=2$ diversity receiver is analogous to the optimum receiver structure for a signal with random phase in Gaussian noise [2.109].

Two-branch receivers have two disadvantages.

(1) 90° hybrids needed for two-branch receivers are inherently lossy devices: the loss of a symmetric 90° hybrid cannot be smaller than 2.3 dB [2.110], and all known implementations have a 3 dB loss.

(2) Receiver imperfections (such as polarization misalignment [2.111], local oscillator intensity noise [2.112] and reflection [2.113] gave a stronger effect on two branch receivers than on receivers with $N>2$, and lead to larger sensitivity penalties.

Three-branch receivers ($N=3$) require more electronic signal processing hardware and a larger local oscillator power than two-branch receivers. In return, three branch receivers offer two advantages.

(1) 120° hybrids needed for three-branch receivers have no intrinsic loss and can be implemented as low-loss 3×3 couplers.

(2) Three-branch receivers are more tolerant to receiver imperfections than two-branch receivers.

For three-branch receivers, ideal symmetric lossless 3×3 couplers can, in principle, be used as 120° hybrids. Unfortunately, practical 3×3 couplers can suffer from several imperfections, including loss, non-uniform distribution of input power among the output ports, and polarization dependence. As a result of these imperfections, practical 3×3 couplers are generally not guaranteed to act as 120° hybrids. One can, however, find selected 3×3 couplers that are sufficiently close to the ideal performance to be used as 120° hybrids [2.114].

Four-branch receivers ($N=4$) require four photodetectors which generate four photocurrents; $i_1 = A \cos \phi$; $i_2 = A \sin \phi$; $i_3 = -A \cos \phi$; and $i_4 = -A \sin\}$. Consider the quantities $i_{13} \equiv i_1 - i_3 = 2A \cos \phi$ and $i_{24} \equiv i_2 - i_4 = 2A\sin\phi$, which can be easily generated by the back-to-back connection of the photodiodes (1 with 3, and 2 with 4, respectively), the so-called balanced configuration. If i_{13} and i_{24} are squared and added together, the result is independent from the phase noise, as in other phase-diversity receivers. Thus, a balanced four-branch receiver combines the best features of both phase-diversity and balanced receivers: it protects the system against the phase noise, the intensity noise of the local oscillator, and reflections in the local oscillator path. Such a receiver needs four photodetectors, but only two sets of RF hardware (amplifiers, filters, and squarers); in other words, it has four optical branches and two electronic branches. A hybrid for such a receiver has no intrinsic loss, so that both signal and LO powers are used efficiently. However, a four-branch receiver requires a larger LO power than either two-branch or three-branch receivers. The currently available knowledge suggests that receivers with larger number of branches ($N>5$) do not offer any additional advantages judging from (2.104), their performance is expected to be worse than that of their counterparts with $N<5$.

OPTICAL HYBRIDS FOR POLARIZATION-DIVERSITY HETERODYNE RECEIVERS

A bulk-optics hybrid can be used in a polarization-diversity receiver. The polarization controller on the LO path is adjusted to produce equal LO powers at the output E_1 and E_2; it does not have to be readjusted as the signal polarization varies in time. The polarization controller on the signal path is not needed and can be omitted.

A fiber hybrid can be more attractive in practical applications. The two fiber optical polarization analysers are orthogonal to each other, and the polarization controller on the LO path is adjusted to produce equal LO powers at the outputs E_1 and E_2. The main advantage of the first approach is that the losses can be allocated at will between the signal and the LO paths; if a 3 dB coupler is used, then both signal and LO suffer a 3 dB loss.

The main advantage of the second approach is that the entire professing occurs in the single mode fiber domain, so that insertion losses and alignment problems associated with bulk optics are eliminated; if a 3 dB coupler and polarization analysers are used, then both signal and LO suffer a 3 dB loss. The output currents generated by the two detectors are identical in both cases; for the ASK format, the currents are

$$i_1 = \sqrt{2LRd}\ \sqrt{P_S P_{LO}}\ \cos\theta\ \cos(\omega t + \phi) \tag{2.102a}$$

$$i_2 = \sqrt{2LRd}\ \sqrt{P_S P_{LO}}\ \sin\theta\ \cos(\omega t + \phi + \psi) \tag{2.102b}$$

where the signal was for simplicity assumed to be linearly polarized with an angle θ with respect to polarization of E_1, and ψ is an arbitrary angle. The currents i_1 and i_2 are squared and added together. If the intermediate frequency and the IF bandwidth are much larger than the signal bandwidth stemming from the modulation, phase noise and polarization fluctuations, then the lowpass part of the result is [2.104]:

$$i_{TOT} = \text{lowpass}\ \{i_1{}^2 + i_2{}^2\} = L^2 R^2 dP_S P_{LO} \tag{2.103}$$

where lowpass $\{x\}$ denotes the lowpass part of x. Note that the result is independent both of polarization (of θ) and of the phase noise (of θ); it can be shown that (2.103) is valid for an arbitrary polarization of the signal.

A hybrid polarization diversity receiver does not have to have any intrinsic loss whatsoever, and L can, in principle, be equal to unity. Indeed some implementations

[2.115] have a 0 dB intrinsic loss, of course the actual implementation loss is larger than 0 dB.

2.6.2 Diversity receivers

POLARIZATION DIVERSITY RECEPTION

In practical coherent optical communication systems, it is a crucial step to cope with the fluctuations of state of polarization (SOP) in the received signal.

In the polarization diversity optical receivers, a lightwave signal with an arbitrary state of polarization (SOP) is split into two orthogonal polarization components, which are separately mixed with a local oscillator lightwave signal having constant SOP and then detected separately. After that the detected signals are processed and combined electronically.

In such a receiver, two types of electronic combining method can be used. One is the IF combining scheme [2.85]. In this scheme, two IF signals obtained from two orthogonal polarizations are combined after phase adjustment, and then the combined IF signal is demodulated. The other is the baseband combining scheme. In this scheme two IF signals are demodulated independently, and then the demodulated (baseband) signals are added together. This type of polarization diversity receiver for the DPSK scheme was proposed in [2.116]. A similar approach of this type but with a phase-diversity scheme was proposed independently in [2.120]. Recombination at the IF stage is difficult because the phases of the two IF signals must be well matched, while recombination at the baseband stage after demodulation is promising because this does not require strict phase matching. Since the baseband combining scheme needs no phase adjustment at the IF stage, it seems more practical than the IF combining scheme. However, it is difficult to extract a stable IF signal for AFC.

A polarization-diversity reception using dual-balanced receivers [2.117] and tested in a 1.2 Gbit/s optical DPSK transmission experiment is presented as an example of this scheme [2.118]. The system configuration for the 1.2 Gbit/s transmission experiment is shown in Fig. 2.21 [2.118]. The polarization diversity receiver is enclosed by dotted lines. The received signal and local laser output were divided by separate polarization beam splitters (PBS) into two orthogonally polarized components. Each PBS has an extinction ratio of more than 20 dB and an insertion loss of about 2 dB. Signal and local laser components of the same polarization were mixed in the polarization-maintaining fiber couplers and with insertion loss of less than 0.4 dB. The output pair from the co-couplers was detected by a pair of dual-balanced optical receivers. The signals generated by the dual-balanced optical receivers were demodulated separately in a delay-line demodulator using a double balanced mixer and then summed to generate the polarization-insensitive baseband signal.

The causes of receiver sensitivity degradation in the polarization-diversity receiver and the power penalties are discussed. First, the thermal noise of the front-end circuit becomes severe because the local laser power is shared by two balanced receivers and is reduced by the insertion loss of the polarization beams splitter and optical

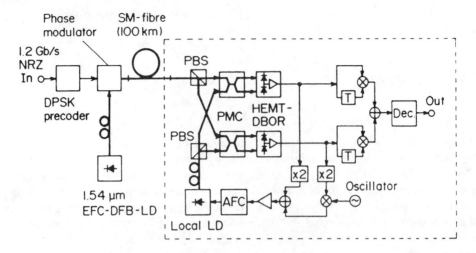

Fig. 2.21 DPSK transmission using polarization diversity receiver. (From [2.118] Chikama *et al.* (1990) *J. Lightwave Technol.* **LT-8**(3), 309–321. Reproduced by permission. ©1990 IEEE.)

coupler. In the polarization-diversity experiment, the effective local laser power was +0.6 dBm, considering the local laser output power of +8.5 dBm, the insertion loss of the optical circuits was 2.4 dB, and the quantum efficiency of the dual-balanced optical receiver was 0.55. On the other hand, the effective local laser power was +3.6 dBm for the experiment without diversity. Considering these parameters, the power penalty was 1.3 dB can be attributed to the lack of local laser power. To reduce the power penalty, a low noise front-end, low insertion-loss optical circuits, and a high-power local laser are needed.

PHASE DIVERSITY RECEPTION

Phase diversity reception (or multiport detection) is based on techniques developed for microwave measurements and was first applied to optical frequencies by Walker and Carroll [2.119]. In optical phase diversity the local oscillator laser is operated at a frequency comparable to the incoming frequency, but the need for phase locking of the optical local oscillator is avoided and only frequency locking is required. Therefore it avoids the major problem with homodyne detection. The receiver sensitivity of phase diversity, at best, is the same as heterodyne detection. So the intensive research in phase diversity techniques generated many interesting results [2.111, 2.112, 2.114, 2.120–2.122].

In particular, it showed that phase diversity reception can tolerate wide laser linewidth, even when the linewidth is of the same order of magnitude as the bit rate. In addition, phase diversity receivers operate in the baseband part of the frequency spectrum. These advantages allow construction of phase diversity receivers from commercially available components: DFB lasers and p–i–n/FET modules. These advantages, of course, are achieved at the expense of greater complexity. In addition, phase diversity receivers require exact optical and electrical

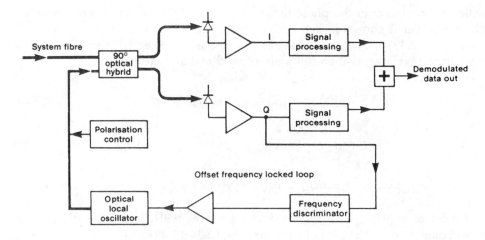

Fig. 2.22   Phase diversity receiver: two phase. (From [2.49] Davis *et al.* (1987) *J. Lightwave Technol.* **LT-5**(4), 561–572. Reproduced by permission. ©1987 IEEE.)

signal processing and show large penalties for deviations from the ideal characteristics [2.111, 2.112].

Since the hybrids must be symmetrical, the signal and LO are split two or three ways. This leads, under shot-noise limited conditions, to a 3 dB (2 times) sensitivity penalty for two-phase processing (compared with homodyne, with a coupler giving negligible loss in the signal path) and a 4.8 dB (3 times) penalty for three-phase processing. At best, therefore, phase diversity sensitivity is the same as for heterodyne, but with the great advantage of baseband operation [2.49].

Phase diversity receivers are possible using both 90° hybrids and other types, such as three-phase 120° hybrids. In the 90° case, the receiver is similar to a Costas loop

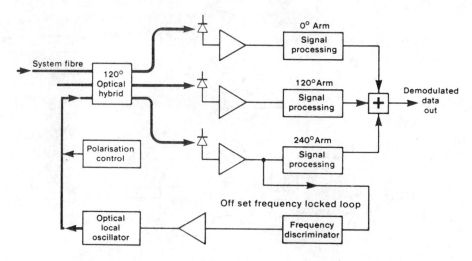

Fig. 2.23   Phase diversity receiver: three phase. (From [2.49] Davis *et al.* (1987) *J. Lightwave Technol.* **LT-5**(4), 561–572. Reproduced by permission. ©1987 IEEE.)

without completion of the phase locked loop. An outline block diagram of a two-phase receiver is shown in Fig. 2.22 [2.49].

Fig. 2.23 [2.49] shows an outline of a three-phase receiver. In this case, the offset frequency envelopes will be mutually phased at 120° intervals.

For two phase

$$i = K_1 M(t) \exp[j(2\pi f_{\text{offset}}t + n\pi/2)] \qquad n = 0,1. \tag{2.104}$$

For three phase

$$i = K_2 M(t) \exp[j(2\pi f_{\text{offset}}t + 2n\pi/3)] \qquad n = 0,1,2. \tag{2.105}$$

SIGNAL PROCESSING OF PHASE DIVERSITY RECEIVERS

The form of signal processing used in each arm will depend on the received modulation format-ASK, DPSK, or FSK (Fig. 2.24 [2.49])

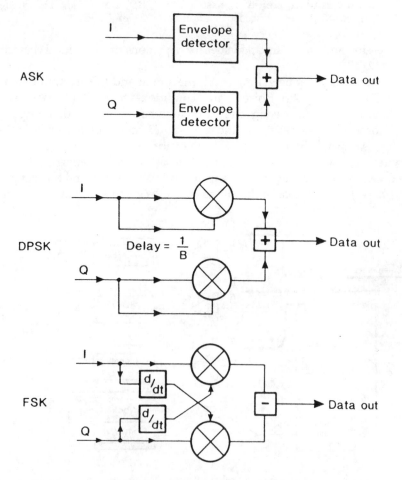

Fig. 2.24 Phase diversity receivers: demodulation. (From [2.49] Davis *et al.* (1987) *J. Lightwave Technol.* **LT-5**(4), 461–572. Reproduced by permission. ©1987 IEEE.)

(1) ASK: Envelop detection (either linear or square law) is used in each channel. The offset frequency can vary quite widely since the only distorting factor is the band limiting of the front-end receive modules and any noise filtering. If the system is dc coupled, varying through zero frequency is permissible. However, one advantage of phase diversity operation is that ac coupling can be used. In this case, a small positive offset frequency ($\approx B/100$) is required to avoid notching out the carrier. The zero notch appears in one sideband and has negligible effect.

(2) DPSK: The well known technique of splitting the signal, delaying one path by one bit, and multiplying is employed in each arm. This effectively demodulates the signal and performs differential decoding in one operation. The bit stream must be differentially encoded at the transmitter. The offset frequency is more critical than with ASK, because a penalty results if the bit rate is not integrally related to the IF frequency as

$$f_{IF} = NB/2 \qquad N = 0, 1, 2 \cdots$$

i.e. if $f_{IF} = NB/4$, $N = 1, 3, 5 \cdots$, the one-bit delayed and the undelayed carriers are in quadrature, and no output results from the multiplier. This penalty is apparent for all IF values, including zero IF.

(3) FSK: Two basic methods are available for demodulating FSK in phase diversity receivers.

(4) Discriminator: A frequency discriminator can be inserted in each channel and the outputs combined. Alternatively, there is an elegant solution for I and Q receivers that has been used at RF [2.123]. As seen in Fig. 2.36 of that reference, differentiation of each signal is required, before cross multiplication with the other channel. It is now known how well this could be implemented for high bit rate systems.

(5) Filter and envelope detectors: Filters for 'mark' and 'space' frequencies (one of which may be the 'zero IF') are inserted in each channel, envelope detected and summed. An extreme case of this is the 'FSK' single filter method [2.27] already reported at optical frequencies, where the deviation is so wide that one frequency only is received.

(6) Other functions: In addition to the demodulation process, other functions such as amplification, AGC, and noise filtering and equalization are needed. With phase diversity systems, these can all be baseband, leading to the possibility of total integrated circuit realization, and the use of existing direct detection hardware.

EXPERIMENTAL PHASE DIVERSITY RECEIVERS

An experimental demonstration of double stage phase diversity scheme using a 100 Mbit/s single filter FSK coherent optical communication system was reported [2.124]. It features phase diversity type homodyne detection followed by phase diversity type frequency up-conversion, giving a heterodyne-like IF signal. It has the advantages of both heterodyning and homodyning. The 90° optical hybrid consists of a $\lambda/4$-plate and polarization beam splitter. The result shows that the system works exactly as expected. A sensitivity 0.6 dB worse than that of a heterodyne receiver is achieved.

A 5 Gbit/s high-speed operation of a three-branch single-filter detection phase diversity homodyne system experiment has been performed [2.125]. A 3×3 optical fiber coupler, which operates as a 120° optical hybrid, has the advantages of low loss and easy polarization control, compared with an optical 90° hybrid. The demodulation circuit for up to 5 Gbit/s phase diversity reception consists of three wide-band phase-matched receivers, wide-band DC-coupled squaring circuits and a summing circuit. The receiver sensitivity for a 10^{-9} error rate is -30.5 dBm at 4 Gbit/s and -27.0 dBm at 5 Gbit/s.

CONCLUSION OF DIVERSITY RECEPTION

Polarization-diversity and phase-diversity receptions are more complex than conventional coherent receivers, and need larger LO powers and more precise signal processing. However, diversity receivers have important advantages.

(1) They are able to tolerate large polarization fluctuations if polarization diversity is employed, large LO intensity noise if a balanced receiver is employed, and wide laser linewidth.
(2) They require a smaller receiver bandwidth and slower photodetectors than heterodyne receivers if phase diversity is employed.

In the long run, the advantages of diversity receivers may outweigh the disadvantages; so, intensive research in this area is likely to continue. Balanced polarization-diversity receivers seem to be particularly attractive, and are a good candidate for further efforts.

REFERENCES

[2.1]   R. A. Linke and A. H. Gnauck (1988) "High-capacity coherent lightwave systems" *J. Lightwave Technol* **LT-6**(11), 1750–1769.

[2.2]   M. C. Teich (1971) "Quantum theory of heterodyne detection" in *Proc. Third Photo-conductivity Conf.*, ed E. M. Prell (New York: Pergamon) pp 1–5.

[2.3]   T. Okoshi (1982) "Heterodyne and coherent optical fiber communications: Recent progress" *IEEE Trans. Microwave Theory Tech.* **MTT-30**(8), 1138–1149.

[2.4]   P. S. Henry (1985) "Lightwave primer" *IEEE J. Quantum Electron.* **QE-21**(12), 1862–1879.

[2.5]   J. P. Gordon (1962) "Quantum effects in communication systems" *Proc. IRE* **50**(9), 1898–1908.

[2.6]   F. E. Goodwin (1967) "A 3.39 μm infrared optical heterodyne communication system" *IEEE J. Quantum Electron.* **QE-3**(11), 524–531.

[2.7]   O. E. DeLange (1970) "Wide-band optical communication systems: Part 2-Frequency-division multiplexing" *Proc. IRE* **58**(10), 1683–1690.

[2.8]   T. Kimura (1987) "Coherent optical fiber transmission" *J. Lightwave Technol.* **LT-5**(4), 414–428.

[2.9]   R. A. Linke and P. S. Henry (1987) "Coherent optical detection: a thousand calls on one circuit" *IEEE Spectrum*, pp 52–57.

[2.10] M. Schwartz *et al.* (1966) *Communication Systems and Techniques* (New York: McGraw-Hill) esp. ch. 7.

[2.11] J. Salz (1985) "Coherent Lightwave Communications" *AT&T Tech. J.* **64**(10), 2153–2209.

[2.12] T. Okoshi *et al.* (1981) "Computation of bit-error rate of various heterodyne and coherent-type optical communication schemes" *J. Opt. Commun.* **2**(3), 89–96.

[2.13] Y. Yamamoto (1980) "Receiver performance evaluation of various digital optical modulation–demodulation system in the 0.5–10 μm wavelength region" *IEEE J. Quantum Electron.***QE-16**, 1251–1259.

[2.14] D. Marcuse (1970] *Engineering Quantum Electronics* (New York: Harcourt Brace) ch. 6.

[2.15] M. Shikada *et al.* (1985) "1.5 μm high bit-rate long span transmission experiments employing a high power DFB-DC-PBH laser diode" *5th IOOC/11th ECOC., Oct. 1985* post-deadline paper.

[2.16] V. W. S. Chan *et al.* (1983) "Heterodyne lasercom systems using GaAs lasers for ISL applications" *IEEE Int. Conf. Commun., Boston, MA, June 19–22, 1983* paper E1–5.

[2.17] T. Okoshi (1986) "Ultimate performance of heterodyne/coherent optical fiber communications" *J. Lightwave Technol.* **LT-4**(10), 1556–1562.

[2.18] T. Okoshi (1979) Feasibility study of frequency-division multiplexing optical fiber communication systems using optical heterodyne or homodyne schemes *Tech. Group. IECE Japan, Feb. 27, 1979,* paper OQE78–139 (in Japanese).

[2.19] S. Kobayashi *et al.* (1982) "Direct frequency modulation in AlGaAs semiconductor lasers" *IEEE J. Quantum Electron.* **QE-18**(4), 582–595.

[2.20] R. C. Alferness (1981) "Guided-wave devices for optical communication" *IEEE J. Quantum Electron.* **QE-17**(6), 946–959.

[2.21] Y. Yamamoto and T. Kimura (1981) "Coherent optical fiber transmission systems" *IEEE J. Quantum Electron.* **QE-17**(6), 919–935.

[2.22] Y. Sakakibara *et al.* (1980) "Single-mode oscillation under high speed direct modulation in GaInAsP/InP integrated twin-guide lasers with distributed Bragg reflectors" *Electron. Lett.* **16**, 455–458.

[2.23] S. Kobayashi *et al.* (1980) "Single mode operation of 500 Mbit/s modulated AlGaAs semiconductor laser by injection locking" *Electron. Lett.* **16**, 746–747.

[2.24] T. Mukai and Y. Yamamoto (1981) "Gain, frequency bandwidth and saturation power of AlGaAs DH laser amplifiers" *IEEE J. Quantum Electron.* **QE-17**(6), 1028–1034.

[2.25] M. Nakamura *et al.* (1978] "Longitudinal mode behavior of mode stabilized AlGaAs injection lasers" *J. Appl. Phys.* **49**, 4644–4648.

[2.26] T. Isobe and M. Tokida (1970) "Power amplification of FM and PM signals with synchronized IMPATT oscillators" *IEEE Trans. Microwave Theory Tech.* **MTT-18**, 906–911.

[2.27] S. Yamakazi *et al.* (1986) "Long span optical FSK heterodyne single-filter detection transmission equipment using a phase tunable DFB laser diode" *Electron. Lett.* **22**, 5–7.

[2.28] C. F. Buhrer *et al.* (1962) "Optical frequency shifting by electrooptic effect" *Appl. Phys. Lett.* **1**, 46–49.

[2.29] W. J. Thaler (1964) "Frequency modulation of He-Ne laser beam via ultrasonic waves in quartz" *Appl. Phys. Lett.* **5**, 29–31.

[2.30] R. Welter *et al.* (1989) "Sixteen-channel coherent broadcast network at 155 Mbit/s" *J. Lightwave Technol.* **7**(10), 1438–1444.

[2.31] K. Emura *et al.* (1990) "An optical FSK heterodyne dual filter detection system for taking advantage of DFB LD applications" *J. Lightwave Technol.* **8**, 243–250.

[2.32] F. K. Reinhart and R. A. Logan (1980) "Electrooptic frequency and polarization modulated injection laser" *Appl. Phys. Lett.* **36**, 954–957.

[2.33] J. E. Ripper *et al.* (1966) "Direct frequency modulation of a semiconductor laser by ultrasonic waves" *IEEE J. Quantum Eectron.* **QE-2**, 603–605.

[2.34] S. Kobayashi *et al.* (1981) "Modulation frequency characteristics of directly optical frequency modulated AlGaAs semiconductor laser" *Electron. Lett.* **17**, 350–351

[2.35] S. Kobayashi *et al.* (1981) "FM signal amplification characteristics of injection locked type and Fabry–Perot type semiconductor laser amplifiers" *Ann. Meeting IECE Japan, Mar. 1981.*

[2.36] S. Kobayashi and T. Kimura (1980) "Injection locking characteristics of an AlGaAs semiconductor laser" *IEEE J. Quantum Electron.* **QE-16**, 915–917.

[2.37] I. P. Kaminow and L. W. Stalz (1975) "Efficient strip-guide modulator" *Appl. Phys. Lett.* **27**, 555–557.

[2.38] W. R. Bennet and J. R. Davey (1965) *Data Transmission* (New York: McGraw-Hill).

[2.39] E. D. Sunde 1969 *Communication Systems Engineering Theory* (New York: Wiley).

[2.40] K. Miyauchi *et al.* (1975) "A trial 4-phase PSK millimeter wave transmission system" *Rev. ECL* **23**, 55–66.

[2.41] H. Yamamoto and K. Kohiyama (1974) "Construction and overall performance of 20 GHz digital radio repeat" *Rev. ECL* **22**, 571–578.

[2.42] M. C. Brain *et al.* (1990) "Progress towards the field deployment of coherent optical fiber systems" *J. Lightwave Technol.* **8**(3), 423–437.

[2.43] J. Mellis *et al.* (1988) "Miniature packaged external-cavity semiconductor laser with 50 GHz continuous electrical tuning range" *Electron. Lett.* **24**, 988–989.

[2.44] T. Imai *et al.* (1990) "Over 300 km CPFSK transmission experiment using 67 photon/bit sensitivity receiver at 2.5 Gbit/s" *Electron. Lett.* **26**, 357–358.

[2.45] S. Ogita *et al.* (1989) "Long-cavity, multiple-phase-shift, distributed feedback laser for linewidth narrowing" *Electron. Lett.* **25**, 629–630.

[2.46] J. M. Kahn (1989) "1 Gbit/s PSK homodyne transmission system using phase-locked semiconductor lasers" *IEEE Photon. Technol. Lett.* **1**, 340–342.

[2.47] J. M. Kahn *et al.* (1989) "High-stability 1.5 μm external-cavity semiconductor lasers for phaselock applications" *IEEE Photon. Technol. Lett.* **1**, 159–161.

[2.48] C. H. Henry (1982) "Theory of the linewidth of semiconductor lasers" *IEEE J. Quantum Electron* **QE-18**, 259–264.

[2.49] A. W. Davis *et al.* (1987) "Phase diversity techniques for coherent optical receivers" *J. Lightwave Technol.* **LT-5**(4), 561–572.

[2.50] H. K. Phillip *et al.* (1983) "Costas loop experiment for a 10.6 μm communications receiver", *IEEE Trans. Commun.* **COM-31**, 1000–1002.

[2.51] D. J. Malyon (1984) "Digital fiber transmission using optical homodyne detection" *Electron. Lett.* **20**, 281–283.

[2.52] T. G. Hodgkinson (1985) "Phase locked loop analysis for pilot carrier coherent optical receivers" *Electron. Lett.* **21**, 1202–1203.

[2.53] L. G. Kazovsky [1986] "Balanced phase locked loops for optical homodyne receivers: Performance analysis, design considerations, and laser linewidth requirements" *J. Lightwave Technol.* **4**(2), 182–195.

[2.54] D. W. Stowe and T. Y. Hsu (1983) "Demodulation of interferometric sensors using a fiber optic passive quadrature demodulator" *J. Lightwave Technol.* **1**(3), 519–523.

[2.55] L. G. Kazovsky (1985) "Decision driven phase locked loop for optical homodyne receivers: Performance analysis and laser linewidth requirements" *J. Lightwave Technol.* **3**(6), 1238–1247.

[2.56] G. J. Foschini *et al.* (1988) "Noncoherent detection of coherent lightwave signals corrupted by phase noise" *IEEE Trans. Commun.* **COM-36**(3), 306–314.

[2.57] J. Garrett and G. Jacobsen (1986) "Theoretical analysis of heterodyne optical receivers for transmission systems using lasers with nonnegligible linewidth" *J. Lightwave Technol.* **LT-4**(3), 323–334.

[2.58] L. G. Kazovsky (1985) "Optical heterodyning versus optical homodyning: A comparison" *J. Opt. Commun.* **6**(1), 18–24.

[2.59] L. G. Kazovsky (1986) "Impact of laser phase noise on optical heterodyne communication systems" *J. Opt. Commun.* **7**(2), 66–78.

[2.60] G. Jacovsen and I. Garrett (1987) "Theory for heterodyne optical ASK receivers using square-law detection and postdetection filtering" *IEE Proc.*, pt. J, **134**(5), 303–312.

[2.61] L. G. Kazovsky and O. K. Tonguz (1990) "ASK and FSK coherent lightwave systems: a simplified approximate analysis" *J. Lightwave Technol.* **8**(3), 338–352.

[2.62] F. G. Stremler (1982) *Introduction to Communication Systems* (Reading, MA: Addison-Wesley).

[2.63] I. Garrett and G. Jacobsen (1986) "Theoretical analysis of heterodyne optical receivers for transmission systems using semiconductor lasers with nonnegligible linewidth" *J. Lightwave Technol.* **4**(3), 323–334.

[2.64] G. Jacobsen and I. Garrett (1987) "Theory for heterodyne optical ASK receivers using square-law detection and postdetection filtering" *IEE Proc.* Pt. J, **134**(5), 303–316.

[2.65] S. D. Personick (1977) "Receiver design for digital fiber-optic communication systems, part 1 and 2" *Bell Syst. Tech. J.* **52**, 843–886.

[2.66] M. Schwarz *et al. Communication Systems and Techniques* (New York: McGraw-Hill) ch 7–8.

[2.67] A. R. DiDonato and M. P. Jarnagin (1962) "A method for computing the circular coverage function" *Math. Comput.* **16**, 347–355.

[2.68] G. Jacobsen and I. Garrett (1985) "A theoretical analysis of coherent optical communication receivers with nonnegligible laser linewidths" *Report R297, Electromagnetics Institute, Technical University of Denmark, Nov. 1984. R17/003/85* (British Telecom Res. Lab.)

[2.69] S. Saito *et al.* (1983) "S/N and error rate evaluation for an optical FSK-Heterodyne detection system using semiconductor lasers" *IEEE J. Quantum. Electron.* **QE-19**, 180–193.

[2.70] J. M. P. Delavaux *et al.* (1989) "Dependence of optical receiver sensitivity, in a 147 Mbit/s heterodyne DPSK system, as a function of optical and electrical parameter variations" *J. Lightwave Technol.* **7**, 138–149.

[2.71] R. Wyatt *et al.* (1983) "1.52 μm PSK heterodyne experiment featuring an external cavity diode laser local oscillator" *Electron. Lett.* **19**, 550–551.

[2.72] R. A. Linke *et al.* (1985) "Coherent lightwave transmission over 150 km fiber lengths at 400 Mbit/s and 1 Gbit/s data rate using DPSK modulation" *EOOC'85, Venice, Italy, 1985* post-deadline paper.

[2.73] A. Gnauck *et al.* (1987) "Coherent lightwave transmission at 2 Gbit/s over 170 km of optical fiber using phase modulation" *Electron. Lett.* **23**, 286–287.

[2.74] K. Iwashita *et al.* (1987) "Optical CPFSK 2 Gbit/s 202 km transmission experiment using a narrow linewidth multielectrode DFB LD" *Electron. Lett.* **23**, 1022–1023.

[2.75] R. E. Tench *et al.* (1987) "Electrooptic phase modulators for coherent fiber communication" *J. Lightwave Technol.* **5**, 492.

[2.76] J. M. P. Delavaux *et al.* (1986) "Performance evaluation of an X-cut Ti:LiNbO$_3$ waveguide traveling wave modulator", *CLEO'86, San Francisco, 1986* paper FN3.

[2.77] R. C. Steel and I. W. Marshal (1987) "Investigation of optical communication systems with DPSK and PSK modulation and the use of a traveling wave laser amplifier in a PSK transmission experiment" *Proc. SPIE* **841**, 322.

[2.78] T. Imai *et al.* (1985) "Optical polarisation control using an optical heterodyne detection scheme" *Electron. Lett.* **21**(2), 52–53.

[2.79] C. S. Brown (1985) "The polarization properties of installed single-mode fiber cables" *SPIE Proc. Coherent Technology in Fiber Optics Systems* **568**, 50–55.

[2.80] L. Giehmann and M. Rocks (1987) "Measurement of polarization fluctuations in installed single mode optical fiber cables" *Opt. Quantum Electron* **19**, 109–113.

[2.81] T. Okoshi (1983) "Review of polarization-maintaining single-mode fiber" *IOOC'83, Tokyo, Japan, 1983* paper 28A4–1.

[2.82] T. Okoshi (1986) "Polarization-maintaining optical fibers", *Opto-Electronics for the Information Age* ed C. Lin (New York: Van Nostrand Reinhold).

[2.83] I. P. Kaminow (1981) "Polarization in optical fibers" *IEEE J. Quantum Electron.* **QE-17**(1), 15–22.

[2.84] R. A. Harmon (1982) "Polarization stability in long lengths of monomode fiber" *Electron. Lett.* **18**(24), 1058–1061.

[2.85] T. Okoshi *et al.* (1983) "Polarization diversity receiver for heterodyne/coherent optical fiber communications" *IOOC'83 Tokyo, Japan, 1983* paper 30C3-2.

[2.86] T. Okoshi (1985) "Polarization-state control schemes for heterodyne or homodyne optical fiber communications" *J. Lightwave Technol.* **Lt-3**(6), 1232–1237.

[2.87] R. Ulrich (1979) "Polarization stabilization on single-mode fiber" *Appl. Phys. Lett.* **35**(11), 840–842.

[2.88] H. C. Lefevre (1980) "Single-mode fiber fractional wave devices and polari controllers" *Electron. Lett.* **16**(20), 778–780.

[2.89] T. Imai *et al.* (1985) "Optical polari control utilising an optical heterodyne detection scheme" *Electron. Lett.* **21**(2), 52–53.

[2.90] Y. Namihira *et al.* (1987) "Polarization fluctuation in optical-fiber submarine cable under 8000 m deep sea environmental conditions" *Electron. Lett.* **23**(3), 100–101.

[2.91] Y. Namihira *et al.* (1987) "Polarization fluctuation of submarine-cabled single-mode optical fibers in ocean installation" *Electron. Lett.* **23**(7), 343–344.

[2.92] R. Noe *et al.* (1988) "Endless polarization control systems for coherent optics" *J. Lightwave Technol.* **6**(7), 1199–1208.

[2.93] R. M. A. Azzam and N. M. Bashare (1977) *Ellipsometry and Polarized Light* (Amsterdam: North-Holland).

[2.94] G. N. Ramachandran and S. Ramaseshan (1962) "Crystal Optics", in *Handbook of Physics* vol 25/1 ed S. Flugge (Berlin: Springer).

[2.95] M. Johnson (1979) "In-line fiber-optical polarization transformer" *Appl. Opt.* **18**(9), 1288–1289.

[2.96] F. Mohr and U. Scholz (1983) "Active polarization stabilization systems for use with coherent transmission systems or fiber-optic sensors" *Proc. ECOC* 313–316.

[2.97] H. Heidcrich *et al.* (1986) "Polarization transformer with unlimited range on Ti:LiNbO$_3$" *Proc. 12th ECOC* **1**, 411–414.

[2.98] H. Heidrich *et al.* (1987) "Integrated optical compensator on Ti:LiNbO$_3$ for continuous and reset-free polarization control" *Proc. 13th ECOC* **1**, 257–260.

[2.99] R. Noe (1987) "Error-tolerant endless polarization control system with negligible signal losses for coherent optical communications" *Proc. 13th ECOC* **1**, 371–374.

[2.100] R. Noe (1986) "Endless polarization control for heterodyne/homodyne receivers" *Proc. Fiber Optics SPIE* **630**, 150–154.

[2.101] R. Noe and G. Fischer (1986] "17.4 Mbit/s heterodyne data transmission at 1.5 μm wavelength with automatic endless polarization control" *Proc. Opto. Paris, France: ESI, 1986*.

[2.102] R. Noe (1986) "Endless polarization control system with three finite elements of limited birefringence ranges" *Electron. Lett.* **22**(25), 1341–1343.

[2.103] C. J. Mahon and G. D. Khoe (1986) "Compensational deformation: New endless polarization matching control schemes for optical homodyne or heterodyne receivers which require no mechanical drivers" *Proc. 12th ECOC* **1**, 267–270.

[2.104] L. G. Kazovsky (1989) "Phase- and polarization-diversity coherent optical techniques" *J. Lightwave Technol.* **7**(2), 279–292.

[2.105] L. G. Kazovsky *et al.* (1987) "ASK multiport optical homodyne receivers" *J. Lightwave Technol.* **5**(6), 770–791.

[2.106] J. G. Foschini *et al.* "Noncoherent detection of coherent optical pulses corrupted by phase noise and additive gaussian noise" *IEEE Trans. Commun.* **36**(3), 306–314.

[2.107] T. Okoshi (1987) "Recent advances in coherent optical fiber communication systems" *J. Lightwave Technol.* **5**(1), 44–52.

[2.108] L. G. Kazovsky (1986) "Impact of laser phase noise on optical heterodyne communications systems" *J. Opt. Commun.* **7**(2), 66–78.

[2.109] H. L. Van Trees (1968) *Detection, Estimation, and Modulation Theory*, vol. 1 (New York: Wiely).

[2.110] L. G. Kazovsky *et al.* (1988) "Wide linewidth phase diversity homodyne receivers" *J. Lightwave Technol.* **6**(10), 1527–1536.

[2.111] T. G. Hodgkinson *et al.* (1985) "Demodulation of optical DPSK using in-phase and quadrature detection" *Electron. Lett.* **21**(19), 867–868.

[2.112] L. G. Kazovsky *et al.* (1987) "Impact of laser intensity noise on ASK two-port optical homodyne receivers" *Electron. Lett* **23**(17), 871–873.

[2.113] L. G. Kazovsky (1988) "Impact of reflections on phase diversity optical homodyne receivers" *Electron. Lett.* **24**(9), 522–524.

[2.114] A. W. Davis and S. Wright (1986) "Coherent optical receiver for 680 Mbit/s using phase diversity" *Electron. Lett.* **22**(1), 9–11.

[2.115] T. E. Darcie *et al.* (1987) "Polarization diversity receiver for coherent FSK communications" *Electron. Lett.* **23**(25), 1369–1370.

[2.116] B. Glance (1987) "Polarization independent coherent optical receiver" *J. Lightwave Technol.* **5**, 274–276.

[2.117] H. Kuwahara *et al.* (1986) "New receiver design for practical coherent lightwave transmission system" *ECOC'86, Barcelona, Spain* pp. 407–410.

[2.118] T. Chikama *et al.* (1990) "Modulation and demodulation techniques in optical heterodyne PSK transmission systems" *J. Lightwave Technol.* **8**(3), 309–321.

[2.119] N. G. Walker and J. E. Carroll (1984) "Simultaneous phase and amplitude measurements on optical signals using a multiport junction" *Electron. Lett.* **20**, 981–983.

[2.120] T. Okoshi and Y. H. Cheng (1987) "Four-port homodyne receiver for optical fiber communications comprising phase and polarization diversities" *Electron. Lett.* **23**(8), 377–378.

[2.121] R. Welter and L. G. Kazovsky (1988) "150 Mbit/s phase diversity ASK homodyne receiver with a DFB laser" *Electron. Lett.* **24**(4), 199–201.

[2.122] R. Schneider and J. Pietzch (1987) "Coherent 565 Mbit/s DPSK transmission experiment with a phase diversity receiver" *Proc. ECOC'87, Helsinki, Finland* **3**, 5–8.

[2.123] I. A. W. Vance (1980) "An integrated circuit VHF radio receiver" *Radio Electron. Eng.* **50**(4), 158–164.

[2.124] T. Okoshi and S. Yamashita (1989) "Double-stage phase-diversity (DSPD experiment using 100 Mbit/s FSK coherent optical communication system" *Electron. Lett.* **25**(22), 1512–1513.

[2.125] K. Emura *et al.* (1989) "5 Gbit/s optical phase diversity homodyne detection experiment" *Electron. Lett.* **25**(6), 400–401.

3

OPTICAL FREQUENCY DIVISION MULTIPLEXING (FDM) TECHNOLOGY

3.1 SIGNIFICANCE OF OPTICAL FREQUENCY DIVISION MULTIPLEXING

Frequency division multiplexing (FDM) technology or wavelength division multiplexing (WDM) technology, by which multiple optical channels can be simultaneously transmitted at different wavelengths through a single optical fiber, is a useful means of making full use of the low-loss characteristics of optical fibers over a wide wavelength region.

This concept first appeared in 1970 according to published reports [3.1]. This research focused on practical applications to communications systems [3.2–3.4]. Since then, research has intensified along with dramatic progress in optical fiber light sources and detectors. In recent years, accelerated research efforts have explored more sophisticated optical fiber components, optoelectronic devices and techniques that allow processing of signals in the optical or hybrid optic–electrical (including mixed microwave–optical) domain. Areas of investigation have included wavelength division multiplexing (WDM) [3.5–3.7], optical switching [3.8], coherent detection [3.9, 3.10], tunable laser diodes [3.11, 3.12] and wavelength filters [3.13, 3.14]. In particular, optical multiplexers/demultiplexers, which are key devices in WDM transmission systems, are currently among the most popular research and development items. Many types of multiplexers/demultiplexers have been proposed and fabricated to date. Optical wavelength division multiplexing (WDM) technology has been utilized to combine several larger channel-spacing systems [3.15, 3.16]. Channel spacing for WDM is about 10–100 nm, i.e. 1.3–13 THz in the frequency domain at the 1.5 μm wavelength region. Multimode semiconductors LDs or LEDs have been used in WDM systems. These WDM techniques will be discussed in detail in the following section.

When channel spacing is narrower, e.g., 0.1–1 nm, corresponding to 13–130 GHz at the 1.5 μm wavelength, this multichannel transmission is called DWDM. As the expression is more applicable to a frequency domain than a wavelength domain for the channel spacing near 10 GHz, so the term optical frequency division multiplexing

(FDM) is used in this case. The developing technology based on the FDM transmission, doped fiber amplifers and precise frequency filters will be one of the most dominant fields in fiber communications.

From a long-term perspective, the channel spacing will be even narrower, so the term dense frequency division multiplexing (DFDM) is more suited to the case of 0.01 nm or less (1.3 GHz or less than 1 GHz for the 1.55 μm range), especially in FDM coherent systems and coherent networks (see Section 3.4).

Table 3.1 summarizes the above different classes of WDM and FDM. In such a FDM or DFDM system, laser diode sources with single-frequency (in DFDM, sometimes narrow-linewidth LDs), frequency stabilized and continuously frequency tunable are the most essential components.

Although FDM or DFDM technology is not fully mature at present, it has been gradually introduced into experimental systems [3.17, 3.18]. It is also expected that this technology will play a major role in future communications systems. Multiplexing and transport techniques for both the interoffice (exchange) network and the subscriber loop must be geared for transmission capacities orders of magnitude greater than those typical in today's telecommunications environment. In addition to providing high-speed transmission capabilities in single mode fiber systems, the FDM network must also incorporate switching, routing, and multiplexing techniques that are appropriate for broad-band services.

In this chapter, the WDM technique will be discussed in Section 3.2, while the FDM technique will be discussed in some detail later. Although microwave subcarrier multiplexing (SCM) is not exactly the same technique in the optical FDM domain, it is still more convenient to be discussed in Section 3.5. Wavelength division photonic switching (really frequency division here) is discussed in Section 3.6. It is quite an important technique in potential broad-band communication networks. Principles and experimental results as well as some multiplexer/demultiplexers are discussed in each section.

3.2 WAVELENGTH DIVISION MULTIPLEXING (WDM)

Most early applications of WDM have utilized relatively simple two-channel devices, such as directional couplers or dielectric thin film (DTF) filters, that multiplex one signal with another at two different wavelengths with a rather large interval. However, there now exist considerably more sophisticated devices capable of multiplexing a much larger number of channels. Grating-based wavelength multiplexers/demultiplexers have achieved accommodation over 20 channels. Even

Table 3.1 Terminology and technology of WDM and FDM

100 nm 13 THz	10 nm 1.3 THz	1 nm 130 GHz	0.1 nm 13 GHz	0.01 nm 1.3 GHz	Multiplexers/demultiplexers
< -------- WDM ----------->					Grating, DTF F–P
		< ---DWDM or FDM---->			couplers + DFBs FFP
				< --DFDM---	waveguide M-Z filter

smaller channel spacings, as low as 0.04 nm (5 GHz), have been achieved [3.17, 3.18] using FDM techniques. The use of coherent receivers for multichannel filtering [3.9, 3.10] would also fall into this chapter. Crosstalk and excess loss are two critical performance parameters. Crosstalk between WDM channels results in system power penalties [3.19], since additional signal power is required at the receiver to overcome noise from interfering channels. WDM loss and crosstalk strongly depend on device structure, number of channels, and channel spacing. Commercial grating-based devices, for example, have typically achieved crosstalk levels of the order of −30 dB and excess loss from 1 to 7 dB. Such low levels of crosstalk rarely result in system penalties greater than a few tenths of a dB; hence, device excess loss usually has a more significant system impact. Grating components utilizing single mode fiber arrays generally suffer higher loss, since the fiber's relatively small core diameter (9 μm compared with 50 μm for multimode fiber) results in larger losses when coupling (refocusing) the light back into the fiber array. Of course, the relative importance of the WDM excess loss depends on a variety of factors, such as system configuration, transmitter power, bit rate, and receiver sensitivity.

3.2.1 Principles of multiplexing/demultiplexing devices for WDM [3.5]

The multiplexing/demultiplexing devices are really important key components for developing the WDM/FDM technologies. Especially in the practical dense frequency division multiplexing systems, a multiplexing/demultiplexing device with in-line, fine and stable spectral resolution, wavelength tuning, low insertion losses, simpler alignment and more compact configuration is an extremely important component for optical communications in future.

ANGULARLY DISPERSIVE DEVICES

Angularly dispersive devices could be used as WDM multiplexers/demultiplexers (MUX/DEMUX) with larger wavelength separation, such as prisms and gratings. In practice, a blazed grating that can efficiently diffract light into a specific diffraction order is primarily and widely used as the angularly dispersive element. For a standard example, the Littrow has been more extensively studied than other types of grating elements. A typical structure for grating-Littrow type demultiplexer [3.21] are shown in Fig. 3.1(a) (conventional lens type) and (b) (GRIN-rod lens type) [3.5]. The GRIN-rod lens type is superior to the conventional lens type in compactness and ease of alignment. Examples of the conventional lens type and GRIN-rod type are listed in [3.20–3.23] and [3.24–3.26], respectively.

The principles and performances for this kind of demultiplexer, referring to the structure shown in Fig. 3.1(a), are discussed here as an example. An input fiber and multiple output fibers are arranged on the focal plane of the lens. Wavelength multiplexed light from the input fiber is collimated by the lens and reaches the diffraction grating. The light is angularly dispersed, according to different wavelengths, and reflected at the same time. Then, the different wavelengths pass

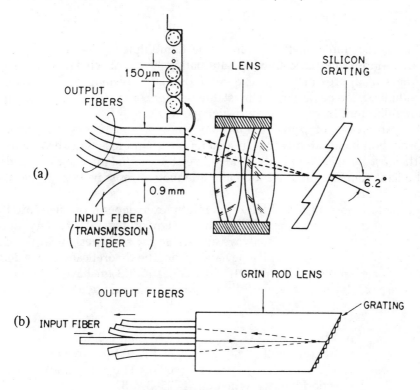

Fig. 3.1 Basic structure for grating-Littrow type DEMUX. (From [3.5] Ishio *et al.* (1984) *J. Lightwave Technol.* **LT-2**(8), 448–463. Reproduced by permission. ©1984 IEEE.)

through the lens and are focused to their corresponding output fibers. Each wavelength is fed to one individual detector through its output fiber. Interchannel wavelength spacing is equal to the 25 nm designed value. The 0.1 dB bandwidths are all 11 nm. Such wide passband width can be obtained due to the difference between the core diameter Δd of the input fiber (core diameter: 60 μm) and the output fibers (core diameters: 130 μm). Bandwidth B_w can be estimated from

$$B_w = (\Delta d \Delta \lambda)/\Delta Z \tag{3.1}$$

where $\Delta \lambda / \Delta Z$ is linear dispersion in the focal plane, which is 25 nm/150 μm in this device. Since Δd is 70 μm, B_w is 11.7 nm from (3.1). This value is nearly equal to the experimental value of 11 nm. This wide bandwidth is necessary in a demultiplexer to prevent insertion loss being influenced by wavelength variation in the LDs.
  The performance of Littrow-type demultiplexers includes:

number of channels	3–20
insertion loss	1–4 dB
wavelength spacing	20–40 nm
crosstalk attenuation	20–30 dB.

DIELECTRIC THIN-FILM FILTER TYPE

The dielectric thin film (DTF) filter type demultiplexer contains high and low refractive-index dielectric films in alternating layers. Each layer has an optical thickness of approximately one quarter or half wavelength. Therefore, a DTF demultiplexer can be designed almost independent of the fiber parameters used and is quite different from angularly dispersive devices.

The parameters of primary importance are passband loss, passband width, and rejection band attenuation. In general, passband loss must be low, passband width must be wide (typically, wider than the oscillation wavelength deviation in the light source), and rejection band attenuation must be large (typically, more than 20–30 dB).

The performance of a multiplexer/demultiplexer depend on dielectric thin-film characteristics to a great extent, so the design of the DTF is very important. The matrix method [3.24] is always applicable for such design. In the case of a bandpass filter (BPF) less the 1% deviation from the theoretical value is demanded [3.25]. For example, BPFs of 0.81, 0.89, 1.2, and 1.3 μm have been developed. These are 23 layer 3 cavity BPFs composed of SiO_2 and TiO_2 films. Their characteristics are:

passband loss	less than 0.2 dB
passband width	400 A (0.8 μm region)
	600 A (1.3 μm region)
rejection band attenuation	25–30 dB.

Typical examples of the DTF demultiplexer are shown in Fig. 3.2 [3.5]. There are the Fig. 3.2(a) GRIN-rod lens type [3.26, 3.28, 3.29] (b) multireflection type [3.30, 3.31], and (c) fiber-end type [3.32]. While these examples are shown as demultiplexers they can also be modified for multiplexing or multiplexing/demultiplexing configurations by reversing the light input at the ports.

The GRIN-rod-lens type has the advantage of compactness, but it is not easy to realize with more than three channels, because this would require GRIN-rod lenses of more than three elements with an off-axis cascade connection between elements, and such an off-axis cascade connection has very critical alignment. The multireflection type has a more complicated structure and has an even more difficult alignment. However, it is applicable to multiplexing or demultiplexing with more than six channels. The fiber-end type has the simplest structure, but it is difficult to fabricate for multiplexing or demultiplexing with more than three channels. This is because it tends to increase insertion loss based on alignment, since it has no collimating lens.

DTF-type devices are now available with the following performance chacteristics:

number of channels	2–6 channels
insertion loss	1–5 dB
crosstalk attenuation	20–70 dB
channel spacing	30–100 nm

(or very wide spacing, for example, 0.8 and 1.3 μm).

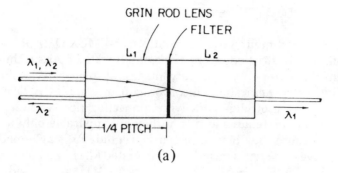

(a)

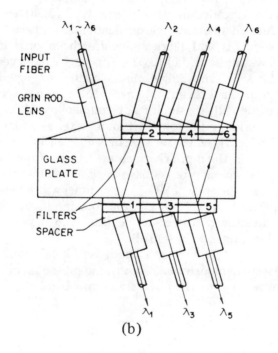

(b)

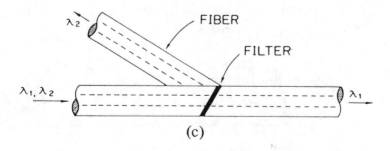

(c)

Fig. 3.2 DTF-type demultiplexer. (From [3.5] Ishio *et al.* (1984) *J. Lightwave Technol.* **LT-2**(8), 448–463. Reproduced by permission. ©1984 IEEE.)

PLANAR WAVEGUIDE TYPE DEVICES

These kinds of optical multiplexers/demultiplexers (MUX/DEMUX) with different configurations are constructed on the basis of integrated optics technology. These configurations have the potential advantages of planar technology, such as reproducibility, batch fabrication, incorporation of complex optical circuits into photomasks, and ease of alignment with connecting fibers. Therefore, they are expected to become more advanced. But they are applicable only to single mode WDM system because they have a planar waveguide for single-mode operation.

The planar waveguide types include the integrated Mach–Zehnder interferometer type [3.33], the elliptical Bragg reflectors type [3.34], the chirped-grating types [3.35], the contradirectional Bragg-coupler type [3.36], the Bragg-grating type [3.37], the selective cascaded-mode-converter type [3.38] and others.

An integrated four-channel multiplexer/demultiplexer based on Mach–Zehnder interferometers elements and fabricated with phosphorus doped SiO_2 planar waveguides on Si was presented in [3.33]. This device has a very low (0.05 dB/cm) waveguide loss for $\lambda = 1.5\,\mu m$. Single mode channel waveguide losses as low as 0.05 dB for $\lambda = 1.5\,\mu m$ have been obtained. The channel spacing is 77 Å. The multiplexer was made with six similar 3 dB couplers based on couplers of 700 μm length. The transmission of the four-port multiplexer as a function of wavelength is shown in Fig. 3.3 [3.33], where the output is normalized with respect to the total output from the device. The input channels are labelled from 1 to 4 and the output channel is C. The transmission peaks have a separation in wavelength of 77 Å and an average crossover efficiency of 90%. The channel width between 1 dB points is 40 Å. The multiplexer transfers power from four input channels into a common output channel with an average fiber-to-fiber loss of 2.6 dB per channel. The average crosstalk of the demultiplexer is −16 dB.

Four-channel WD multiplexers and bandpass filters based on elliptical Bragg reflectors (EBRs) were demonstrated recently and presented in [3.34]. The channel spacing is 50 Å near 1.56 μm. The EBRs are narrow band elliptical mirrors that can

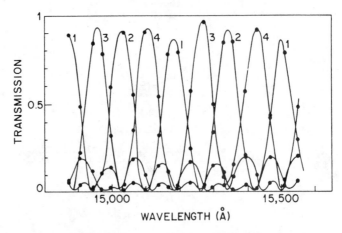

Fig. 3.3 Planar-waveguide demultiplexer. (From [3.33] Verbeek *et al.* (1988) *J. Lightwave Technol.* **LT-6**(6), 1011–1015. Reproduced by permission. ©1988 IEEE.)

refocus light from an input waveguide to any one of a number of output waveguides. Multiplexers with single filtering and double filtering were demonstrated. With single filtering, the fiber-to-fiber insertion loss was 3.0 ± 0.5 dB, and the crosstalk was -20 dB. With double filtering, the insertion loss was 4.0 ± 0.5 dB, and the crosstalk was -30 dB. About 2 dB of this loss was due to coupling between the fibers and the waveguide.

3.2.2 Design and parameters of WDM systems [3.5]

The major parameters for WDM transmission system design are loss, interchannel crosstalk and modal noise of MUX/DEMUX, as well as the emission spectrum and wavelength dependence on temperature of light source. These parameters are closely related to each other because a number of optical signals having different wavelengths share a single fiber whose loss and bandwidth depend on wavelength. Among these parameters, optical interchannel crosstalk is the most important when considering a WDM system design.

In comparison with conventional fiber-optic transmission systems, a WDM transmission system needs additional optical multiplexers/demultiplexers. When multiplexer/demultiplexer insertion loss is considered, the loss budget design for a WDM transmission system is basically almost the same as that for conventional fiber-optic transmission systems. However, other factors must be taken into account.

In optical multiplexers/demultiplexers, optical interchannel crosstalk is related to light-source spectrum width, photodetector sensitivity (which is dependent on wavelength) and so on. The tolerable crosstalk levels are determined by the nature of the transmission signal [3.39–3.41].

In some multimode fiber transmission systems, such as in an analogue transmission system using laser diodes or in a very high speed digital transmission system, an optical multiplexer/demultiplexer caused significant SNR, differential gain and differential phase degradation [3.42]. This is primarily due to fiber propagation mode filtering in fiber-to-fiber coupling through the lenses in the optical multiplexers/demultiplexers [3.43, 3.44].

In general, when an input fiber diameter is smaller than the output core diameter, the modal noise can easily be reduced. Since the photodetector sensing area diameter is more than 100 μm, it is easy for demultiplexer output ports to use large-core-diameter fibers. However, for multiplexer input ports, core diameter reduction decreases the LD-to-fiber coupling.

MULTIPLEXER/DEMULTIPLEXER DESIGN

Multiplexer/demultiplexing loss, interchannel crosstalk and channel spacing for given wavelength-selective element performance and light-source characteristics are discussed in [3.5].

The following terminology for multiplexer/demultiplexer design will be used hereafter:

L_0: sum of multiplexing loss and demultiplexing loss (dB);

L'_0: demultiplexing loss (dB);

S/X: optical crosstalk from adjacent optical channels (dB);

L_e: Effective optical multiplexing/demultiplexing loss for digital transmission systems (dB).

L_e is obtained as

$$L_e = L_0 + \Delta P \qquad \Delta P = \Delta\mathrm{SNR} \cdot (2+x)/(2+2x)$$

$$\Delta\mathrm{SNR} = -20 \log(1 - 2 \times 10^{-(s/x)/10}) \qquad (\mathrm{dB}) \qquad (3.2)$$

where x is the excess noise factor for an avalanche photodiode. Hereafter, $x = 0.4$ is used. $\Delta\mathrm{SNR}$ is the signal-to-noise ratio deterioration caused by crosstalk and ΔP is the loss penalty due to crosstalk.

In order to simplify the argument, the reflection and aberration losses in optical systems are neglected in the following discussion. It is also assumed that the LED and LD emitting spectra are expressed by [3.39]

$$\mathrm{LED} \qquad \Gamma_{\mathrm{LED}}(\lambda) = \left[1 + \left| \frac{2(\lambda - \lambda_0)}{\Delta\lambda_0} \right|^3 \right]^{-1} \qquad (3.3)$$

and

$$\mathrm{LD} \qquad \Gamma_{\mathrm{LD}}(\lambda) = \delta(\lambda - \lambda_0) \qquad (3.4)$$

where λ_0 and $\Delta\lambda_0$ are the centre wavelength and the emitting spectrum halfwidth, respectively.

When fiber baseband width is sufficient for signal bandwidth, the digital system repeater span l (km) is determined by

$$\alpha l = P_0 - P_r - L_e$$

$$L_e = L_0 + \Delta P \qquad (3.5)$$

where P_0 is an optical power coupled to a fiber (dBm), P, is the minimum received power (dBm), and α is fiber loss (dB/km).

In analogue transmission system design, the multiplexer/demultiplexer loss L_e is kept to a minimum.

In analogue transmission system design, the multiplexer/demultiplexer configuration should be chosen so that L_0 is the smallest among multiplexer/demultiplexer configurations in which the crosstalk level is less than the desired level, for example, $-30\,\mathrm{dB}$ for a 4 MHz baseband analogue system.

DTF MULTIPLEXER DESIGN

A DTF multiplexer with multireflection structure which includes optical bandpass filters with different reflectance/transmittance wavelength dependencies is shown in Fig. 3.4 [3.5]. Wide LED emission spectrums or LD oscillation wavelength deviations due to temperature changes create interchannel crosstalk. When the bandwidth of the bandpass filter becomes narrow, crosstalk decreases but multiplexer loss increases.

(1) LED systems. When LEDs are utilized in a WDM transmission system using DTF-type multiplexer, filtering loss L_0 and crosstalk S/X are given as

$$L_0 = 10 \log \, [S]/[O]$$

$$S/X = 10 \log \, [S]/[X] \qquad (3.6)$$

$$[O] = \int_0^\infty \Gamma_{LED}(\lambda) d\lambda \qquad (3.7)$$

$$[S] = \int_0^\infty \Gamma_{LED}(\lambda) T_t(\lambda) [1 - T_t(\lambda + \lambda_s)] [1 - T_t(\lambda + 2\lambda_s)]$$
$$[1 - T_r(\lambda + 2\lambda_s)] [1 - T_r(\lambda + \lambda_s)] T_r(\lambda) d\lambda \qquad (3.8)$$

$$[X] = \int_0^\infty \Gamma_{LED}(\lambda - \lambda_s) T_t(\lambda - \lambda_s) [1 - T_t(\lambda)] [1 - T_r(\lambda + \lambda_s)] (1 - T_t(\lambda + 2\lambda_s))$$
$$[1 - T_r(\lambda + \lambda_s)] T_r(\lambda) d\lambda + \int_0^\infty \Gamma_{LED}(\lambda + \lambda_s) T_t(\lambda + \lambda_s) [1 - T_t(\lambda + 2\lambda_s)]$$
$$[1 - T_r(\lambda + 2\lambda_s)] (1 - T_r(\lambda + \lambda_s) T_r(\lambda) d\lambda \qquad (3.9)$$

$$T_t(\lambda) = \exp [-2 \ln 2 |(\lambda - \lambda_0)/\Delta\lambda_t|^2] \qquad (3.10)$$

$$T_r(\lambda) = \exp [-2 \ln 2 |(\lambda - \lambda_0)/\Delta\lambda_r|^2] \qquad (3.11)$$

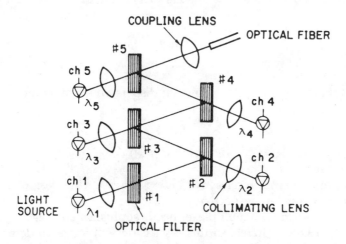

Fig. 3.4 Schematic of multireflection-type MUX=DEMUX. (From [3.5] Ishio *et al.* (1984) *J. Lightwave Technol.* **LT-2**(8), 448–463. Reproduced by permission. ©1984 IEEE.)

where λ_s is channel spacing, $\Delta\lambda$, and $\Delta\lambda_r$ are filter bandwidths of the multiplexing device and of the demultiplexing device, respectively. $T_t(\lambda)$ and $T_t(\lambda)$ express the transmittance of a filter in the multiplexing device and in the demultiplexing device, respectively.

(2) LD systems. When using LDs, it is necessary to consider wavelength deviation due to temperature change and/or wavelength setting error. If LD wavelength shifts $\Delta\lambda_d$ from the given wavelength λ_0 while the adjacent channel is at the longer wavelength side, LD oscillates at $\lambda=\lambda_0+\lambda_s-\Delta\lambda_d$ and, in another adjacent channel, LD wavelength is $\lambda=\lambda_0-\lambda_s+\Delta\lambda_d$. This is the worst case for the given wavelength deviation $\Delta\lambda_d$. In this case, L_0 and S/X are obtained by replacing $\Gamma_{LED}(\lambda)$ and $\Gamma_{LED}(\lambda\pm\lambda_s)$ in (3.6)–(3.8) by $\delta(\lambda+\Delta\lambda_d)$ and $\delta[\lambda\pm(\lambda_s-\Delta\lambda_d)]$, respectively.

ANGULAR DISPERSION DEMULTIPLEXER DESIGN

A schematic structure of an angular-dispersive (grating or prism) optical demultiplexing circuit is shown in Fig. 3.5 [3.5]. The light signals from the input fiber are collimated. After striking the dispersive element, the propagation angles of input WDM signals correspond to their wavelengths. They are then focused on the receiver plane by a Fourier transform lens having focal length f_2 and each light beam incident to the centre of its channel window (a photodiode or a fiber terminal). Thus light signals incident to an adjacent channel window create crosstalk. The collimated beam diameter D is

$$D=2f_1(NA)/n_M \qquad (3.12)$$

where (NA) is the numerical aperture of the input fiber, n_M is the reflective index of the medium between the dispersive element and the lens, and f_1 is the focal length of the collimating lens, the angular dispersion for a Littrow-mounted grating is given by

$$\frac{d\theta}{d\lambda}=2\frac{\tan\theta}{\lambda}\approx2\frac{\tan\theta_B}{\lambda} \qquad (3.13)$$

where θ_B is the blaze angle. Prism angular dispersion for minimum deviation angle is given by

$$\frac{d\theta}{d\lambda}=\frac{B}{n_M D}\frac{dn}{d\lambda} \qquad (3.14)$$

where n, B, and D are the refractive index, length of the base of the prism, and incident-beam diameter, respectively [3.37].

A grating- or prism-type multiplexer needs sophisticated optical systems for low insertion loss [3.45]. Therefore, this discussion will focus on demultiplexers.

From (4.12), positional dispersion for a Littrow-type grating demultiplexer is given by

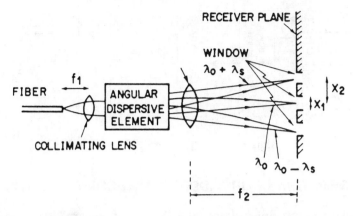

Fig. 3.5 An angular-dispersive DEMUX circuit. (From [3.5] Ishio *et al.* (1984) *J. Lightwave Technol.* **LT-2**(8), 448–463. Reproduced by permission. ©1984 IEEE.)

$$\Delta x = \int_{\lambda 0}^{\lambda_0 + \delta\lambda} f_2 \, \frac{\tan \theta}{\lambda} \, d\lambda \simeq 2f_2 \tan \theta_B \int_{\lambda 0}^{\lambda_0 + \delta\lambda} \frac{1}{\lambda} d\lambda$$

$$(3.15)$$

$$= 2f_2 \tan \theta_B \ln \left(1 + \frac{\delta\lambda}{\lambda_s} \right).$$

From (3.12) and (3.14) positional dispersion for a prism-type demultiplexer is given by

$$\Delta x = \frac{Bf_2}{2f_1(\text{NA})} \int_{\lambda_0}^{\lambda_0 + \delta\lambda} \frac{dn}{d\lambda} \, d\lambda \, .$$

$$(3.16)$$

(1) LED systems. When LEDs are used as light sources, the wide emission spectrum creates crosstalk. The demultiplexing loss and crosstalk for a grating type demultiplexer are obtained as

$$L_0' = -10 \log [S]/[O]$$

$$(3.17)$$

$$S/X = 20 \log [S]/[X]$$

$$(3.18)$$

$$[O] = \int_0^\infty \Gamma_{\text{LED}}(\lambda) d\lambda$$

$$(3.19)$$

$$[S] = \int_{\lambda_0 - \delta\lambda}^{\lambda_0 + \delta\lambda} \Gamma_{\text{LED}}(\lambda) d\lambda$$

$$(3.20)$$

$$[X] = \int_{\lambda_0 - \delta\lambda}^{\lambda_0 + \delta\lambda} \Gamma_{\text{LED}}(\lambda - \lambda_s) + \Gamma_{\text{LED}}(\lambda + \lambda_s) d\lambda$$

$$(3.21)$$

$$\delta\lambda = \lambda_0 \exp \left[\frac{X_1(\text{NA})}{2Dn_M \tan \theta_B} \right] - \lambda_0.$$

$$(3.22)$$

(2) LD system. When using LDs, an angular dispersive demultiplexer improves performance. It becomes demultiplexing loss free and crosstalk free if proper channel spacing λ_s and sufficient window dimensions x_1 are chosen to receive a light signal deviating within a certain wavelength range.

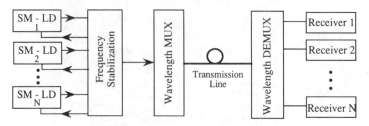

Fig. 3.6 Optical FDM transmission.

3.3 FDM NETWORK IN BROADBAND TELECOMMUNICATION

The capabilities and principles of optical FDM technology are illustrated in Fig. 3.6.

An 18-channel FDM achieved the very high bit-rate distance product, 2.07(Tb/s)km. Both the DFB lasers and the FDM components used in these techniques were packaged devices obtained from manufacturers. The wavelength multiplexer and demultiplexer were grating-base devices. The multiplexer was fully single mode, with excess losses of 6–7 dB per channel. The demultiplexer utilized multimode fiber on its output ports; the relatively large core diameter of the multimode fiber resulted in significantly lower excess loss (<2 dB/channel) than that incurred in the multiplexer. The FDM channels were spaced at 2 nm intervals (this example is not so densely spaced, and could even be treated as WDM network) between 1527 nm and 1561 nm. The individual FDM/WDM passbands were approximately 0.3 nm FWHM, necessitating temperature stabilization of each DFB laser to within ≈1°C. Each laser was modulated at 2 Gbit/s and provided a minimum −3 dBm average (modulated) output power. The multiplexed light was transmitted through 57.5 km of standard single-mode fiber. At the demultiplexer output, crosstalk in each FDM channel due to the cumulative optical power from all other channels was below −27 dB; no crosstalk-related bit error rate (BER) degradations were observed.

The above experiment illustrates the use of multichannel FDM in one of its most well known applications, namely, as a means of greatly increasing transmission capacity on point-to-point fiber links. However, in addition to such use in point-to-point links, the unique characteristics of FDM allow fundamentally new network configurations whose properties may play critical roles in future broadband systems for both interoffice and subscriber-loop applications.

DISTRIBUTION OF BROADCAST SIGNALS [3.46, 3.48]

We select a so-called LAMBDANET (a trademark of Bellcore) architecture, illustrated in Fig. 3.7 [3.46], as an example of an FDM network designed for distribution of broadband signals in the interoffice environment [3.47, 3.48]. Each node in the network transmits on a unique frequency, and the transmitted light from all N nodes is combined in a passive star coupler. Since each output fiber contains $1/N$ of each of the N inputs, broadcast distribution is thus achieved. Each node may receive information from all other nodes, independently and simultaneously, by using an N-channel wavelength demultiplexer and N receivers.

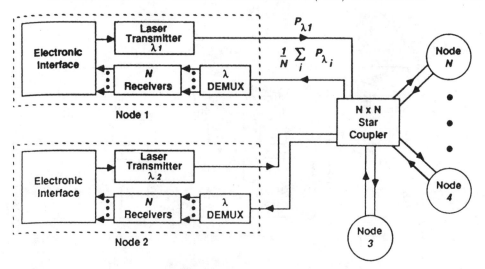

Fig. 3.7 LAMBDANET FDM network. (From [3.46] Wagner and Kobrinski (1989) *IEEE Commun. Mag.* **27**(3), 22–29. Reproduced by permission. ©1989 IEEE.)

The high-capacity broadcast nature of the LAMBDANET design makes it particularly suitable for interoffice video distribution. As suggested by Fig. 3.7, each node could be a telephone Central Office (CO) having its own DFB laser and assigned transmission wavelength. Each office then serves as a Local Access and Transport Area wide broadcast point for any video vendor in its vicinity.

The LAMBDANET FDM network configuration was demonstrated experimentally [3.48] by connecting the 18 DFB lasers to an 18×16 star coupler. The lasers were modulated at 1.5 Gbit/s, and the outputs of the star were transmitted through spans of single mode fiber. Coupling and loss variations within the star coupler produced corresponding variations in the maximum achievable transmission distance from the output ports of the star (42 km for the weakest port to 57.8 km for the strongest port). Adding the bit-rate-distance product achieved on each link, it was found that this experimental network afforded a point-to-multipoint capacity of approximately 21.5(Tb/s)km node. With such capacity, a single head end could deliver 180 digital video channels (150 Mbit/s uncompressed NTSC or compressed high definition) to each of 16 central offices located an average of 45 km from the head end. Eventual upgrading to higher capacities can be achieved through the addition of more wavelengths, possibly through the use of echelon grating demultiplexers that feature multiple sets of FDM passbands [3.49].

ROUTING OF SWITCHED SIGNALS [3.46, 3.52]

In addition to providing point-to-multipoint broadcast connectivity, future interoffice networks also must afford point-to-point connectivity for switched services. To provide such connectivity for today's narrow-band interoffice traffic, hubbed networks have been proposed as economic alternatives to fully interconnected mesh topologies [3.50]. In a hubbed network, traffic is transmitted to a central switching

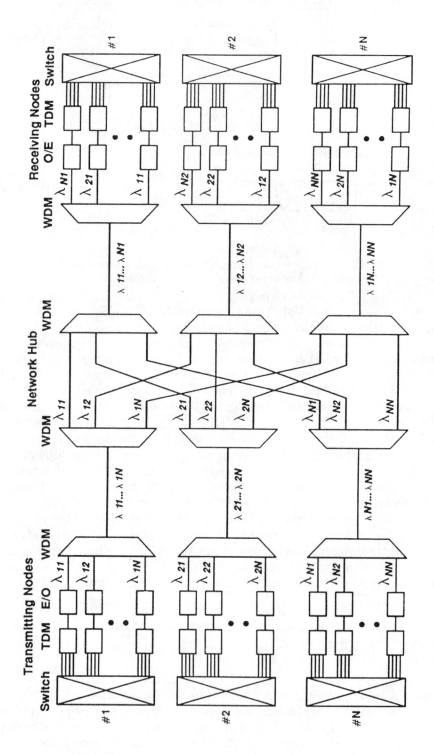

Fig. 3.8 Crossconnection scheme for FDM interoffice network. (From [3.46] Wagner and Kobrinski (1989) *IEEE Commun. Mag.* 27(3), 22–29. Reproduced by permission. ©1989 IEEE.)

location (hub) that electronically crossconnects and routes the traffic according to destination. If such electronic hubbing is used for broadband services, for which the required throughput may be in excess of 100 Gbit/s per end office, it implies massive high-speed electronic switching, multiplexing, and demultiplexing at the hub site [3.51].

Fig. 3.8 illustrates an interoffice network [3.52, 3.53] that employs multichannel FDM techniques to retain the desirable features of a hubbed topology while eliminating the need for high-speed electronic processing at the hub. For clarity, Fig. 3.8 shows each node (end office) with its transmitter portion on the left and receiver portion on the right. Prior to transmission, switch outputs are groomed according to their destination, and all signals destined for a given node are time-multiplexed to a single higher-bit-rate data stream. These high-speed streams modulate lasers having unique wavelengths corresponding to the destination nodes. The optical signals are then wavelength-multiplexed and transmitted over a single fiber to the hub site. At the hub, wavelength demultiplexers separate the optical signals, and the signals are regrouped (via passive cross-connection) according to destination. Since each of the optical streams intended for a given destination was assigned a different wavelength, they can be remultiplexed via FDM and sent to that destination over a single fiber. The FDM channels are then wavelength-demultiplexed, detected, and time-demultiplexed so that the inputs to the local switch are the original, lower bit-rate signals.

At first glance, it may appear that full connectivity of N nodes would require $O(N^2)$ different wavelengths and N different wavelength multiplexing/demultiplexing devices. However, the two conditions required by this architecture—namely, that all wavelengths transmitted from a given node $k(\lambda_{kj}, j=1, 2, \ldots, N)$ be different and that all wavelengths destined for a given node $m(\lambda_{jm}, m=1, 2, \ldots, N)$ be different—can be satisfied with only N wavelengths. As a result, all FDM devices can be identical, considerably simplifying implementation of the architecture.

FDM DISTRIBUTION SYSTEMS WITH OPTICAL FILTERS

Optical frequency division multiplexing distribution systems providing more than ten frequency multiplexed optical signals, separated by of the order of GHz, distribute signals to plural receivers, where one of the signals is selected by a frequency selection switch (FS-SW). Light sources with different carrier frequencies, each of which is independently modulated, are optically multiplexed by a multiplexer and coupled to a single mode fiber. The multiplexed signals are distributed to subscribers by a star coupler. At the receiving equipment, multiplexed signals are selected by the FS-SW and directly detected by a photodetector. An FS-SW is constructed by cascade connection of Mach–Zender interferometer-type periodic filters. The frequency selection is performed by changing the phase shift value of the optical waveguide of each filter.

Frequency spacing among optical sources is about several GHz, so several ten or several hundred frequencies can be multiplexed is a broad, low-loss wavelength region of silica fiber, 0.8–1.7 μm, is fully available. Such a distribution system is also possible to select optical frequencies by changing the local optical source frequency using heterodyne detection [3.54]. This detection requires placing a frequency stabilized optical source with a frequency tuning function at each receiver.

For high density frequency division multiplexing (such as a LAMBDANET FDM network), an angle-tuned in-line Fabry–Perot etalon filter for optical channel selection was proposed and demonstrated; their relative advantages have been discussed in recent years [3.55].

The FDM distribution using a Mach–Zender interferometer filter is reported [3.17].

3.4 COHERENT OPTICAL FDM TECHNOLOGY [3.66]

3.4.1 FDM coherent star network

A coherent detection, wavelength division multiplexed, optically star-coupled network offers several advantages in interconnecting large numbers of users [3.59–3.62]: (1) FDM allows the use of the wide available optical bandwidth [3.63]; (2) coherent detection yields high receiver sensitivity and high-frequency selectivity; and (e) the star configuration efficiently distributes the optical power through the network. By combining these features, such a FDM coherent star network can provide a large throughput (number of users × bit rate).

The minimum channel spacing needed to achieve this result is limited by the interference generated by the heterodyne detection process [3.64, 3.65]. In a large multichannel system, channel selection with a feasible IF frequency is done by positioning the LO frequency near the desired channel. A large enough frequency interval is thus needed between channels to avoid interference from the image frequency since the optical mixer is equally sensitive to inputs above and below the LO frequency (one input is the image frequency of the other). The resulting heterodyne process interleaves, in the IF domain, the channels on the low-frequency side of the LO signal with the channels on the high-frequency side of this signal. As a result, co-channel interference is generated in the IF domain where the channels are closer.

Additional interference due to frequency beat between adjacent channels is suppressed by the use of a balanced mixer. (For FSK signals, interference from direct detection terms arises only from frequency-beat between adjacent channels [3.64].) For an IF channel spacing yielding an acceptable level of co-channel interference, the minimum optical channel spacing D is obtained when the IF channels are spaced equally. The quantity D is then equal to twice the IF channel spacing. The two lowest IF values yielding this result are $IF_1 = D/4$ and $IF_2 = 3D/4$. The first of these values corresponding to a LO frequency near the desired channel is shown in Fig. 3.9(a) [3.66]. Unfortunately, the resulting IF frequency is too low for FSK signals to suppress interference between the filtered IF signal and the demodulated baseband signal. As a result, the IF must be increased and the channel spacing is augmented accordingly. This problem is avoided by the use of a second IF value. In this case, the LO frequency is near the adjacent channel (Fig. 3.9(b)). Thus, the desired channel is heterodyned as the second lowest IF signal. The increase of co-channel interference (due to the presence of two adjacent channels instead of one) can be offset by increasing slightly the channel spacing. Therefore, the minimum channel spacing

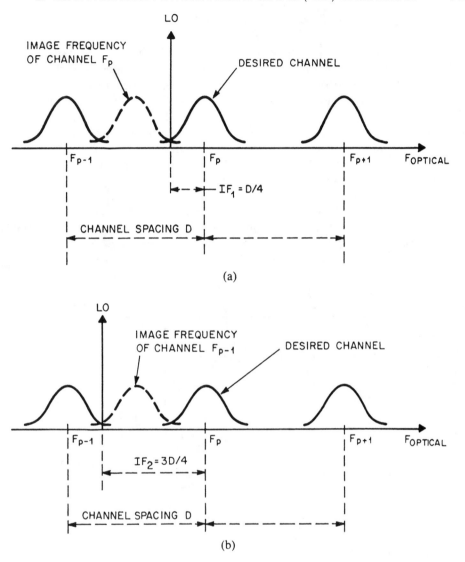

Fig. 3.9 Optimum channel spacing in coherent EDM. (From [3.66] Glance *et al.* (1988) *J. Lightwave Technol.* **LT-6**, 1770–1781. Reproduced by permission. ©1988 IEEE.)

is provided by the second IF value. The use of this relatively large IF frequency may require the use of a front-end receiver [3.67, 3.68] tuned to the IF to achieve shot-noise limited performance for a high bit-rate system.

The minimum achievable channel spacing also depends on the modulation format. The optimum result is given by FSK with a modulation index of one [3.64]. This type of modulation presents several other advantages: it allows direct modulation of the laser current, and provides the means to stabilize the frequency of the optical sources.

DESCRIPTION OF FDM COHERENT NETWORK

The optical sources used for transmission are fast frequency tunable external cavity lasers [3.69]. These lasers provide a narrow-line single-frequency signal at 1.28 μm, which can be frequency tuned over about 4000 GHz as well as frequency modulated up to 100 MHz. Modulation is done by FSK at 45 Mbit/s with a modulation index of about one by a random NRZ bit stream with $2^{15}-1$ pattern length. The three sources transmit at optical frequencies spaced by a frequency interval of 300 MHz. The three optical signals are combined by a 4×4 fiber star coupler. The schematic of the FDM network used for making the measurements is shown in Fig. 3.10 [3.66].

Each of the four output fibers of the coupler carries the three FDM signals. The signal from one of these fibers is combined by means of a 3 dB fiber coupler with the LO signal provided by a conventional external-cavity laser [3.70, 3.71]. The polarization of the transmitted signals are manually adjusted to match the polarization state of the LO signal. The combined signals from the two outputs of the coupler feed a

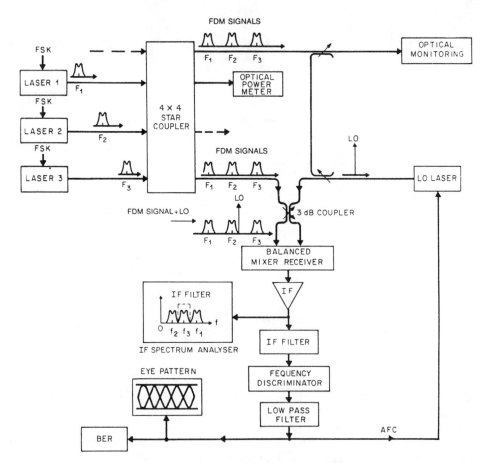

Fig. 3.10 FDM network used for measurements. (From [3.66] Glance *et al.* (1988) *J. Lightwave Technol.* **LT-6**, 1770–1781. Reproduced by permission. ©1988 IEEE.)

balanced mixer receiver [3.72] which heterodynes the received signal to an IF frequency of 225 MHz. The optical power of the LO signal at the photodetectors is about 0.7 dBm. As a result, the shot noise due to the LO signal dominates the thermal noise of the receiver. This is verified by observing, using a spectrum analyser, that the IF noise increases by 12 dB in the presence of the LO signal.

The use of a balanced mixer is essential in an FDM star network. This type of mixer eliminates the interference (channel–cross-channel as well as intrachannel) arising from the direct-detection terms [3.64]. It also allows a more efficient use of the optical power of the LO source and thus reduces the degradation caused by the shot noise originating from the received signal. In a FDM star network, the shot noise generated by the entire collection of the received FDM signals is equal to the shot noise due to a single transmitting optical source minus the excess loss of the network. Thus for a system excess loss of, say, 8–10 dB, the degradation associated with this shot-noise contribution will be a few tenths of a decibel. The IF signal is amplified and then filtered by an IF filter 100 MHz wide centred at 225 MHz. Demodulation is accomplished by frequency discriminator, and the resulting baseband signal in filtered by a low-pass filter having a 3 dB cutoff of 35 MHz. Selection of the desired channel is achieved by tuning the LO frequency to the value that centres the wanted channel within the bandwidth of the IF filter.

The resulting IF is maintained by an automatic frequency control (AFC) circuit controlling the optical frequency of the LO signal (also Fig. 3.10). The signal from one of the three remaining output fibers from the star coupler is used, after combining with a fraction of the LO signal, to monitor the four optical signals (the three FDM signals and the LO signal). This is done using a spectrometer and a scanning Fabry–Perot etalon. Another output fiber is utilized to measure the received signal. The measurement takes into account the slight difference of received signal between this fiber and the fiber connected to the receiver.

IF CHANNEL SELECTION

In order to operate at an IF frequency above the frequency of the baseband signal, the desired optical channel is heterodyned by LO signal which is tuned to the edge of an adjacent optical channel. Thus, the channel selected by the IF filter is the second lowest IF channel. This arrangement minimizes the frequency spacing needed between the optical channels and separates the demodulated baseband signal from its IF signal. For a modulation rate of 45 Mbit/s, the minimum optical frequency spacing preventing co-channel interference in the IF domain is found to be about 260 MHz [3.73], to provide a protective margin, we selected a channel spacing of 300 MHz. The resulting frequency spacing between the interleaved IF channels is 150 MHz.

CO-CHANNEL INTERFERENCE

Channel spacing is deliberately reduced in order to determine co-channel interference. The power penalty caused by co-channel interference is 1 dB when the selected channel is interfered in the IF domain by a second channel spaced by 100 MHz, received with the same optical power and modulated at the same bit rate of 45 Mbit/s but with a different random bit stream.

The co-channel interference increases very fast when the frequency spacing is reduced to 90 MHz. It completely disappears when the frequency spacing is increased to 130 MHz. Similarly, no co-channel degradation is observed when the selected channel is interleaved by two adjacent channels spaced by 150 MHz. This result agrees well with the theoretical results reported in [3.64].

SYSTEM THROUGHPUT

The system throughput (number of users × bit rate) can be estimated from the above results for a network operating around 1.55 μm (the minimum fiber loss window). This requires the largest possible star coupler compatible with the receiver sensitivity. For example, let us assume that a user transmits 0 dBm of optical power into his fiber connected to a star coupler interconnecting 131 072 (2^{17}) subscribers. The excess loss suffered by the signal propagating through the star coupler is proportional to the 17 stages of 3 dB couplers constituting the star.

3.4.2 Frequency stabilization of FDM network

The optical signals of a densely spaced FDM system must be confined to a comb of equally spaced frequencies. A Fabry–Perot resonator offers the ideal solution for solving this problem. Such a device provides a set of equally spaced resonances which can be used to assign the frequencies of the system. The result is achieved by frequency-locking each optical source to one of the resonances. A simple fiber Fabry–Perot having a low finesse can be used for this application.

FREQUENCY STABILIZATION OF FDM SIGNAL ORIGINATING FROM THE SAME LOCATION

Experimental circuit

The three-channel system described above was used for demonstration of this frequency stabilization circuit. As previously, each laser is FSK modulated at 45 Mbit/s by an independent random NRZ bit stream with a modulation index of about one. The frequency stabilization circuit is connected to one of the outputs of the 4×4 optical coupler (Fig. 3.11, [3.66]).

The optical part of the frequency stabilization circuit consists of a single-mode fiber Fabry–Perot resonator providing a comb of resonances equally spaced by about 500 MHz with a 3 dB bandwidth of about 50 MHz. One end of the regular section of fiber is connected to one of the outputs of the 4×4 optical coupler; the other end illuminates a photodetector.

When an optical source drifts from the peak of a Fabry–Perot resonance, the photodiode detects a baseband signal having the same pattern as the FSK bit stream

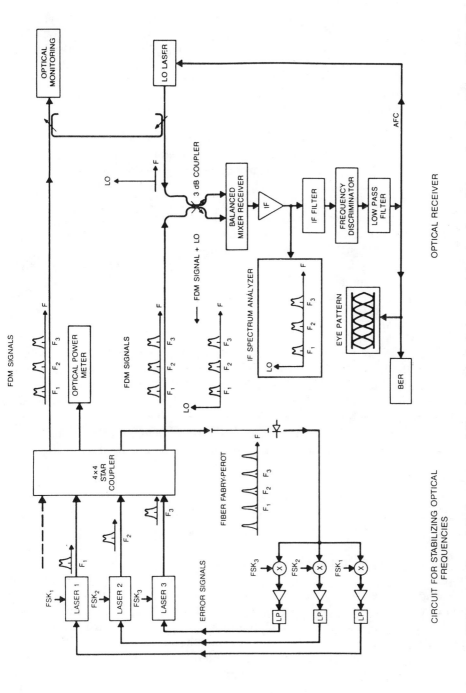

Fig. 3.11 FDM network used for frequency stabilization of optical signals originating from the same location. (From [3.66] Glance *et al.* (1988) *J. Lightwave Technol.* **LT-6**, 1770–1781. Reproduced by permission. ©1988 IEEE.)

modulating the optical source. This signal is used as the error signal to frequency-lock the optical sources. The polarity of the detected pattern relative to that used to modulate the laser depends on which side of the resonance the frequency drift occurred. The error signal [3.67] can be obtained by multiplying the FSK bit stream by the detected signal and filtering the resulting amplified product by a low-pass filter. The error signal is then used to lock the laser frequency to the selected Fabry–Perot resonance.

FREQUENCY STABILIZATION OF FDM SIGNALS ORIGINATION FROM DIFFERENT LOCATIONS

The problem of the previous stabilization circuit can be resolved by frequency-locking each laser to a separate tunable fiber Fabry–Perot resonator. The set of resonances is made the same for all the lasers by synchronizing the Fabry–Perot resonators. One signal is supplied by the laser, the other signal originates from a central optical source frequency-stabilized by its own Fabry–Perot resonator. This optical signal is frequency modulated by a known frequency tone supplied to each separate station. Thus, two error signals are obtained from the detected output of each local Fabry–Perot resonator. One error signal is used to frequency-lock the laser, the other error signal is utilized to frequency-lock the local Fabry–Perot to the central Fabry–Perot acting as a master for the whole system.

Experimental circuit

The above scheme was demonstrated with the circuit shown in Fig. 3.12 [3.66] which allows the use of the three available optical sources to generate three equally spaced channels. The optical signals are modulated by FSK.

As previously described, each optical source is frequency-locked to a resonance of its Fabry–Perot resonators. This is achieved (for a set of resonances) by using a fraction of the power of one optical signal to lock the Fabry–Perot resonators of the two other sources. This is achieved the following way: one optical source, used as a reference, feeds a fraction of its signal into its tunable fiber Fabry–Perot resonator at a frequency adjusted near one of the resonances. A photodiode detects the output of this device. An error signal is generated by multiplying the FSK bit stream by the detected signal and filtering the resulting amplified product by a low-pass filter. This error signal is used to lock the laser frequency to the selected Fabry–Perot resonance. This resonance is then used as the master frequency to synchronize the Fabry–Perot resonator of the other optical sources as described below.

The two other optical sources have the same frequency-locking circuit as described above except that the optical signal feeding the Fabry–Perot resonator includes a fraction of the reference signal (Fig. 3.12). In this case, the photodiode detects a mixture of two FSK baseband signals: one due to the frequency drift of the laser relative to the selected Fabry–Perot resonance; the other due to the drift of the Fabry–Perot resonance relative to the reference frequency. Two error signals are thus generated by dividing the photodetected signal into two branches, each feeding a balanced mixer. The second input of one of the balanced mixers is fed by the FSK baseband signals modulating the laser; the second input of the other balanced mixer

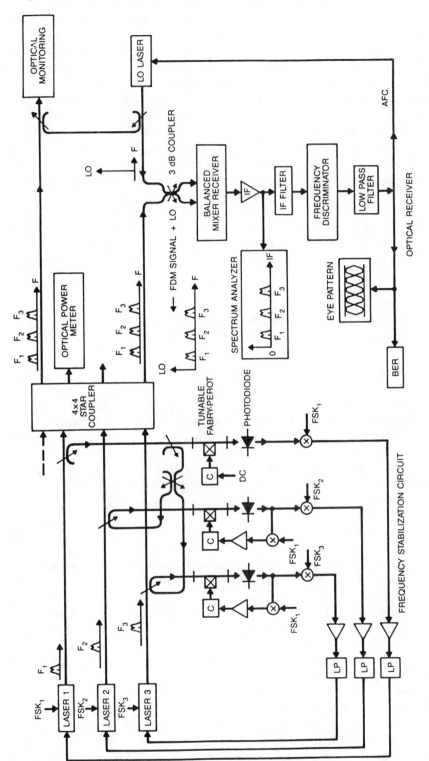

Fig. 3.12 FDM network used to synchronize the FP resonators of the two other channels. (From [3.66] Glance *et al.* (1988) *J. Lightwave Technol.* **LT-6**, 1770–1781. Reproduced by permission. ©1988 IEEE.)

is driven by the FSK baseband signal modulating the reference signal (the three FSK signals have independent random patterns). Two balanced mixers act as correlators yielding, after amplification and filtering, the two desired independent error signals: one due to the frequency drift of the laser relative to the selected Fabry–Perot resonance, the other originating from the drift of the Fabry–Perot resonance relative to the reference frequency. The first of these error signals locks the laser frequency, and the second error signal controls the Fabry–Perot resonator.

With this set up, the Fabry–Perot resonators are synchronized over a set of resonances whose number depends on the tolerance with which the length of the Fabry–Perot cavities are fabricated. In the present case, this length is equal to $30\,cm \pm 0.3\,mm$. The corresponding free spectral range is $346 \pm 0.35\,MHz$. Consequently, starting from the reference resonance, each following resonance of a slave Fabry–Perot is shifted relative to the corresponding master Fabry–Perot by a maximum of $\pm 0.35\,MHz$ or about $1/1000$ of a free spectral range. The cumulative frequency shift over one hundred resonances is thus $35\,MHz$, or $1/10$ of a free spectral range. This defines approximately the number of synchronized resonances usable for a system with the above tolerance.

3.5 MICROWAVE SUBCARRIER MULTIPLEXING MULTICHANNEL OPTICAL COMMUNICATION

Recently, significant interest in microwave optical links has been keeping pace with the fast progress occurring in high-speed optoelectronic devices. Potential applications include antenna remoting, communications systems, radar systems, and phased-array signal distribution. Also, considerable interest has arisen regarding the use of fiber-optic analogue multichannel video transmission systems for modulated VHF carriers.

The use of microwave subcarriers in optical communication systems represents an attractive approach to the design of wide-band distribution systems including combinations of video services, as well as voice and data. The attractive features of subcarrier multiplexing (SCM) in optical communication systems are: (1) to provide a way to exploit the multigigahertz bandwidth potential of single mode optical fiber and lightwave components using well established and commercially available technology; (2) to evolve in step with expected changes in video technology such as high definition television, new digital compression techniques, and simultaneous transmission of SCM and baseband signals; (3) to offer a highly flexible way to design broad-band systems which can simultaneously accommodate both analogue and digital modulation of the microwave subcarriers without relying on time division multiplexing (TDM); (4) in particular, SCM techniques do not require an absolute frequency stabilization scheme, which is necessary for conventional multicarrier (i.e. multiple laser) coherent FDM systems.

In a conventional frequency division multiplexed (FDM) architecture, information for each video channel is transmitted by a separate optical carrier. Alternatively, more efficient multichannel transmission can be achieved with one optical carrier using subcarrier multiplexing (SCM) techniques. SCM networks provide an attractive approach for utilizing the wide bandwidth of single mode fiber and the electro-optic components, while taking advantage of commercially available microwave electronics. In addition, the bandwidth allocation is very flexible. Both digital and analogue multichannel direct-detection SCM systems have been developed [3.74–3.80], and research on coherent systems has also been reported [3.81, 3.82].

3.5.1 Microwave transmitter (multiplexer) and optical link [3.74–3.81]

The schematic blocks of the SCM system, shown in Fig. 3.13, are the SCM transmitter, the optical link, and the SCM receiver. This section discusses the design of the microwave transmitter and optical link.

SUBCARRIER GENERATION AND MODULATION

Each microwave subcarrier frequency is generated by a voltage-controlled oscillator (VCO) and multiplexed using microwave power combiners as illustrated in Fig. 3.14 [3.74] for one of the 20 channels. The VCO is modulated by adding the ac signal to a dc voltage in a bias tee and applying the composite voltage to the VCO tuning port. The dc voltage term is adjusted to set the proper subcarrier frequency, and the ac voltage term is adjusted to set a frequency deviation of 100 MHz for the 100 Mbit/s data. Isolators are used at the VCO outputs to prevent the intermodulation distortion resulting from VCO interaction through the power combiners. Low-pass filters are used following the isolators for the channels that could produce in-band second harmonics. Finally, combinations of single carrier amplifiers and attenuators are used to provide the microwave power required to establish the desired optical modulation depth m.

The 20 digital data sources are generated by voltage dividing the data from a bit error rate test set. The data is first divided in a 1×4 divider and then four 1×5

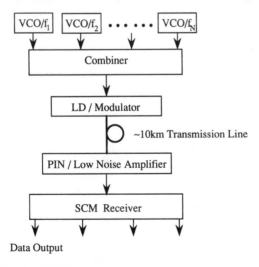

Fig. 3.13 Schematic block of SCM system.

dividers. Each signal is amplified following a voltage divider to establish the appropriate frequency deviation at each channel. The channel being monitored for digital performance is modulated directly from a second data output port on the bit error rate test set.

The output of the transmitter with the modulation signal off is shown in Fig. 3.15(a) [3.74] and with the modulation signal on in Fig. 3. 15(b). The transmitted output signal is very clean, with all distortion terms at least 50 dB below the carrier. All of the channels are independent since each carrier is independent and at a random phase to all the other channels.

The microwave subcarrier frequencies are selected to minimize adjacent channel interference. The subcarriers are placed between 2.1–5.9 GHz with 200 MHz carrier spacing and a frequency deviation of 100 MHz for a 100 Mbit/s signal. This particular subcarrier frequency placement provides a 100 MHz guard band between each subcarrier. All of the harmonic frequencies and second-order intermodulation products will fall in the centre of the guard band minimizing the noise in the receiver 120 MHz signal band.

OPTICAL LINK

The optical link consisting of a high-speed 1.3 μm InGaAsP laser [3.77], 12 km of single mode fiber, and a InGaAsP p–i–n photodetector [3.78]. The complete optical receiver consists of the photodetector followed by a bias tee and a 50 Ω low-noise amplifier (LNA) as the optical receiver. A 2–6 GHz isolator is used between the bias tee and LNA to reduce reflections caused by the impedance mismatch of the photodetector and 50 Ω amplifier. The noise figure (NF) of the LNA is 2.5–3.5 dB over the 2–6 GHz bandwidth.

The system performance is determined by the overall carrier-to-noise ratio (CNR). The total system noise consists of the LNA thermal noise, the diode laser relative intensity noise (RIN), and intermodulation distortion products (IMPs). The first

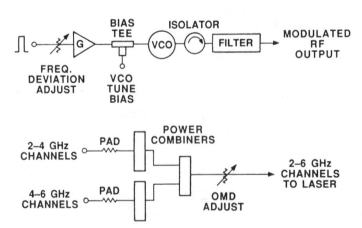

Fig. 3.14 The SCM microwave transmitter. (From [3.74] Hill and Olshansky (1990) *J. Lightwave Technol.* **LT-8**(4), 554–560. Reproduced by permission. ©1990 IEEE.)

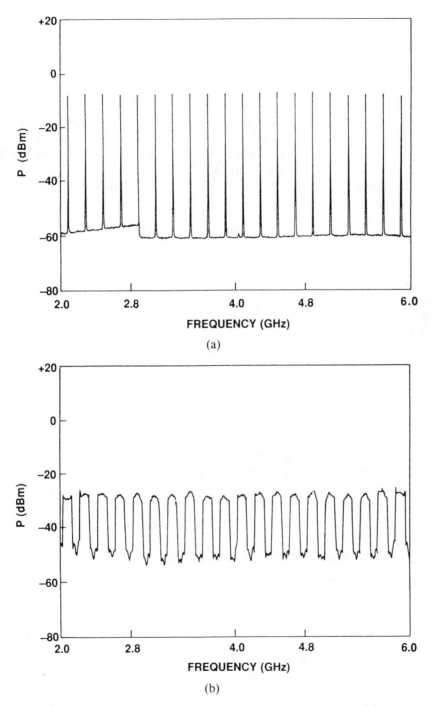

Fig. 3.15 The microwave output spectrum for the 20-channel SCM transmitter. (From [3.74] Hill and Olshansky (1990) *J. Lightwave Technol.* **LT-8**(4), 554–560. Reproduced by permission. ©1990 IEEE.)

CNR term, (CNR_1), is the ratio of the carrier power to the sum of the optical receiver thermal noise and the RIN and is given [3.79] as

$$\text{CNR}_1 = \frac{\frac{1}{2}(m\ I_{dc})^2 R_L}{NF\ kT\ B + \text{RIN}\ I_{dc}^2\ R_L B} \qquad (3.23)$$

where I_{dc} is the dc photocurrent, $R_L = 50\ \Omega$ is amplifier impedance. kT is the thermal noise, and B is the microwave receiver filter bandwidth. A CNR of 16 dB is required for a 10^{-9} bit error rate (BER). The diode laser performance characteristics impact the magnitude of I_{dc}, RIN, and noise due to the IMPs.

The magnitude of the intrinsic laser RIN [3.80] is determined to be the ratio

$$\text{RIN} = \left(\frac{I_n}{I_{dc}}\right)^2 \qquad (3.24)$$

where I_n is the noise curent per $(\text{Hz})^{1/2}$, and I_{dc} is the detected photocurrent. The actual RIN may be higher than (3.24) unless care is taken to minimize [3.81–3.83] optical reflection back into the laser.

The magnitude of CNR_1 can be improved by increasing the modulation depth m or by increasing the photocurrent up to a point. For any RIN and amplifier noise figure, the RIN term in (3.23) will begin to dominate the noise at some photocurrent. For large RIN or large I_{dc}, (3.23) reduces to

$$\text{CNR}_1 = \frac{m^2}{2\ \text{RIN}\ B} \qquad (3.25)$$

and the CNR_1 can only be increased by increasing m. The modulation depth can only be increased to a point as well. As the modulation depth increased, the noise contributed by the IMPs will increase causing a degradation in the total CNR.

3.5.2 Microwave receiver and experimental results

MICROWAVE RECEIVER

Figure 3.16 [3.74] shows schematically the SCM receiver. The SCM receiver consists of a subcarrier frequency selection circuit and a 2.67 ns delay line discriminator which demodulates the FSK signal and closely resembles a typical FM radio receiver.

The frequency selection is accomplished by up-shifting the desired carrier to a 6.5 GHz first intermediate frequency (IF) and a mixer and tunable VCO. Up-conversion is used to avoid typical mixer images. The first IF is then passed through a bandpass filter with a 3 dB bandwidth B of 120 MHz and down-converted to the second IF of 1.2 GHz with an additional mixer and fixed frequency VCO. The bandwidth of the bandpass filter also represents the receivers noise bandwidth of

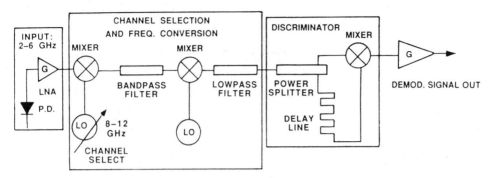

Fig. 3.16 The SCM optical and microwave receiver block diagram. (From [3.74] Hill and Olshansky (1990) *J. Lightwave Technol.* **LT-8**(4), 554–560. Reproduced by permission. ©1990 IEEE.)

120 MHz. While not provided in this SCM demon-stration, one or both of the receiver VCOs could be phase locked to track any subcarrier frequency drift at the transmitter.

A low-pass filter is used before the discriminator to remove the first IF and second local oscillator frequencies. The second IF frequency is selected to match the zero crossing point of the discriminator frequency–voltage transfer function with the most linear slope over a 100 MHz bandwidth. The magnitude of the discriminator output voltage over the 100 MHz bandwidth is a linear function of the input power over a range of about −5 to +2 dBm. An output voltage of ±50 mV is achieved with −2 dBm at the input. A linear-amplifier–variable-attenuator combination preceded the discriminator to compensate for the mixer conversion loss and to ensure the discriminator is used in its linear input power range. A better choice for the amplifier may be a limiting amplifier to minimize AM noise and maintain a constant power to the discriminator.

EXPERIMENTAL RESULTS

A field experiment of a microwave multicarrier transmission system for satellite Earth stations is an example for applications [3.83]. The system was investigated because the high capacity and low attenuation of optical fiber made it ideal for extending the reach of existing satellite-based networks, serving as a link between Earth stations and terrestrial networks. Motivation for transporting satellite downlink multicarriers at microwave frequencies directly via optical fiber stems from considerations of simplicity and economics. Unlike previously reported regional optical fiber networks [3.84] that transported down-converted satellite signals at IF frequencies, in the method described here there is no need for signal processing equipment at the Earth station antenna site.

The complete system configuration is shown in Fig. 3.17(*a*) [3.83]. The downlink satellite signal from an Earth station antenna is amplified and used to directly modulate a high-speed InGaAsP LD which has a 6 GHz bandwidth at a bias level of $3 I_{th}$ (threshold current $I_{th} = 15$ mA at room temperature). Figure 3.17(*b*) shows

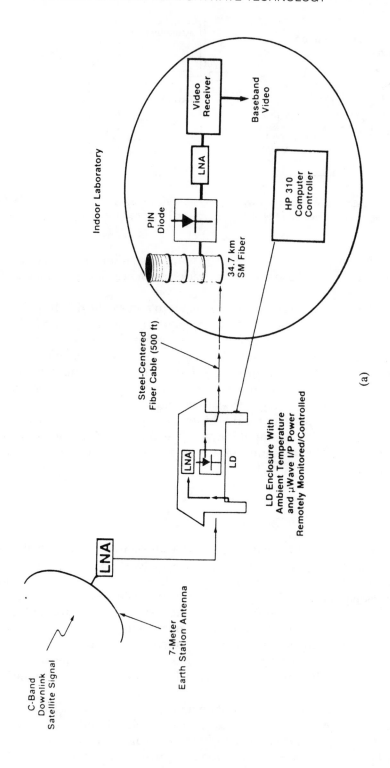

(a)

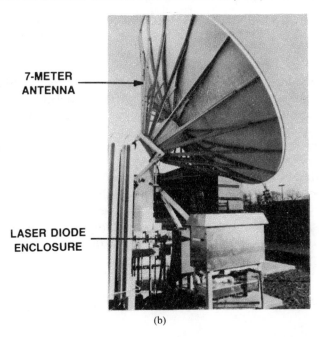

(b)

Fig. 3.17 Microwave multicarrier transmission system. (From [3.83] Way *et al.* (1987) *J. Lightwave Technol.* **LT-5**(9), 1325–1332. Reproduced by permission. ©1987 IEEE.)

the 7 m antenna and the LD enclosure. The optical signal was detected by an InGaAs p–i–n diode which has a measured bandwidth of 6 GHz. The satellite microwave signal from the p–i–n diode was then amplified by a C-band low noise amplifier (LNA). No other signal processing at the Earth station site is necessary.

After having discussed the signal power and the various major contributions to noise power, the system CNR (carrier-to-noise ratio) per unit bandwidth could be summarized as in the following equation [3.83]:

$$(\text{CNR})^{-1} = \frac{2\,(F-1)\,k\,T}{m^2\,I_D^2\,R_D} + \frac{2\,\text{RIN}_T}{m^2}$$

$$+ \frac{2\,P_{\text{IM}}}{R_D G\,(I_D\,m)^2}. \tag{3.26}$$

Calculated results based on this equation will be used to compare with measured results.

The measured C-band carrier to noise ratio CNR (per 36 MHz bandwidth) after transmission over 0 and 35 km of fiber are shown in curves (*c*) and (*e*) in Fig. 3.18 [3.83], respectively. Both curves are closely matched to a calculated curve (*d*) in the same figure. Curve (*d*) was obtained by using the measured RIN_T and OMI values and equation (3.26). Note that RIN$_T$ and OMI values were measured after the 35 km fiber link. Baseband video quality of a demodulated FM-TV was experimentally characterized. The measured baseband SNR of an FM-TV signal at 3840 MHz for a 0 and 35 km transmission distance are shown in curve (a) and (b), respectively. When the CNR is 16.5 dB, the measured SNRs (the SNR for video is commonly defined as the ratio of the total luminance signal of the output baseband video signal to the rms value of noise) are close to the predicted value of 56 dB (with a 2 dB range). The measured video differential gain (DG) and differential phase (DP) after 35 km transmission as functions of received optical power are shown in curves (f) and (g), respectively.

The CNR and SNR degradations observed in curves (b) and (e), as compared with curves (a) and (c), Fig. 3.18 [3.83], respectively, are due to (1) the additional reflection noise caused by eight more pairs of biconic connectors with interconnected eight spools of fibers, and (2) the limited bandwidth of long length of fibers. Uncleaned connectors can severely degrade the video signal quality.

3.6   WAVELENGTH DIVISION (WD) PHOTONIC SWITCHING

Photonic switching is expected to play a key role in optical, one-link 'transparent' networks over optical fiber transmission highways. The optical one-link networks have design flexibilities to meet requirements for future optical telecommunication networks providing multimedia services, especially for video information services. In the 1990s, photonic switching research trends will be divided into two directions: applying photonic switching technologies to practical telecommunications networks, and applying photonic processing technologies to photonic switching systems. Photonic switching systems have advantages over conventional electronic switching systems for use in exchanging broad-band signals. Broad-band networks providing various of services, such as video telephony and video broadcasting, have been receiving increasing attention in recent years. To realize such networks, photonic switching technologies are attractive, both for transmission and switching. Photonic switching fall into three categories, namely, space division (SD), time division (TD), and frequency division or wavelength division (WD) switching as mentioned in Section 1.2–4 (the terms wavelength division switching and wavelength switch are already used conventionally and we will use them even when sometimes the channel spacing is quite narrow). In extending the photonic switching network to a large capacity switching system, the SD switching network will encounter problems, wherein it requires a rapidly growing number of optical cross points and a huge number of optical fibers for inlet/outlets. The photonic TD and WD switching networks have the potential capability of giving a good solution for these problems [3.85, 3.86].

As compared with the TD switching network, the WD switching network has some advantages. One is bit-rate independence for individual wavelength channels. Then, various speed broadband signals can be exchanged without difficulty. The other advantage is that there is no necessity for high-speed operation in switching control

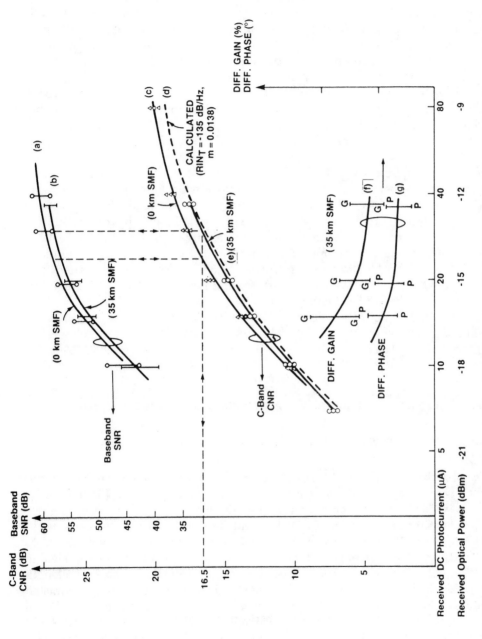

Fig. 3.18 The measured parameters of microwave multicarrier system. (From [3.83] *Way et al.* (1987) *J. Lightwave Technol.* **LT-5**(9), 1325–1332. Reproduced by permission. ©1987 IEEE.)

circuits. In the TD switching network, ultrahigh speed operations are required, not only for the switching network but also for control circuits, in order to exchange a time division multiplexed broad-band signal. For example, if 32-channel 150 Mbit/s signals are time division multiplexed, the operational speed of the control circuits, as control memories, reaches as high as 4.8 Gbit/s. On the other hand, the WD switching network is controlled by the tuning of wavelengths. High-speed tuning is not required for circuit switching networks. Therefore, conventional low-speed electronic circuits, whose operation speed is micro- or milliseconds can be used as control circuits. Moreover, the WD switching system has the potential capability for extension to a wide-area network in partnership with the wavelength division multiplexed (WDM) transmission system. The WDM transmission also has bit-rate independence for individual channels. Therefore, this wide-area network will be able to provide an optical bit-rate independent connection between subscribers.

From this viewpoint, an optical broad-band network architecture, using photonic WD switching systems and WDM optical transmission systems, was proposed [3.87]. Various photonic WD switching systems have already been proposed and demonstrated. Two kinds of wavelength switch, which accomplish the wavelength interchange, and a wavelength selective space switch have been demonstrated, using acousto-optic deflectors and distributed feedback laser diode (DFB LDs) as tunable wavelength filters [3.88, 3.89]. Also proposed and demonstrated are the passive wavelength routing systems [3.90, 3.91] and WDM passive-star systems [3.92–3.96].

Coherent optical detection technologies can realize dense FDM optical transmission, due to their high frequency selectivity. Coherent optical technologies are also important for achieving a large-scale photonic switching system [3.97–3.99].

3.6.1   Wavelength division (WD) switching techniques [3.100]

PRINCIPLES OF WD SWITCHING SYSTEMS

A wavelength switch (λ switch), which can interchange wavelengths from an input WDM, signal, is a basic component in the WD switching network [3.101–3.103], as a time switch is a basic component in the TD switching network. There are two kinds of switches, as shown in Fig. 3.19(a) and Fig. 3.19(b) [3.100]. In the λ switch shown in Fig. 3.19(a) an input WDM signal is split and parts thereof are led to individual variable wavelength filters, each of which extracts a specific wavelength signal from the input signal. The output optical signal from the tunable wavelength filter is then applied to an optically controlled modulator and is intensity modulated onto a pre-assigned-wavelength light carrier, which is extracted from the WDM wavelength reference light by a fixed wavelength filter. Through this process, input wavelengths λ_a, λ_b, $\cdots$, and up to λ_z, in Fig. 3.19, are converted into wavelengths λ_1, λ_2, $\cdots$, and up to λ_n, respectively, without any loss in intensity modulated information, which results in wavelength interchanging. Moreover, by controlling multiple tunable filters to select the same wavelength signal, multiple wavelength carriers can be modulated according to the same information. A multicast function can be achieved in this way. If the multicast function is not necessary, attenuation, caused by optical signal splitting in the switch shown in Fig. 3.19(*a*) [3.100], can be avoided by using another type of wavelength filter which can separate a specific wavelength signal from the WDM signal, as shown in Fig. 3.19 (*b*).

Multistage switching networks are necessary to construct large capacity switching systems. A λ^m switching network [3.100] is proposed for this purpose. As shown in Fig. 3.20 [3.100], a wavelength demultiplexer (DMUX) and multiplexer (MUX)

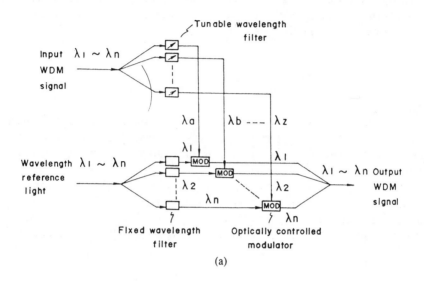

(a)

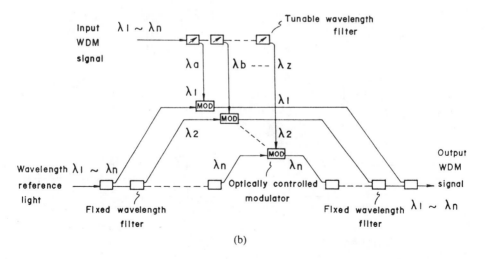

(b)

Fig. 3.19 Wavelength switch. (From [3.100] Suzuki *et al.* (1990) *J. Lightwave Technol.* **LT-8**(5), 660–666. Reproduced by permission. ©1990 IEEE.)

are mounted in interstage connections. Between individual DMUX output ports and individual MUX input port, an optical path is provided, thereby giving each λ switch potential connectivity to every nextstage switches. As shown in Fig. 3.20, the λ^3 switching network has n input/output links, each of which is n-channel wavelength division multiplexed. Therefore, line capacity for the λ^3 network is expressed as n^2.

In InP tunable wavelength filters, their selected wavelengths are generally controlled by their refractive indexes. A 1% refractive index control possibility in InP materials has been reported, using the free-carrier plasma effect [3.104] or the non-linear optical properties in multiple quantum well (MQW) structures [3.105]. Hence, a 1% tuning range, 150 Å in the case of the 1.5 μm region, in the tunable wavelength filter is expected to be achieved in the future.

EXPERIMENT ON WD SWITCHING

An experiment on a WD switching system using InGaAsP/InP phase-shift-controlled DFB LD tunable wavelength filters [3.106] was performed to test the feasibility of a switching system based on today's optical semiconductor tunable wavelength filter technology.

Figure 3.21 [3.100] shows the experimental photonic WD switching system block diagram considered in this section. This system consists of a wavelength multiplexer, λ switch, and wavelength demultiplexer. The wavelength multiplexer consists of LD modulators [3.107], which intensity modulate light carriers supplied

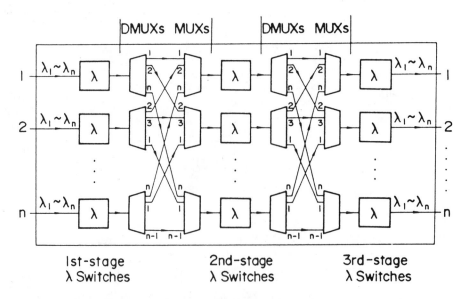

Fig. 3.20 λ^3 switching network. (From [3.100] Suzuki *et al.* (1990) *J. Lightwave Technol.* **LT-8**(5), 660–666. Reproduced by permission. ©1990 IEEE.)

from wavelength reference light sources according to input CH 1-n signals, and an optical combiner. In the λ switch, an input WDM signal is split and parts are led to individual tunable wavelength filters, each of which extracts a specific wavelength signal. The output signal from the tunable wavelength filter is then converted to an electronic signal by an optical–electronic converter. A pre-assigned-wavelength light carrier is intensity modulated by a LD modulator according to the electronic signal from the optical–electronic converter. The wavelength demultiplexer consists of an optical splitter, fixed wavelength filters, and optical–electronic converters.

The number of WD channels n in this photonic switching network is mainly determined by the tunable filter performance. The phase-shift-controlled DFB LD tunable filter has the advantages of wide wavelength tuning range and narrow transmission bandwidth with constant gain and bandwidth over the tuning range. Tunable filter optical gain and spontaneous emission noise characteristics determine the required filter input power. Calculated filter input peak power values, taking the filter spontaneous emission into account, are -31.5 dBm and -38.5 dBm to satisfy the 10^{-10} error rate for a 200 Mbit/s signal, on condition that the tunable filter optical gain is 15 and 75, respectively. Assuming 3 dBm modulator output peak power and 5 dB coupling loss, between 15 and 75 optical gains could be assigned to the combiner and splitter maximum loss, while the operation margin is 3.5 dB. Therefore, it is concluded that the n values can reach 8 and 16, respectively, where 0.5 dB single stage optical coupler excess loss in the combiner and splitter is counted.

With improvement in filter optical gain and replacement of the optical combiner by a low-loss wavelength multiplexer, n can increase. For example, required filter input peak power will be reduced to -43 dBm using a filter with 200 optical gain.

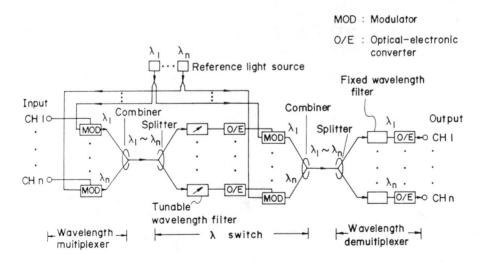

Fig. 3.21 Experimental WD switching system. (From [3.100] Suzuki *et al.* (1990) *J. Lightwave Technol.* **LT-8**(5), 660–666. Reproduced by permission. ©1990 IEEE.)

As a result, a 32.5 dB loss value can be assigned to the wavelength multiplexer and optical splitter. Therefore, n can reach 100 by using the wavelength multiplexer with 10 dB loss and the optical splitter with 22.5 dB loss.

3.6.2   Coherent wavelength division photonic switching system [3.116]

The future broad-band network should support a variety of services, such as voice, high-speed data, high-speed still picture, and video services. A high-speed packet switching system will support a wide variety of services. However, a broad-band circuit switching system would still be attractive for supporting broad-band/high-speed communication services, even in the future broad-band integrated networks [3.108]. The coherent WD switching system is very attractive for application in such a broad-band circuit switching system. For the circuit switching application, extremely high-speed operation are not required for wavelength switching devices. In this case, some kind of computer-controlled LO wavelength tuning scheme [3.109, 3.110] can be applied for random access for the WD channels. In addition, coherent switch technology can be applied to a time division [3.111] or packet switching system [3.112, 3.113] with fast wavelength switching operation in the LOs. Fast wavelength switching experiments have already been demonstrated using multi-section DFB LDs or DBR LDs [3.111, 3.114, 3.115]. Further development is still required in fast wavelength tuning control schemes.

A coherent wavelength division photonic switching system has the following two excellent features: (1) low crosstalk switching for dense WDM signal, and (2) large line capacity capability.

The key element for the coherent photonic WD switching system is the coherent wavelength switch (coherent switch) which is used to accomplish wavelength interchange in a coherent WD system. In the coherent switch, WD channel selection is accomplished in IF signal level, not in optical signal level, utilizing IF filters with a steep cutoff characteristic and high outband rejection. Therefore, low crosstalk switching for dense WDM signals is possible in comparison with conventional WD switching systems using tunable wavelength filters.

ARCHITECTURES OF A COHERENT WD PHOTONIC SWITCHING SYSTEM

Fig. 3.22 [3.116] shows the structure for the proposed coherent photonic WD switching system, consisting of a wavelength multiplexer, wavelength demultiplexer, and a coherent λ switch. In this system, the frequency shift-keying (FSK) modulation format is considered because an FSK system is most practical regarding laser linewidth and modulation. In Fig. 3.22, lightwave signal paths are indicated by solid lines and electrical signal paths by dashed lines.

The traffic generated by each user (s_1, s_2, $\cdots$, s_n) is wavelength division multiplexed in the wavelength multiplexer, using the equally spaced comb of

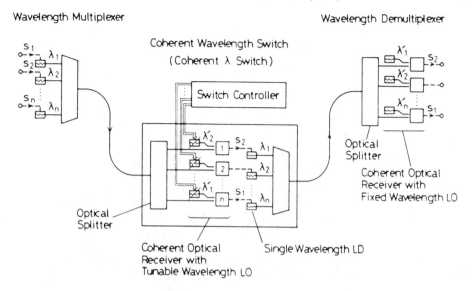

Fig. 3.22 Coherent WD switching system. (From [3.116] Fujiwara *et al.* (1990) *J. Lightwave Technol.* **LT-8**(3), 416–422. Reproduced by permission. ©1990 IEEE.)

wavelength λ_1, λ_2, $\cdots$, λ_n. The wavelength multiplexer may be located close to the coherent switch or close to the users and terminates the users. The WDM signal from the wavelength multiplexer is transmitted to the coherent switch by a single-mode optical fiber (SMF) tranmission line.

In the λ switch, an input WDM signal is split and parts thereof are led to individual coherent optical receiver with a tunable wavelength local oscillator (LO). At every coherent optical receiver, the desired signal channel can be selected and demodulated by tuning the LO wavelength with a control signal from the switch controller. Each demodulated electrical signal from the coherent optical receiver is applied to a single-wavelength laser diode (DFB or DBR LD) to create an output optical FSK signal. The wavelengths of LDs are pre-assigned also to be λ_1, λ_2, $\cdots$, λ_n, as shown in Fig. 3.22.

This process accomplishes wavelength interchange. By way of example, it can be seen in Fig. 3.22 that LO wavelengths of the coherent optical receivers 1 and n are tuned to λ_2' and λ_1' to select wavelength channels λ_2 and λ_1, respectively. The wavelengths λ_1', λ_2', $\cdots$, λ_a' differ from the signal wavelengths λ_1, λ_2, $\cdots$, λ_n by the microwave intermediate frequency (IF), respectively.) As a result, wavelength interchange from λ_2 to λ_1 and λ_1 to λ_n is performed using the coherent optical receiver 1 and n. Any kind of one-to-one wavelength interchange is possible. Moreover, a multicast function can be achieved by controlling all coherent optical receivers to select the same wavelength channel.

On the outbound side of the coherent λ switch, the wavelength interchanged optical signals are multiplexed to form and output WDM signal and sent to the wavelength demultiplexer. There, a switched signal for each user is obtained using a coherent optical receiver with fixed wavelength LO.

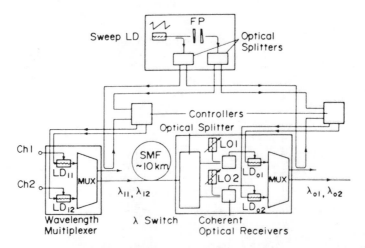

Fig. 3.23 Coherent WD photonic switching. (From [3.116] Fugiwara *et al.* (1990) *J. Lightwave Technol.* **LT-8**(3), 416–422. Reproduced by permission. ©1990 IEEE.)

EXPERIMENTS ON COHERENT WD PHOTONIC SWITCHING

A two-channel wavelength-synchronized switching experiment was carried out to demonstrate the switching function of the coherent switch. Fig. 3.23 [3.116] shows the experimental switching system diagram. The experimental switching system consists of a wavelength multiplexer, a coherent λ switch, and a wavelength controller for wavelength synchronization.

The 'reference pulse method' was applied for wavelength synchronization. Both the direct output from the sweep LD and the output via the Fabry–Perot resonator (F–P) were distributed to the wavelength multiplexer and the λ switch. LD wavelengths were controlled so that the generation time for the beat pulses between the transmitter LD lights and sweep LD light coincide with that for the F–P output pulses [3.110], both in the wavelength multiplexer and the λ switch. As a result, each wavelength of the corresponding LDs in the wavelength multiplexer and the λ switch was stabilized to the same respective resonant frequency of the F–P resonator. Through this process, wavelength synchronization is achieved between the input and output WDM highways for the switch. In the experiments, the optical frequency discrepancies, between corresponding LDs in the wavelength multiplexer and the λ switch, have been controlled to less than 2.7% of the frequency spacing (8 GHz) [3.99].

1.55 μm wavelength tunable DBR LDs were used as the transmitters, LOs, and sweep LD. Transmitted and switched signals are 280 Mbit/s optical FSK signals with a frequency deviation of 1 GHz. Frequency separation for the WDM signals was set to be 8 GHz, which is sufficient to avoid interchannel crosstalk up to 400 Mbit/s [3.111]. Coherent optical receivers in the λ switch consist of balanced receivers and FSK single filter detection systems [3.112] with 400 MHz–1.4 GHz passbands. Beat spectral linewidths, between two DBR LDz, were around 30 MHz, which is sufficiently narrow for FSK single filter detection [3.112]. SMF couplers were used as wavelength multiplexer and optical splitters.

REFERENCES

[3.1]   O. E. Delange (1970) "Wideband optical communication systems, Part 2—Frequency division multiplexing" *Proc. IEEE* **58**, 1683.

[3.2]   W. Tomlinson (1977) "Wavelength multiplexing in multimode optical fibers" *Appl. Opt.* **18**(8), 2180–2194.

[3.3]   H. Ishio and T. Mike (1977) "A preliminary experiment on wavelength division multiplexing transmission using LED" *Proc. IOOC'77, Tokyo, Japan* **C7-3**.

[3.4]   K. Nosu and H. Ishio (1979) "A design of optical multi/demulti-plexers for optical wavelength-division multiplexing transmission" *Trans. IECE* **62-B**, 1030–1036.

[3.5]   H. Ishio *et al.* (1984) "Review and status of wavelength-division multiplexing technology and its applications" *J. Lightwave Technol.* **LT-2**(8), 448–463.

[3.6]   G. Winzer (1984) "Wavelength multiplexing components—A review of single-mode devices and their applications" *J. Lightwave Technol.* **LT-2**, 369–378.

[3.7]   C. A. Brackett (1988) "Multiwavelength photonic networks" *Proc. CLEO'88, Anaheim, CA* paper WA1.

[3.8]   Y. Silberberg *et al.* (1988) "Digital optical switch" *Proc. OFC'88, New Orleans, LA* paper THA3.

[3.9]   I. Stanley (1985) "A tutorial review of techniques for coherent optical fiber transmission systems" *IEEE Commun. Magazine* **23**, 37–53.

[3.10] T. Kimura (1987) "Coherent optical fiber transmission" *J. Lightwave Technol.* **LT-5**, 414–428.

[3.11] I. Mito (1988) "Recent advances in frequency tunable DFB-DBR lasers" *Proc. OFC'88, New Orleans, LA,* paper THK1.

[3.12] J. M. Cooper *et al.* "Nanosecond wavelength switching with a double-section distributed feedback laser" *Proc. CLEO'88, Anaheim, CA* paper WA4.

[3.13] A. R. Chraplyvy *et al.* (1988) "Network experiment using an 8,000 GHz tuning-range optical filter" *Electron Lett.* **24**, 1701–1702.

[3.14] C. Lin *et al.* (1988) "Wavelength-tunable optical channel transmission experiment" *Electron. Lett.* 1215–1217.

[3.15] I. Kobayashi *et al.* (1980) "A 4 channel wavelength-division multiplexing transmission system" *Proc. FOC'80 San Francisco Sept. 1980* 27–30.

[3.16] J. Popocics (1982) "Optical fiber multiservice subscriber connection system: SAFO" *IEEE Trans. Commun* **COM-30**, 2215–2220.

[3.17] H. Toba *et al.* (1986) "A conceptual design on optical frequency-division-multiplexing distribution systems with optical tunable filters" *IEEE J. Sel. Areas Commun.* **SAC-4**, 1458–1467.

[3.18] H. Toba *et al.* (1987) "5 GHz-spaced, eight-channel, guided-wave tunable multi/demultiplexer for optical FDM transmission systems" *Electron Lett.* **23**, 788–789.

[3.19] A. M. Hill and D. B. Payne (1985) "Linear crosstalk in wavelength-division-multiplexed optical fiber transmission systems" *J. Lightwave Technol.* **LT-3**, 643–651.

[3.20] Y. Fujii *et al.* (1980) "Optical demultiplexer using a silicon echelette grating" *IEEE J. Quantum Electron.* **QE-16**, 165–169.

[3.21] R. Watanabe *et al.* (1980) "Optical grating multiplexer in the 1.1–1.5 μm wavelength region" *Electron. Lett.* **16**, 108–109.

[3.22] K. Kobayashi and M. Seki (1980) "Microoptic grating multiplexers and optical isolators for fiber-optic communications" *IEEE J. Quantum Electron.* **QE-16**, 11–12.

[3.23] M. Seki *et al.* (1982) "20 channel micro-optic grating demultiplexer for 1.1–1.6 μm band using a small focusing parameter graded-index rod lens" *Electron. Lett.* **18**, 257–258.

[3.24] A. Thelen (1969) "Design of multilayer interference filters" in *Physics of Thin Films* vol. 5, ed G. Hass and R. E. Thun (New York: Academic) pp. 47–86.

[3.25] J. Minowa and Y. Fujii (1983) "Dielectric multilayer thin film filters for WDM transmission systems" *J Lightwave Technol.* **LT-1**, 116–121.

[3.26] J. Minowa and Y. Fujii (1984) "High performance bandpass filter for WDM transmission" *Appl. Opt.* **23**(2), 139.

[3.27] S. Sugimoto *et al.* (1978) "Wavelength division two-way fiber-optic transmission experiments using micro-optic duplexer" *Electron. Lett.* **14**, 15–17.

[3.28] T. Tanaka *et al.* (1982) "Fiber-optic multifunction device using a single GRIN-rod lens for WDM transmission systems" *Appl. Opt.* **21**, 3423–3429.

[3.29] Y. Fujii *et al.* (1983) "Practical two-wavelength multi- demultiplexer: Design and performance" *Appl. Opt.* **22**, 3090–3097.

[3.30] K. Nosu *et al.* (1979) "Multireflection optical multi/demultiplexer using interference filters" *Electron. Lett.* **15**, 414–415.

[3.31] S. Ishikawa *et al.* (1980) "Multi-reflection wavelength-division multi/demultiplexers with the wavelength tunability" *ECOC'80, U.K.*, 298–301.

[3.32] G. Winzer *et al.* (1981) "Single-mode and multimode all-fiber directional couplers for WDM" *Appl. Opt.* **20**, 3128–3135.

[3.33] B. H. Verbeek *et al.* (1988) "Integrated four-channel Mach–Zehnder multi/demultiplexer fabricated with phosphorous doped SiO_2 waveguide on Si" *J. Lightwave Technol.* **LT-6**(6), 1011–1015.

[3.34] C. H. Henry *et al.* (1990) "Four-channel wavelength division multiplexers and bandpass filters based on elliptical Bragg reflectors" *J. Lightwave Technol.* **LT-8**, 748–755.

[3.35] A. Livanos *et al.* (1977) "Chirped-grating demultiplexers in dielectric waveguides" *Appl. Phys. Lett.* **30**, 519–521.

[3.36] P. Yhe and F. Taylor (1980) "Contradirectional frequency-selective couplers for guided-wave optics" *Appl. Opt.* **19**, 2848–2855.

[3.37] T. Suhara *et al.* (1982) "Monolithic integrating and photodiodes for wavelength demultiplexer" *Appl. Phys. Lett.* **40**, 120–122.

[3.38] R. Alferness and L. Buhl (1980) "Electro-optic waveguide TE-TM mode converter with low drive voltage" *Opt. Lett.* **5**, 473–475.

[3.39] H. Ishio and K. Nosu (1978) "On channel spacing and interchannel crosstalk in WDM transmission" in *Proc. Nat. Conf. IECE Japan* Paper no. 2002.

[3.40] K. Nosu *et al.* (1980) "Optical access loop network design" *IECE Japan* paper CS80–4.

[3.41] A. Tomita (1983) "Crosstalk caused by stimulated Raman scattering in single mode wavelength division multiplexed systems" *Opt. Lett.* **8**, 412–414.

[3.42] R. Watanabe *et al.* (1983) "Design and performance of multi/demultiplexer for subscriber loop system" *Proc. IOOC'83* **30C1-2**.

[3.43] R. E. Epworth (1978) "The phenomenon of modal noise in analog and digital optical fiber system" *ECOC'78, Genoa, Italy*.

[3.44] K. Peterman and G. Arnold (1982) "Noise and distortion characteristics of semiconductor lasers in optical fiber communication systems" *IEEE J. Quantum Electron.* **QE-18**, 543–555.

[3.45] K. Nosu *et al.* (1979) "Multireflection optical multi/demultiplexer using interference filters" *Electron. Lett.* **15**, 414.

[3.46] S. S. Wagner and H. Kobrinski (1989) "WDM Applications in Broadband telecommunication networks" *IEEE Commun. Magazine* **27**(3), 22–29.

[3.47] M. P. Vecchi *et al.* (1988) "High-bit-rate measurements in the LAMBDANET Multiwavelength optical star network" *Proc. OFC'88, New Orleans, LA* paper WO2.

[3.48] H. Kobrinski *et al.* (1987) "Demonstration of high capacity in the LAMBDANET architecture; A multiwavelength optical network" *Electron. Lett.* **23**, 824–826.

[3.49] D. H. McMahon *et al.* (1987) "Echelon grating multiplexers for hierarchically multiplexed fiber-optic communication networks" *Appl. Opt.* **26**, 2188–2196.

[3.50] M. Kerner *et al.* (1986) "An analysis of alternative architectures for the interoffice network" *IEEE J. Sel. Areas Commun.* **SAC-4**, 1404–1413.

[3.51] M. Eiger (1987) "New approaches for the broadband interoffice network" *Proc. Globecom'87, Tokyo* paper 37.2.

[3.52] H. Kobrinski (1987) "Crossconnection of wavelength-division- multiplexed high-speed channel" *Electron Lett.* **23**, 975–977.

[3.53] G. R. Hill (1988) "A wavelength routing approach to optical communication networks" *Proc. Infocom'88, New Orleans, LA*.

[3.54] E. J. Bachus *et al.* (1984) "Digital transmission of TV signals with a fiber-optic heterodyne transmission system" *J. Lightwave Technol.* **LT-2**, 381–384.

[3.55] A. Frenkel and C. Lin (1989) "Angle-tuned etalon filters for optical channel selection in high density wavelength division multiplexed systems" *J. Lightwave Technol.* **7**, 615–624.

[3.56] H. Toba *et al.* (1985) "450 Mbit/s optical frequency-division-multiplexing transmission with an 11 GHz channel spacing" *Electron Lett.* **21**, 656–657.

[3.57] Y. Itaya *et al.* (1983) "Longitudinal mode spectra of 1.5 μm GaInAsP/InP distributed feedback lasers" *IOOC'83, Tokyo, Japan, 1983*, 154–155.

[3.58] H. C. Lefevre (1980) "Single-mode fiber fractional wave devices and polarization controllers" *Electron. Lett.* **16**, 778–780.

[3.59] S. Yamazaki *et al.* (1990) "A coherent optical FDM CATV distribution system" *J. Lightwave Technol.* **8**(3), 396–405.

[3.60] B. S. Glance *et al.* (1988) "WDM coherent optical star network" *J. Lightwave Technol.* **LT-6**(1), 67–72.

[3.61] E. J. Bachus *et al.* (1986) "Ten-channel coherent optical fiber transmission" *Electron. Lett.* **22**, 1002–1003.

[3.62] A. R. Chraplyvy and R. W. Tkach (1986) "Narrow-band tunable optical filter for channel selection in densely packed WDM systems" *Electron. Lett.* **22**, 1084–1085.

[3.63] P. S. Henry *et al.* (1987) "Introduction to lightwave systems" in *Optical Fiber Telecommunications*, ed S. E. Miller, and I. P. Kaminow (New York: Academic).

[3.64] L. G. Kazovsy (1987) "Multichannel coherent optical communications systems" *J. Lightwave Technol.* **LT-5**, 1095–1102.

[3.65] G. P. Agrawal (1987) "Evaluation of crosstalk penalty in multichannel ASK heterodyne optical communication systems" *Electron. Lett.* **23**(17), 906–908.

[3.66] B. S. Glance *et al.* (1988) "Densely spaced FDM coherent star network with optical signals confined to equally spaced frequency" *J. Lightwave Technol.* **6**, 1770–1781.

[3.67] J. X. Kan *et al.* (1987) "Transformer-tuned frontends for heterodyne optical receivers" *Electron. Lett.* **23**, 783–787.

[3.68] G. Jacobsen *et al.* (1987) "Improved design of tuned optical receivers" *Electron. Lett.*, **23**, 787–788.

[3.69] B. Glance *et al.* "Fast frequency-tunable external-cavity laser" *Electron. Lett.* **23**, 98–99.

[3.70] R. Wyatt and W. J. Delvin, Jr. (1983) "10 kHz linewidth 1.5 μm InGaAsP external cavity laser with 55 nm tuning range" *Electron. Lett.* **19**, 110–112.

[3.71] N. A. Olsson and J. P. van der Ziel (1987) "Performance characteristics of 1.5 μm external cavity semiconductor lasers for coherent optical communication" *J. Lightwave Technol.* **LT-5**, 510–515.

[3.72] B. L. Kasper *et al.* (1986) "Balanced dual-detector receiver for optical heterodyne communication at Gbit/s rates" *Electron. Lett.* **22**, 413–415.

[3.73] B. Glance *et al.* (1987] "Densely spaced WDM coherent optical star network" *Electron. Lett.* **23**, 857–876.

[3.74] P. M. Hill and R. Olshansky (1990) "A 20-Channel optical communication system using subcarrier multiplexing for the transmission of digital video signals" *J. Lightwave Technol.* **LT-8**(4), 554–560.

[3.75] P. Hill and R. Olshansky (1988) "Twenty channel FSK subcarrier multiplexed optical comunication system for video distribution" *Electron. Lett.* **24**, 892–893.

[3.76] T. Darcie *et al.* (1988) "Bidirectional multichannel 1.44 Gbit/s lightwave distribution system using subcarrier multiplexing" *Electron. Lett.* **24**, 649–650.

[3.77] W. Way and C. Castelli (1988) "Simultaneous transmission of 2 Gbit/s digital data and ten FM-TV analogue signals over 16.5 km SM fibre" *Electron. Lett.* **24**, 611–613.

[3.78] R. Olshansky *et al.* (1988) "Simultaneous transmission of 100 Mbit/s at baseband and 60 FM video channels for a wideband optical communication network" *Electron. Lett.* **24**, 1234–1235.

[3.79] R. Olshansky and V. Lanzisera (1987) "60 channel FM video subcarrier multiplexed optical communication system" *Electron. Lett.* **23**, 1196–1198.

[3.80] P. Rosher *et al.* "Broadband video distribution over passive optical network using subcarrier multiplexing techniques" *Electron. Lett.* **25**, 115–117.

[3.81] R. Gross and R. Olshansky (1990) "Multichannel coherent FSK experiments using subcarrier multiplexing techniques" *J. Lightwave Technol.* **8**(3), 406–415.

[3.82] A. van Bochove *et al.* (1989) "A coherent optical system for analogue multichannel video transmission" in *Proc. ECOC'89, Brighton, U.K.* 541–544.

[3.83] W. I. Way *et al.* (1987) "A 1.3 μm 35 km fiber-optic microwave multicarrier transmission system for satellite earth stations" *J. Lightwave Technol.* **LT-5**(9), 1325–1332.

[3.84] J. J. Prisco (1986) "Fiber-Optic regional area networks in New York and Dallas" *IEEE J. Selected Areas Commun.* **SAC-4**, 750–757.

[3.85] J. E. Midwinter (1987) "Photonic switching components, current status and future possibilities" *Photonic Switching* (New York: Springer-Verlag).

[3.86] J. E. Midwinter (1988) "Photonic switching technology, component characteristics versus network requirements" *IEEE J. Lightwave Technol.* **LT-6**, 1152–1159.

[3.87] S. Suzuki and K. Nagashime (1987) "Optical broad-band communication network architecture utilizing wavelength- division switching technology" in *Tech. Dig. Topical Meeting on Photonic Switching (Nevada) Mar. 18–20, 1987* 21–23.

[3.88] Y. Shimazu *et al.* (1987) "Wavelength-division-multiplexing optical switch using acoustooptic deflector" *J. Lightwave Technol.* **LT-5**, 1742–1747.

[3.89] M. Nishio *et al.* (1988) "Eight-channel wavelength-division switching experiment using wide-tuning range DFB LD filters" *Tech, Dig. ECOC'88, Brighton, U.K.* pt. 2, 59–52.

[3.90] N. A. Olsson and W. T. Tsang (1984) "An optical switching and routing system using frequency tunable cleaved-coupled-cavity semiconductor lasers" *IEEE J. Quantum Electron* **QE-20**, 332.

[3.91] H. Kobrinski (1987) "Crossconnection of wavelength-division-multiplexed high-speed channels" *Electron. Lett.* **23**, 974–976.

[3.92] M. S. Goodman *et al.* (1986) "Application of wavelength division multiplexing to communication network architecture" *ICC'86 Conf. Rec., Toronto, Ont., Canada, June 22–25, 1986* **2**, 931–933.

[3.93] H. Kobrinski *et al.* (1987) "Demonstration of high capacity in the LAMBDANET architecture: A multiwavelength optical network" *Electron. Lett.* **23**, 824–826.

[3.94] D. B. Payne and J. R. Stern (1986) "Transparent single-mode fiber optical networks" *J. Lightwave Technol.* **LT-4**, 864–869.

[3.95] B. S. Glance *et al.* (1988) "WDM coherent optical star network" *J. Lightwave Technol.* **6**, 67–72.

[3.96] K. Y. Eng *et al.* (1989) "An FDM coherent optical switch experiment with monolithic tunable laser covering a 1,000 GHz range" *Tech. Dig. Topical Meeting of Photonic Switching (Salt Lake City, UT) Mar., 1989*, PD10.

[3.97] M. Fujiwara *et al.* (1987) "Optical switching in coherent lightwave systems" *Tech. Dig. Topical Meeting on Photonic Switching (Nevada) Mar., 1987* 27–29.

[3.98] M. Fujiwara *et al.* (1989) "Application of coherent optical transmission technologies to photonic switching networks" *Trans. Inst. Electron. Inf. Commun. Eng.* **E72**, 55–62.

[3.99] M. Fujiwara *et al.* (1988) "A coherent photonic wavelength-division switching system for broadband networks" *Tech. Dig. ECOC'88, Brighton, U.K., Sept., 1988* pt. 1, 139–142.

[3.100] S. Suzuki *et al.* (1990) "A photonic wavelength-division switching system using tunable laser diode filters" *J. Lightwave Technol.* **LT-8**(5), 660–666.

[3.101] K. Tai *et al.* (1987) "Nonlinear optical logic etalon at today's fiber communication wavelength" *OSA Topical Meeting on Photonic Switching, Mar. 1987.*

[3.102] M. Ikeda (1983) "Switching characteristics of laser diode λ switch" *IEEE J. Quantum Electron.* **QE-19**(2).

[3.103] K. Wakita *et al.* (1986) "Anisotropic electroabsorption and optical modulation in InGaAs/InAlAs multiple quantum well structures" *IEEE J. Quantum Electron.* **QE-22**.

[3.104] K. Kishino *et al.* (1982) "Wavelength variation of 1.6 μm wavelength buried heterostructure GaInAsP/InP laser due to direct modulation" *IEEE J. Quantum Electron* **QE-18**(3), 343.

[3.105] H. Yamamoto *et al.* (1985) "Electric-field-induced refractive index variation in quantum-well structure" *Electron. Lett.* **21**(13), 579.

[3.106] T. Numai *et al.* (1988) "1.5 μm tunable wavelength filter using phase-shift controllable DFB LD with wide tuning range and high constant gain" *Proc. ECOC'88*, pt. 1, 243.

[3.107] M. Nishio *et al.* (1989) "An experiment on photonic wavelength- division and time-division hybrid switching" *Proc. 2nd Topical Meeting on Photonic Switching, 1989* pp. 98–100.

[3.108] H. Suzuki *et al.* (1986) "Very high speed and high capacity packet switching for broadband ISDN" *Conf. Rec. ICC'86, Toronto, Ont., Canada, June, 1986*, 749–754.

[3.109] S. Yamazaki *et al.* (1989) "Random access optical heterodyne receiver for coherent FDM Broadcasting systems" *Electron. Lett.* **25**, 507–508.

[3.110] B. Glance *et al.* (1989) "Densely spaced FDM optical coherent system with near quantum-limited sensitivity and computer-controlled random access channel selection" *Electron. Lett.*, **25**, 883–885.

[3.111] N. Shimosaka *et al.* (1989) "Photonic wavelength-division and time division hybrid switching system utilizing coherent optical detection" *ECOC'89, Gothenburg, Sweden, Sept. 1989*, paper WeA15-2.

[3.112] K. Y. Eng (1987) "A photonic knockout switch for high-speed packet neworks" *Proc. GLOBECOM'87, Tokyo, Japan, Nov. 1987* 1861–1865.

[3.113] E. Arthurs *et al.* (1988) "HY-PASS: An optoelectronic hybrid packet switching system" *IEEE J. Select. Areas Commun.* **SAC-6**, 1500–1510.

[3.114] J. M. Cooper *et al.* (1988) "Nanosecond-tunable double-section DFB laser for dynamic wavelength addressing applications" *Electron. Lett.* **24**, 1237–1239.

[3.115] H. Kobrinski *et al.* (1989) "Simultaneous fast wavelength-switching and direct data modulation using a 3-section DBR laser with a 2.2 nm continuous tuning range" *Tech. Dig. OFC'89, Houston, TX, Feb. 6–9, 1989* paper PD3.

[3.116] M. Fujiwara *et al.* (1990) "A coherent photonic wavelength-division switching system for broad-band networks" *J. Lightwave Technol.* **LT-8**(3), 416–422.

[3.117] C. Clos (1953) "A study of non-blocking switching networks" *Bell Syst. Tech. J.* **32**, 407–424.

[3.118] M. Fujiwara and M. Sakaguchi (1990) "Line capacity expansion schemes in photonic switching" *IEEE LCS Magazine* 47–53.

4

HIGH-SPEED AND TIME DIVISION MULTIPLEXING OPTICAL TRANSMISSION

4.1 INTRODUCTION TO HIGH-SPEED AND TDM TRANSMISSION

In the last two decades, huge information communication capacities have been demanded by human society due to activity in economics and business, the exciting developments in scientific research and outer-space exploration, large-scale and computer-controlled industrial production, and advanced high technology in military use. The capacity characteristics of optical fiber transmission systems have expanded substantially in response to the requirements of huge information data services [4.1]. This is because a single mode fiber is an ideal medium for long haul trunk transmission systems, which require low loss and wide bandwidth. Although the fundamental and practical limits of characteristics such as laser power, transmitter/receiver speed, transmitter spectral purity, receiver sensitivity, fiber dispersion, and fiber loss can be probed at the device level, the efficiency with which a given component trades off one characteristic against another is best evaluated in the system environment. One figure of merit that measures the degree of optimization is the bit-rate × distance product. Another is the efficiency of the use of the total fiber bandwidth. Much effort has been expended to overcome the repeater spacing dispersion limit in developing a practical large capacity system. This limit results from spectral broadening of light sources and chromatic dispersion of single mode fiber [4.2].

Directly modulated single-longitudinal mode semiconductor lasers exhibit dynamic frequency chirp which can cause a performance penalty for multi Gbit/s system operating in the 1.55 μm wavelength region with conventional single mode optical fiber. The natural linewidth of semiconductor lasers operation under CW conditions is of the order of 10^8 Hz and is too narrow to cause a dispersion penalty for bit rates < 10 Gbit/s; however, under amplitude modulation the linewidth dynamically broadens or chirps, and it is this dynamic linewidth that will set the ultimate limit on the performance of single mode transmission systems in dispersive fibers using directly modulated lasers. The dynamically broadened linewidth of a single-longitudinal mode laser is typically in the range of 0.2–0.5 nm and the fiber chromatic dispersion coefficient is 15–19 ps/(km nm).

Single mode fibers are important for high bandwidth transmission because of the absence of intermodal dispersion. Operation at the zero first-order chromatic dispersion wavelength, low-loss single mode fibers provide large information capacity over long distances. It is important to know the theoretical maximum transmission bandwidth as well as the practical limit to the maximum information capacity. Several papers [4.3, 4.6] address the problem of the maximum information bandwidth and discuss pulse propagation at the minimum chromatic dispersion wavelength λ_0, where the first-order dispersion vanishes and the second-order dispersion dominates. It has been shown that, right at λ_0, the propagating pulse experiences oscillatory pulse distortion and its rms width is no longer linearly related to the spectral width of the light source. A more extensive theoretical treatment, including the rms pulsewidth calculation and numerical evaluation of pulse shapes for various conditions, is given in [4.6].

Mode-locking techniques have been developed to generate and manipulate pico- and subpicosecond optical pulsed in laser diodes. Ultrafast optical detectors and electronic components are reported with speeds approaching 100 GHz. Thus, future fiber-optic systems will likely be capable of operating at data rates approaching and even exceeding 50 Gbits. At these data rates the modulation bandwidth is so large that, even for an ideal source without chirp or phase noise, fiber dispersion broadens the optical pulses and thus limits transmission. The deleterious influence of linear fiber dispersion may be minimized by operation at the minimum group velocity dispersion (GVD) wavelength (λ_0). Dispersion is not eliminated completely, however, since higher order dispersion terms in fibers are not zero. At data rates approaching 100 Gbit/s, second-order dispersion limits the transmission distance even at λ_0. Dispersion compensation techniques will be needed for optical communication systems that require exceptionally large bandwidths, such as those that employ address encoding and decoding by frequency domain modulation.

Optical time division multiplexing (OTDM) is a useful technique for increasing the bit rate of lightwave systems beyond the bandwidth capabilities of the drive electronics and promises very high-bit-rate transmission. Optical transmission based on electronic multiplexing has been reported at data rates as high as 8 Gbit/s, but at these bit rates the electronic circuits are difficult to implement. Optical time division multiplexing extends time multiplexing into the optical regime, where the availability of short optical pulses and fast electro-optical switches enables very high bit rate operation while employing moderate bandwidth electronics [4.7, 4.8]. Thus OTDM offers a solution to the electronic speed bottleneck that is encountered in multi Gbit/s electronic multiplexing and demultiplexing. An experimental four-channel optical time division multiplexed transmission system at a bit rate of 16 Gbit/s was reported [4.9]. In this experiment, data at 16 Gbit/s have been transmitted over 8 km of fiber with a bit error rate below 10^{-9}.

In practice, for current systems in use, bit rate transmitted is limited by the speed of electronics used for modulation and detection, or by excess linewidth noise in the transmitter laser. Even if these practical limitations could be overcome, so that signals could be transmitted and detected at any desired bit rate and the linewidth of the transmitter could be made arbitrarily narrow, there would still be limitations on the data rate that could be carried in a fiber. This limitation is caused by group velocity dispersion and any optical signal on the carrier wave must inevitably introduce new frequency components on either side of the carrier frequency. These components then travel through the fiber with slightly different group velocities and are carried away from one time slot towards the surrounding slots in the bit sequence. Hence, as the distance along the fiber increases, neighbouring symbols begin to interfere with each other. Some interference between neighbouring symbols in a bit sequence can be tolerated; and the extent of this tolerance in the transmission system depends on the details of the modulation and detection schemes used. However, for any given modulation and detection scheme, the group velocity dispersion in the transmission fiber would appear to impose a length-dependent limit on the maximum bit rate that can be carried in a single channel. It was suggested some time ago [4.10, 4.11] that 'soliton pulses' could be used as the basic symbols in an optical communication system in order to overcome this dispersion limitation. If pulses of the correct form are transmitted through an optical fiber in the wavelength region of anomalous (negative group velocity) dispersion, then the effect of the non-linear refractive index can counteract exactly the effect of dispersion. Such a pulse is called a soliton and would, in principle, propagate without any change in shape, although in practice loss in the fiber would allow some pulse broadening.

Soliton pulses appear to have many desirable properties that would make them suitable for use as bit symbols in an optical communication system. The most important of these properties are: (1) solitons are 'stable' in the sense that small changes in the pulse shape or amplitude around the exact form for a soliton [4.12] and small amounts of chirp or phase noise in the launched signal [4.13] lead to only small changes in the effective soliton that propagates through the fiber; and (2) in other respects solitons behave exactly as would linear pulses, so long as they are launched sufficiently far apart for the pulse envelopes not to overlap initially. Multiplexed streams of soliton bit sequences of different frequencies pass through each other without any interference.

It was shown that a practical Gbit/s system operation at 1.55 μm wavelength with over 100 km repeater spacing required 1.55 μm zero dispersion fiber in order to be free from impairments caused by chirping and side-mode oscillation of the DFB-LD [4.14]. High-speed optical transmitters utilizing an LD direct modulation scheme or external modulation scheme have been studied and multi Gbit/s optical transmission experiments have been demonstrated [4.15–4.17]. Moreover, high-speed operation of 16 Gbit/s direct modulation of semiconductor lasers was reported [4.18], and a high-speed external modulator using LiNbO$_3$ with greater than 20 GHz bandwidth was also developed [4.19]. The next generation of optical transmission system also should not be degraded by chromatic dispersion of single mode fiber. A multi Gbit/s transmission system can thus be made practical.

4.2 MULTI GBIT/S OPTICAL MODULATION

4.2.1 Direct modulation

High-speed lightwave modulation has been desired for large-capacity optical communications. The simplest method developed is the direct intensity modulation and direct detection system. This conventional method was performed at 16 Gbit/s several years ago [4.18]. Direct intensity modulation, however, has the serious disadvantage of a broad optical spectrum because of wavelength chirping. Time-resolved wavelength measurements on single-frequency pulsed lasers reveal that the wavelength shows a transient shift to the blue as the laser turns on and then to the red side of the equilibrium wavelength as the laser turns off. The effect of modulation-induced laser wavelength shifts an optical transmission through dispersive fibers is shown to be a serious limiting factor at bit rates greater than 1 Gbit/s due to the chirping-induced waveform distortion. Different proposals and experimental results have been presented for reducing the wavelength chirping in direct-modulation multi Gbit/s transmission [4.20–4.22]. In order to avoid dynamic wavelength shift of light sources, some ways of modulation using an external modulator, such as a Ti:LiNbO$_3$ directional coupler modulator or a multiple quantum well optical modulator instead of directly modulating a semiconductor laser, were also introduced [4.16, 4.23].

MODULATION INDUCED TRANSIENT CHIRPING IN LASER DIODES [4.24]

Directly modulated single-longitudinal mode semiconductor lasers exhibit a dynamic shift of the peak emission wavelength due to carrier-induced changes in the refractive index of the active region. This linewidth broadening or chirping is characterized by a shift toward shorter wavelengths during the leading edge of a pulse (blue-shift) and a shift toward longer wavelengths during the trailing edge of a pulse (red-shift) [4.24]. These excursions are readily explained in terms of relaxation oscillations in the carrier density which is temporarily driven out of equilibrium by a change in injection current [4.25]. Wavelength excursions last for hundreds of picoseconds or about one half of the relaxation resonance period. The combination of the laser chirping and group velocity dispersion of conventional single mode optical fibers can cause a performance penalty for high-speed systems operation in the 1.55 μm wavelength region, as the leading and trailing edges of a pulse spread away from the pulse centre [4.26–4.32]. The degradation in the system performance depends upon the laser structure, modulation conditions, fiber length, and receiver filtering.

Time resolved laser spectra were obtained with the apparatus depicted in Fig. 4.1 [4.24]. The pulsed laser output was passed through a spectrometer and focused onto a fast p–i–n detector. The resulting electronic pulse stream was amplified with a bandwidth of 7.8 GHz and sampled by an oscilloscope sampling unit. The sampling instant was triggered by the pulse generator and could be positioned to any time during the pulse cycle using the sampling unit in its manual mode. A laser spectrum was obtained in this way for sampling times spaced 100 ps apart. The spectrometer wavelength resolution of 0.7 Å limited the time resolution of the measurement to about 100 ps.

The above technique was applied to a cleaved-coupled-cavity (C^3) ridge guide laser [4.33, 4.34] and two types of distributed feedback (DFB) lasers, with results shown in Fig. 4.2 [4.24]. The data of Fig. 4.2 show an initial shift in wavelength at turn-on to about 2 Å below the steady state value. The intensity rises to nearly its final value at least 100 ps before the line returns to its equilibrium wavelength.

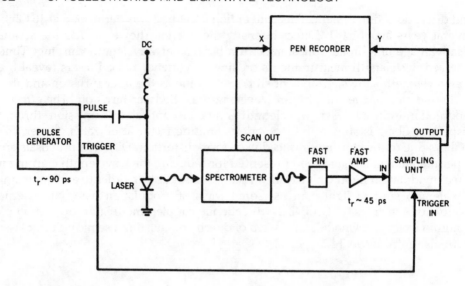

Fig. 4.1 A block of the chirp measurement. (From [4.24] Linke (1985) *IEEE J. Quantum Electron.* **QE-21**, 593–597. Reproduced by permission. ©1985 IEEE.)

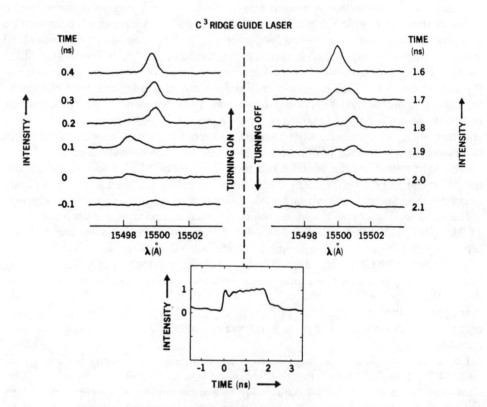

Fig. 4.2 Wavelength shift induced by high-speed modulation. (From [4.24] Linke (1985) *IEEE J. Quantum Electron.* **QE-21**, 593–597. Reproduced by permission. ©1985 IEEE.)

It is this power at the non-equilibrium wavelength which caused signal degradation in a transmission system. Spectra for the interval when the pulse is fully on are not shown because the laser wavelength and intensity remain nearly constant. A shift toward longer wavelengths is seen at the end of the pulse. Again, significant power is radiated at the non-equilibrium wavelength as the light flux falls. The wavelength excursions obtained for pre-bias points at or below the threshold current were found to be much more extreme with maximum deviations of up to 6 Å.

The observation of the chirp time dependence processes on the LD injection current, the wavelength excursions and the light output intensity is expressed in Fig. 4.3 [4.24]. It is consistent with the well known relaxation oscillations in the carrier density observed in transient solutions of laser rate equations [4.25]. When the injection current is suddenly increased, the carrier density rises before the light output can increase to compensate for the higher current (see Fig. 4.3). The result is a temporary jump in carrier density which leads to a temporary reduction in the index refraction in the active region [4.36]. This, in turn, shortens the optical path length of the laser cavity and the wavelength shifts initially to the blue. Strong damping terms [4.37] prevent oscillations from continuing for more than a few periods in practical injection lasers. Similarly, at turn-off the carrier density briefly falls below its equilibrium value and the wavelength shifts to the red. The magnitude of the shift depends on a number of factors, including: (1) the magnitude and speed of the current change, (2) the laser pre-bias condition and (3) the oscillation damping rate. It can be shown that a simple relationship exists between wavelength excursions and the time dependence of the optical power output from any directly modulated semiconductor laser [4.38]. In practice, excursions as large as $\Delta\lambda=0.2$–0.8 nm are observed [4.39–4.43]. The fact that relaxation oscillations are reduced by increased laser bias accounts at least in part for the fact that optimum bias currents in communications systems experiments at 1.5 μm are found to be above the threshold current [4.44, 4.45] even though this results in a reduction of the signal available.

The peak typically seen in the laser small-signal modulation transfer function is a manifestation of the relaxation oscillation quasi-resonance [4.25] which gives rise to the ringing in the injected carrier density and hence to the transient chirp. It is clearly desirable from the point of view of chirp reduction for the laser structure to provide damping of this resonance. Parasitic capacitance shunting the laser could also reduce the amplitude of the resonance but only at the expense of serious degradation of the modulation response which is not observed [4.37, 4.46].

One can obtain a simple model which illustrates the communication implications of this chirp by assuming that the blue-shifted and the red-shifted portions of the emission last for a time, t_c, equal to one half the relaxation oscillation period

$$t_c \approx \frac{1}{2}\,\frac{1}{f_c} \approx \pi \left[\frac{1}{\tau_s \tau_{ph}} \left(\frac{J}{J_{th}} - 1 \right) \right]^{-\frac{1}{2}} \tag{4.1}$$

where f_c is the relaxation oscillation frequency, τ_s and τ_{ph} are the carrier and photon lifetimes, and J and J_{th} are current density and threshold current densities, respectively [4.25]. For a signal transmitted optical pulse of duration $1/B$, where B is the bit rate, the fraction of energy carried by blue-shifted light is approximately t_cB. The same fraction is red shifted. When this bit is transmitted through a

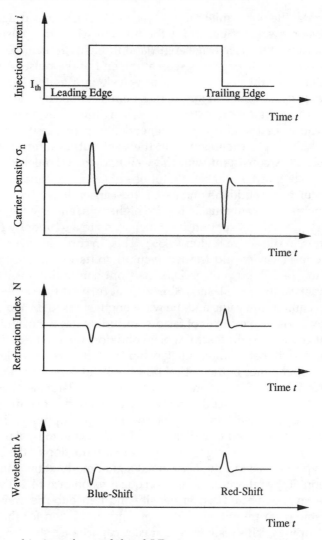

Fig. 4.3 Transient chirping of a modulated LD.

standard fiber with dispersion D (ns/km nm), the blue-shifted power will move to earlier times and the red-shifted power will move to later times with respect to the center of the pulse (for wavelengths longer than the zero dispersion wavelength, $\approx 1.3\,\mu$m). The amount of power which moves outside the symbol interval will increase approximately linearly with distance along the fiber until all the chirped power has left the bit interval, i.e., when $LD = t_c/\Delta\lambda$. The resulting power penalty (i.e., the increase in power required to maintain a given signal-to-noise ratio) can be approximated by assuming that the signal remaining is the difference between the powers in and out of the initial symbol interval [4.47].

Signal $= 1 - 4DLB\Delta\lambda$, so

$$\text{Penalty (decibels} = 10\log\left(1/1 - 4DLB\Delta\lambda\right). \tag{4.2}$$

Equation (4.2) is valid only for $LD < t_0 / \Delta\lambda$. After this point the penalty no longer increases with increased dispersion since all the chirped power has left its original symbol interval and no further degradation occurs.

The power penalties obtained from (4.2) are in good agreement with systems experiment results at 420 Mbit/s and 1 Gbit/s [4.45] performed with the lasers characterized in Fig 4.2.

REDUCTION OF FREQUENCY CHIRPING AND DYNAMIC LINEWIDTH

Directly modulated single-longitudinal mode semiconductor lasers exhibit dynamic frequency chirp or linewidth broadening which can cause a performance penalty for high-speed systems operating with conventional single mode optical fiber and reduce the potential span-rate system product. The dynamically broadened linewidth of a single-longitudinal mode laser is typically in the range of 0.2–0.5 nm and the fiber chromatic dispersion coefficient is 15–19 ps/km nm.

One method used to overcome this problem is to bias the laser above threshold at a point where the spectral linewidth is a minimum, however, this then leads to an extinction ratio power penalty, as an example, 5 dB over 130 km at 2 Gbit/s has been reported [4.48]. Techniques for reducing the frequency chirp, below that of a solitary distributed feedback (DFB) laser, include shaping of the laser injection current waveform [4.49, 4.50], coupling the laser to an external high-Q resonator [4.51–4.53], injection locking [4.54–4.57] and direct phase-shift and self-homodyne intensity modulation (DPSH-IM) [4.58].

SHAPING WAVEFORM OF THE LASER INJECTION CURRENT

A reduction in dynamic linewidth has been predicted theoretically by using a small-amplitude prepulse of duration equal to the relaxation oscillation period [4.49]. Numerical solutions of the multimode diode laser rate equations are used to demonstrate that dynamic line broadening can be significantly reduced by shaping the leading edge of the current pulse. By introducing a small amplitude prepulse, the rms carrier density oscillation can be reduced by a factor of three. Over a 100 km transmission length the power penalty resulting from dynamic line broadening can be reduced from 3.7 to 1.0 dB. More complex shaping of the current pulse can produce theoretically even greater reductions.

A 200 μm cavity length laser with normal front facet reflectivity ($R_1 = 0.3$) and high rear facet reflectivity ($R_2 = 1.0$) is considered as a specific example of this pulse shaping technique [4.49]. It is assumed that the laser is biased at threshold and is driven with a current pulse having an amplitude I_p which would produce 3 mW of CW power. The response of the laser is calculated both for the standard drive current pulse shown in Fig. 4.4(a) (the current pulse is taken to have exponential rise and fall times characterized by a 0.1 ns time constant, and the pulse duration T is taken as 1 ns) and for the optimized drive pulse shown in Fig. 4.4(a) (the parameters of which are chosen by varying the relative amplitude of the prepulse $x_0 = \delta I / I_p$ and its duration t_0 to minimize the rms carrier density oscillation σ_n), with σ_n defined as

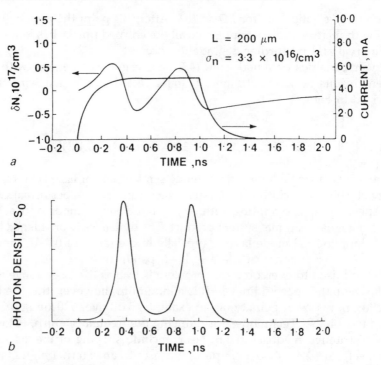

Fig. 4.4 Carrier and photon densities at standard drive current. (From [4.49] Olshansky and Fye (1984) *Electron. Lett.* **20**, 928–929. Reproduced by permission of IEE.)

$$\sigma_n{}^2 = \int_0^t dt S_0(t) [n(t) - \bar{n}(t)]^2 / \int_0^t dt S_0(t) \qquad (4.3)$$

where $\bar{n}(t)$ is the value of the carrier density $n(t)$ averaged over the bit interval T, and $S_0(t)$ is the photon density in the central lasing mode. The overall amplitude of the main current pulse is adjusted so that the total photon density per bit which results from the optimized drive pulse is the same as that obtained with the standard drive pulse.

The responses of the carrier densities and photon densities arising from the standard and optimized drive current pulses are shown in Figs. 4.4(*a*), 4.4(*b*) and Figs. 4.5(*a*), 4.5(*b*) [4.49]. Comparison of Figs. 4.4(*a*) and 4.5(*a*) shows that the optimized current pulse significantly reduces the amplitude of the carrier density relaxations. The value of σ_n has been reduced from $3.3 \times 10^{16}/\mathrm{cm}^3$ to $1.2 \times 10^{16}/\mathrm{cm}^3$. The values of the parameters x_0 and t_0, found by a standard minimization routine, are 0.21 and 0.52 ns, respectively. The dynamic line broadening $\delta\lambda$ can be calculated from the relation

$$\delta\lambda = \lambda \Gamma \sigma_n 1/N \; dN/dn$$

where N is the refractive index and dN/dn is $-2.8 \times 10^{-20} \, \mathrm{cm}^3$. One finds that the standard drive pulse produces an rms broadening $\delta\lambda$ of 1.4 Å, and that this is reduced to 0.5 Å by the optimized pulse shape.

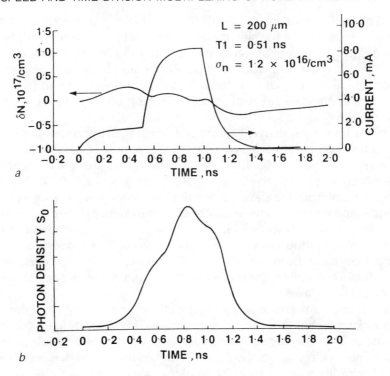

Fig. 4.5 Carrier and photon densities in reduced chirping. (From [4.49] Olshansky and Fye (1984) *Electron. Lett.* **20**, 928–929. Reproduced by permission of IEE.)

The physical mechanism responsible for the reduction in relaxation oscillations is clear. The amplitude of the initial current pulse is reduced to minimize the large overshoot in carrier density which normally occurs at turn-on. The position at t_0 to catch the carrier density oscillation before it undershoots the steady-state value.

The dynamic linewidth of 1.5 μm ridge waveguide DFB lasers is shown to be reduced experimentally by the pulse of the laser modulating waveform. Pulse shaping is provided by a second-order network with a resonant characteristic equal and opposite to that of the laser resonance, thus giving a flat overall laser-network transfer function. The variation of laser small-signal resonance with bias current was compensated by means of a variable capacitor, such that the transient overshoot was minimized for any particular value of bias current. Typical component values for the pulse shaper were $R = 10\,\Omega$, $L = 7\,\text{nH}$ and $C = 0.9$–$2\,\text{pF}$. The corresponding reduction in spectral width is demonstrated in [4.50]; it shows a clear reduction in spectral width using the pulse shaper from 1.4 to 0.55 Å FWHM for a 500 ps pulse.

INJECTION LOCKING TECHNIQUE

The injection locking technique is well known for improving the stability of microwave oscillators and laser oscillators [4.61]. In semiconductors lasers, specifically, injection locking has been used: (*a*) for suppression of intensity fluctuation noise and relaxation oscillations [4.60], (b) for achieving single-longitudinal mode operation by suppressing the side modes in an otherwise multilongitudinal mode laser [4.61], (c) for optical FM signal amplification [4.62]. Also narrow linewidths for injection locked lasers modulated at sub Gbit/s rates have been reported [4.63]. The theory of injection locking of classical oscillators is well known [4.64] and the first experimental investigation of injection locking of semiconductor lasers was reported in [4.65].

A chirp-free transmission over 82.5 km at 2 Gbit/s with injection locked ridge type DFB lasers both as master and slave lasers was reported [4.66]. The experimental arrangement was shown in Fig. 4.6 [4.66]. The two lasers are optically coupled using 0.65 NA mmicroscope objectives; and a YIG crystal isolator with 30 dB of isolation prevents coupling from the slave laser to the master laser. The lasing wavelengths of the master and slave lasers were matched by temperature tuning of the two lasers. For some master–slave laser pairs it was found that, when locking occurred, self-sustained oscillatons around 7 GHz were also induced. This is believed to be caused by residual coupling from the slave to the master laser. However, by careful adjustments of the coupling, the temperature and elimination of specular reflections, this problem can be avoided.

An important parameter for assessing the effectiveness chirp reduction with the injection locking technique is the locking range. This is the range over which the natural frequency of the slave oscillator can be tuned while still locked to the master oscillator. The locking range thus gives a measure of how large chirps can be eliminated with injection locking. The locking range is given by [4.65]

$$\Delta f = (P_{in}/P_1)^{1/2} 1/2 \pi \tau_p \tag{4.4}$$

where P_{in} is the injected power and P_1 the slave laser power, both power levels being measured inside the laser cavity. The photon lifetime τ_p, can be expressed as

$$\tau_p = [(c/n)(\alpha + 1/L\ln(1/R))]^{-1} \tag{4.5}$$

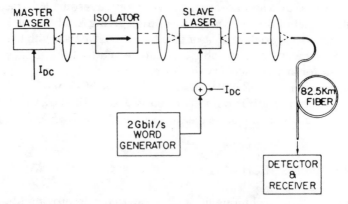

Fig. 4.6 Chirp-free LD with injection locking. (From [4.66] Olsson *et al.* (1985) *J. Lightwave Technol.* **LT-3**, 63–66. Reproduced by permission. ©1985 IEEE.)

where c is the speed of light, n is the refractive index, α is the internal loss, L is the laser length, and R is the facet reflectivity. The value of τ_p for typical laser parameters is 2–5 ps. The injected power required to eliminate a 2 Å (25 GHz) chirp is estimated as $P_{in} = 0.2$–1.2 mW using $P_1 = 2$mW and $\tau_p = 2$–5 ps in (4.4).

In other words, a substantial amount of injected power is required for elimination of typical chirps in single-longitudinal mode lasers. To increase the locking range or reduce the required injected power for a given locking range, it is advantageous to apply an anti-reflection coating to the facets of the slave laser. This will both reduce the photon lifetime and decrease the intracavity power, for a given output power, and hence increase the locking range. The cavity length of the slave laser can also be reduced to decrease the photon lifetime.

4.2.2 External modulation

Frequency chirping in semiconductor lasers under high-speed direct modulation due to carrier density and refractive index change can lead to significant dynamic line broadening. It has been pointed (4.67, 4.43] that such dynamic line broadening in single-longitudinal mode (SLM) semiconductor lasers may become the limiting factor in high-bit-rate long-distance single mode fiber transmission even with dynamic SLM lasers [4.68]. The obvious solution is to avoid frequency chirping completely by using an external modulator with the laser in CW operation. Using the external modulator, such as a multiple quantum well optical modulator [4.69] or a Ti:LiNbO$_3$ external modulator [4.70], the spectral width of the transmitted signal is essentially that minimum value determined by the information bandwidth. Thus, the effect of chromatic dispersion is minimized and the use of the fiber capacity is maximized.

An experimental transmission system with Ti:LiNbO$_3$ external modulator is shown in Fig. 4.7 [4.16]. Four 2 Gbit/s pseudorandom digital data streams of length $2^{15} - 1$, which are obtained from a commercially available 2 Gbit/s BER test set, are combined to form the 8 Gbit/s non-return-to-zero (NRZ) signal using circuits built from discrete GaAs FETs. The length of the single mode, almost-pure-silica core fiber span is 68.3 km with a total loss of 14.8 dB and dispersion of 19 ps/km/nm at 1.53 μm wavelength. The transmitter is a hybrid combination of an InGaAsP/InP injection laser and a Ti:LiNbO$_3$ directional coupler modulator. The laser is of the channel–substrate buried heterostructure design and was cleaved to form two coupled resonators. The drive to the laser is not modulated. In this manner, the laser provides a high power (≈ 10 dBm into air per facet), single frequency, and relatively narrow linewidth emission at 1.53 μm wavelength.

Light from the laser is coupled into a microlensed fiber that is in turn connected to one port of the directional coupler modulator. The crossover waveguide of the Ti:LiNbO$_3$ coupler serves as the output and is connected to the fiber span. Data is encoded by switching the Ti:LiNbO$_3$ modulator between the port attached to the fiber span and the second output port using the amplified signal from the 8 Gbit/s multiplexer. The design of the Ti:LiNbO$_3$ waveguide switch is summarized elsewhere [4.70]. The relevant characteristics of the present modulator are an active length of 1 cm, a switching voltage of 10.6 V, dc extinction ratio of >28 dB, a total

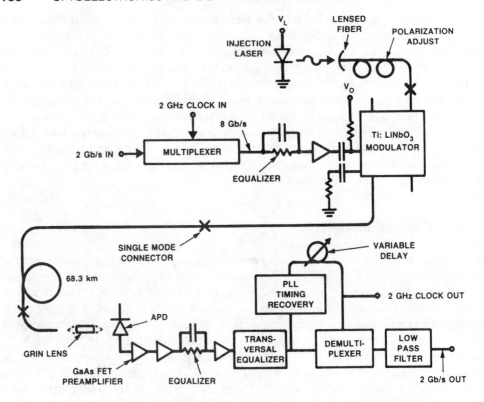

Fig. 4.7 No chirp transmission with external modulator. (From [4.16] Korotky *et al.* (1987) *J. Lightwave Technol.* **LT-5**, 1505–1509. Reproduced by permission. ©1987 IEEE.)

optical insertion loss of 4 dB, an optical reflectivity of ≈ -26 dB, and a small-signal 3 dB electrical bandwidth of ≈ 4–5 GHz. With the use of an external modulator, the observed dispersion penalty of 1 dB and the B^2L product of 4400 (Gbit/s)^2 km attained are in good agreement with the estimated dispersion limit [4.71].

In high-speed systems, low probability phenomena, such as modulation fluctuation and spectrum fluctuation of lasers, cause degradation of system quality, and more advanced methods of analysis which can accurately take into account these rare phenomena accurately are required.

The significance of the eye pattern analysis method applied to high-speed optical transmission systems has already been confirmed. Among them, an eye pattern analysis method, called the 'error rare pattern method', was presented [4.72] recently.

4.3 OPTICAL TIME DIVISION MULTIPLEXING TECHNIQUES

Optical time division multiplexing (OTDM) is a powerful technique to construct high bit-rate transmission by time-multiplexing optically several lower bit-rate data streams, then demultiplexed it optically at the receiving terminal of the system. This approach moves the demand for high-speed performance away from electronic devices to optical and optoelectronic components. It is highly expected that optical and optoelectronic components, such as fiber-couplers or optical waveguide components, have much more broader bandwidths and higher speed than electronic counterpart one.

4.3.1 Principles of optical time division multiplexing and demultiplexing [4.76]

The basic principle of time division multiplexing and demultiplexing is that each of the baseband data streams is allocated a series of time slots on the multiplexed channel. A multiplexer (MUX) assembles the higher bit-rate bit stream from the baseband streams and a demultiplexer (DEMUX) reconstructs bit streams at the original lower bit rate by separating bits in a multiplexed stream. The techniques for this process are well established for electrical time division multiplexing and demultiplexing but are only now emerging in optical systems.

In an electrically time-multiplexed system, Fig. 4.8 multiplexing is carried out in the electrical domain before the electrical-to-optical (E/O) conversion. Demultiplexing is carried out after the optical-to-electrical (O/E) conversion. For n baseband channels, each of bit rate B, the multiplexed bit rate is nB. Potential electronic bottlenecks occur in the WUX and the E/O converter, and in the O/E converter and the DEMUX, where the electronics must operate at the full multiplexed

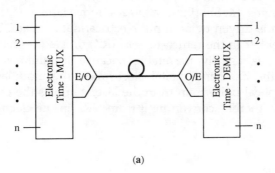

(a)

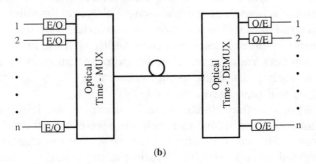

(b)

Fig. 4.8 Time division multiplexing system.

bit rate. These bottlenecks arise from (a) speed limitations of digital integrated circuits, (b) speed limitations of high-power and low-noise linear amplifiers used to drive the laser or modulator in the E/O converter and in the O/E converter, (c) limited modulation bandwidths of lasers and modulators. These problems have so far limited the maximum bit rate for electrically multiplexed systems to ≈ 10 Gbit/s [4.16, 4.74].

In the optically multiplexed system, Fig. 4.8(b), the electronic bottlenecks are removed by moving the E/O and O/E converters (i.e. the transmitters and receivers) into the baseband channels. Multiplexing is carried out after the E/O conversion and demultiplexing is carried out before the O/E conversion. All electronics associated with signal processing operate only at the baseband-bit rate. Note that a control signal is needed to drive the demultiplexer. In general, this control signal could be either electrical or optical, depending on the demultiplexer technology. At present, the most practical optical demultiplexers are based on electro-optic switches, which use electrical control signals. It could be proved that the bandwidth of this electrical control signal need not be large for demultiplexing in an OTDM system.

The operation of time-multiplexing several lower bit-rate baseband channels onto a higher bit-rate channel can be divided into three subfunctions: sampling, timing, and combining.

SAMPLING FUNCTION

This takes samples of the incoming baseband data stream, thereby identifying the value of each incoming bit. Short optical pulses from a laser are incident on an optical modulator, which is driven by an input electrical data stream. The electrical data stream could be either in the return-to-zero (RZ) or the non-return-to-zero (NRZ) format, but NRZ is usually preferable because it minimizes the bandwidth requirements of the baseband digital electronics, the modulator, and its drive amplifiers. The optical pulse train from the laser samples the electrical input data via the modulator, thereby converting it from NRZ in the electrical domain to RZ in the optical domain.

TIMING FUNCTION

This ensures that the samples are available at the correct time slots on the multiplexed channel. The timing scheme for a general n-channel optical time division multiplexed system is shown in Fig. 4.9. The n optical signals incident on the combiner are RZ pulse trains with repetition rates B and pulsewidths T (measured at the baseline). The incoming streams are temporally offset from one another by delays D. Data are encoded on each pulse train, before combining, so in general some of the individual pulses will have zero amplitude.

Fig. 4.9 [4.76] shows that the RZ signal format provides low system crosstalk. Since each baseband signal is always nominally zero except in its allotted time slot on the multiplexed bit stream it cannot interfere with other channels. In practice, the pulse stream will have a finite on/off ratio and the resulting baseline light signal between pulses will cause a component of crosstalk. To avoid possible overlap

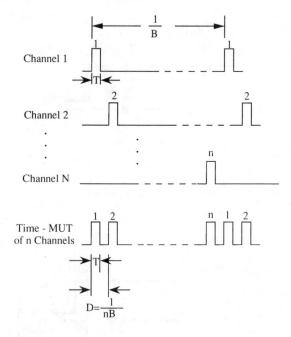

Fig. 4.9 Timing processing in OTDM system.

problems it is usually necessary to ensure that the pulses are somewhat shorter than the one-bit period of the mutliplexed bit stream. Since shorter pulses occupy a wider optical spectrum, reducing the pulse width may increase the dispersion penalty, even if the lasers operate at or near the wavelength of zero first-order chromatic dispersion in the fiber [4.75]. Thus, choosing the optimum pulsewidth may entail a compromise between system crosstalk and pulse spreading caused by fiber dispersion.

COMBINING FUNCTION

This assembles all the sampled baseband data streams to generate the higher bit-rate multiplexed data stream.

Block schematics showing two possible configurations for the E/O converters and combiner for an n-channel OTDM system are presented in Fig. 4.10 [4.76]. The first configuration, Fig. 4.10(a), uses n optical pulse generators, all driven by the same master clock. These pulse generators could be mode-locked or gain-switched semiconductor lasers. The pulse streams are delayed with respect to one another using delay elements either in the electrical clock paths [4.76] or in the optical signal paths [4.77]. Electrical delays would usually be preferred because they are easily made adjustable. Data encoding and sampling can be carried out using optical modulators such as $LiNbO_3$ waveguide electro-optic devices [4.78] at the outputs of the pulse generators. A potential disadvantage of the transmitter–combiner arrangement in Fig. 4.10(a) is that the wavelengths of all lasers need to be closely

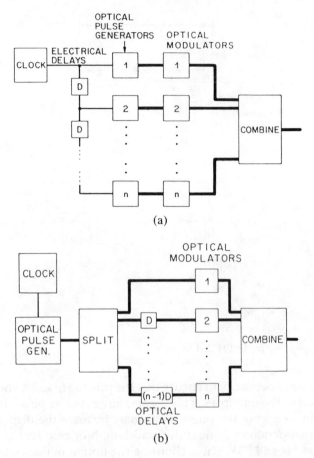

Fig. 4.10 Combining processing in OTDM system. (From [4.76] Tucker (1988) *J. Lightwave Technol.* **LT-6**, 1737–1749. Reproduced by permission. ©1988 IEEE.)

matched to avoid pulse overlap at the receiver end of the fiber caused by different propagation times.

The second OTDM transmitter configuration, Fig. 4.10(*b*), uses a single optical pulse generator [4.77]. The output of the generator is split passively into *n* channels, which are then encoded with data and properly delayed with respect to one another. This arrangement requires only one laser that cannot be achieved with other optical multiplexing methods such as wavelength division multiplexing. The optical combiner in Fig. 4.10(*a*) and (*b*) can, in general, be either passive or active. A passive combiner would incorporate devices such as fiber directional couplers, while active combining would use active devices such as optical switches [4.79]. The passive combiner option has the advantage of simplicity but its losses can become large because each 3 dB directional coupler introduces a 3 dB combining loss. The losses in an active combiner should be smaller because the loss per device is potentially low. An active combiner has the potential of reducing crosstalk, by decreasing pulse overlap caused by finite on/off ratios and leading and trailing tails on the pulses. In a passive combiner, a tail extending into the time slot of a neighbouring bit would be unattenuated, but in an active combiner it would be reduced by the (time-dependent) switching function.

The demultiplexer is the most critical element of an OTDM system. Its purpose is to direct each bit of the arriving multiplexed bit stream to the appropriate O/E converter, as shown in Fig. 4.8(*b*). The basic building block of an optical demultiplexer is a 1×2 switch. Typically, optical switches such as Ti:LiNbO$_3$ directional coupler devices are fabricated as 2×2 crosspoints, which can serve as 1×2 switches by simply terminating one port.

For systems with more than two baseband channels, larger demultiplexer switching networks can be constructed by interconnection 1×2 building blocks [4.80].

4.3.2   High-bit-rate transmission experiments using OTDM

Some experimental results on multi-channel OTDM transmission have been presented in recent years [4.8, 4.9, 4.81, 4.82].

An experimental 16 Gbit/s optical time division multiplexed transmission system is described in [4.9]. In this experiment, data have been transmitted over 8 km of fiber with a bit error rate below 10^{-9}. An optical multiplexing and demultiplexing ratio of $1 : 4$ and a baseband channel bit rate of 4 Gbit/s to attain an aggregate transmission bit rate of 16 Gbit/s. The principal diagram of 16 Gbit/s optical time division multiplexing is illustrated schematically in Fig. 4.11 [4.9]. Four optical transmitters are driven by a common 4 GHz clock which also controls timing of the word generators. Each electrical phase-shifter in the clock path provides a quarter-bit-period time delay between the mode-locked pulses from the four laser transmitters as shown. Additional phase-shifters (not shown) provide synchronization of the modulators with the mode-locked pulses. The transmitters use 1.3 μm mode-locked semiconductor lasers diode, which produce 15 ps optical pulses and provide the low-duty-cycle pulse streams necessary for time multiplexing. Mode-locked LD provide an excellent means of generating short pulses at high repetition rates while using only narrowband electronics. In addition, they offer the potential of good spectral purity—a key requirement for low-distortion, high-bit-rate transmission over long fiber lengths. Data are encoded using Ti:LiNbO$_3$ waveguide intensity modulators, giving return-to zero (RZ) transmitter outputs at 4 Gbit/s. The four mode-locked lasers have a common 4 GHz clock, which also controls timing of the word generators.

The multiplexer in Fig. 4.11 was a passive power combiner comprising a tree structure of 3 dB couplers. In principle, the multiplexer may be an active element, made up of Ti:LiNbO$_3$ switches. If more than four channels were to be multiplexed, there would be an advantage in using an active multiplexer integrated on a single chip. At the receiver end of the system, the 16 Gbit/s RZ signal is decomposed to its 4 Gbit/s baseband components by the demultiplexer (DEMUX). The demultiplexer is comprised of Ti:LiNbO$_3$ directional coupler switches [4.79] and associated drive electronics. The demultiplexed 4 Gbit/s RZ signals are detected using a 4 Gbit/s avalanche photodiode and GaAs FET combination [4.88].

There are two levels of demultiplexing. In the first level, a Ti:LiNbO$_3$ travelling-wave reversed β directional coupler switch is driven by a sinusoid at 8 GHz and

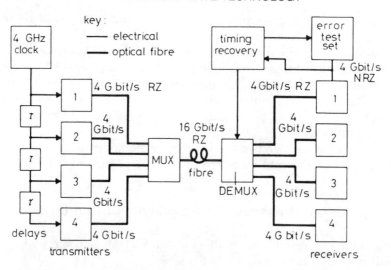

Fig. 4.11 16 Gbit/s optical TDM systems. (From [4.9] Tucker *et al.* (1987) *Electron. Lett.* **23**, 1270–1271. Reproduced by permission of IEE.)

demultiplexed the incoming 16 Gbit/s data stream into two 8 Gbit/s signals. In the second level, the 4 Gbit/s baseband channels are demultiplexed from the intermediate 8 Gbit/s signals using switches driven with 4 GHz sinusoids. The widest electronic bandwidth in the demultiplexer and receiver (and also in the transmitter) is only ≈ 2.5 GHz.

4.4 OPTICAL SOLITON TRANSMISSIONS

In the negative group velocity dispersion (GVD) region of optical fibers, solitons can be generated by balancing self-phase modulation with negative GVD. A soliton is a good information carrier because of its short duration and high stability. Solitons in optical fibers were first proposed and the stability of the solitons was demonstrated by numerical calculations in 1973 [4.10]. In 1980, solitons were observed in optical fibers [4.84]. Recently, there have been many numerical investigations of soliton propagation in connection with the transmission capacity of optical fiber communication systems [4.85–4.87].

For long-distance optical transmission systems, optical amplifiers must be used to compensate for fiber loss. Stimulated Raman scattering (SRS) in optical fiber is proposed for soliton transmission since it operates as a distributed gain medium [4.88]. The distributed amplifier is very suitable for soliton transmission because it provides a nearly loss-free transmission line. Soliton transmission over 6000 km with Raman amplifiers by circulating a 50 ps soliton pulse in a 42 km long fiber loop was demonstrated [4.89].

Currently, Er^{3+}-doped fiber amplifiers operating in the 1.5 μm region are of great interest for optical communication because of their high gain and low insertion loss. They operate as lumped gains. It has been shown that solitons can be amplified and transmitted with Er^{3+}-doped fiber [4.90, 4.91]. A simple and interesting soliton transmission method with a lumped amplifier was reported for 10 km repeater spacing with an $N=1$ soliton [4.92]. This configuration, the lumped amplifier system, is very simple and easy to construct. However, the repeater spacing is small, and so it can be described in terms of perturbation theory.

4.4.1 Optical soliton propagation in fibers [4.93]

The propagation of a non-linear pulse in a fiber with loss is described by the non-linear Schrödinger equation (NLS). In its simplest form, the non-linear propagation of a slowly varying wave envelope $U[z,t]$

$$-i\partial u/\partial z + i\Gamma u + \tfrac{1}{2}(\partial^2 u/\partial s^2) + |u|^2 u = 0. \qquad (4.6)$$

When the loss is ignored ($\Gamma=0$), the equation has a steady state and periodic solutions. The initial condition is

$$u = A \text{ sech } (s) \qquad s = t/\tau \qquad (4.7)$$

which is well known as the lowest order soliton when $\tfrac{1}{2} < A < 3/2$. The soliton period Z_{sp} is the length of non-linear medium at which the pulse shape repeats. Z_{sp} is given by

$$Z_{sp} = \frac{\pi^2 c \tau^2}{\lambda^2 |D|} \qquad (4.8)$$

where D is the GVD, λ is a vacuum wavelength, τ is the normalizing time ($\tau = t_{FWHM}/1.76$ for each shape pulse), and c is the speed of light in a vacuum. The peak power of the pulse required to generate an exact $N=1$ soliton pulse ($A=1$) is given by

$$P_1 = \frac{\lambda}{4 n_2 Z_{sp}} \pi w^2 \qquad (4.9)$$

where n_2 is the non-linear index of refraction (which for silica is 3.2×10^{-20} m/W) and w is the mode field radius.

Because of fiber loss (γ), for example -0.22 dB/km at a wavelength of 1.55 μm, the steady-state soliton solution cannot be maintained. The relation between γ (in dB/km), Z_{sp} (in m), and value Γ is given by

$$\Gamma = \frac{2}{\pi} Z_{sp} \frac{\gamma}{20\,000} [\log_{10} e]^{-1}. \qquad (4.10)$$

Equation (4.6) was solved numerically [4.93] by using the beam-propagation method [4.94]. In the calculation, periodic boundary conditions are used. When the fiber loss is small, the perturbation theory indicates that the soliton pulse width broadens in inverse proportion to its amplitude.

Figure 4.12 [4.93] shows results for the relation between the soliton pulse width and its peak amplitude. The cases of $\Gamma = -0.01$ and -1.0 are illustrated in Fig. 4.12(a) and (b), respectively. The upper half of the figure shows pulse shape (amplitude) change along the propagation distance, assuming sech(t) as an input pulse (i.e. exact $N=1$ soliton). The lower half of the figure shows the product of amplitude and pulse width. For Fig. 4.12(a) ($\Gamma = -0.01$), this product remains almost constant throughout propagation, which means that simple perturbation theory is valid and the pulse propagates as an $A=1$ soliton over the whole distance. On the other hand, for a large Γ of -1.0 given in Fig. 4.12(b), the amplitude decreases very rapidly, and therefore the pulse width remains almost constant. This pulse cannot propagate as an $A=1$ soliton because of its small amplitude.

In optical transmission systems over long distances, periodic amplification is required to compensate for fiber loss. There are two types of amplifiers which can be used for this purpose. One is a distributed amplifier and the other is a lumped amplifier. The Raman amplifier is a typical distributed amplifier. That is, conventional optical fiber itself becomes an amplifying medium. In the Raman amplifier, pump intensity decreases due to fiber loss and pump only when the pump depletion and/or fiber loss is small. It is pointed out that decreases in pump intensity produce a periodic perturbation to the solitons [4.88] and either gain or loss dominates at any given position. That is, the transmission loss of the soliton can only be compensated for by the overall periodic Raman gain [4.95]. When Γ is small, the soliton pulse behaves as described in Fig. 4.12(a), and the pulse can propagate as an $A=1$ soliton over many multiples of Z_{sp}. Pulse transmission over more than 6000 km have been demonstrated [4.89].

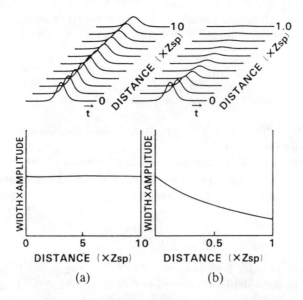

Fig. 4.12 Soliton pulse shape change with distance. (From [4.93] Kubota and Nakazawa (1990) *IEEE J. Quantum Electron.* **QE-26**, 692–700. Reproduced by permission. ©1990 IEEE.)

4.4.2   Demonstration of fiber soliton transmission

An area of increasing interest in optical communications using solitons in fibers has emerged as a result of the recent demonstration of successful soliton transmission over a distance of 6000 km [4.89]. Interest in an optical soliton system has grown also in view of bit-rate limitations in a linear system which originate from dispersion (for a PCM system) [4.96] or non-linearity (for a coherent system) [4.97]. The recent success of optical soliton propagation has shown their potential for future ultra-high-speed, long-distance optical communication systems.

Two major merits of the use of optical solitons for communication purposes are that a soliton propagates free of distortion in the presence of a fiber (group) dispersion and that the fiber loss, which tends to spread the soliton width due to the weakened non-linearity, can simply be compensated by doped fiber amplifier or/and Raman gain in the fiber systems. Optical solitons, injected from a semiconductor laser diode (such as a DFB-LD) and travel through a fiber with loss compensated by periodically pumped fiber amplifiers or Raman gain amplifiers, can provide an all-fiber high bit-rate communication system over very long distances.

SOLITON TRANSMISSION WITH LASER DIODES AND DOPED FIBER AMPLIFIERS

To generate optical solitions at a high repetition rate over Gbit/s in fiber communication systems, laser diodes (LD) are promising light sources because of their advantages of high-speed modulation and small size. To keep the soliton transmission in distances over more than several hundred, or even several thousand kilometers, an Er^{3+}-doped fiber amplifier or/and Raman amplifier is expected to be the most promising candidate for future ultra-high-speed long-distance optical soliton communication. Several optical soliton experiments using laser diodes and with Er^{3+}-doped fiber amplifiers and/or Raman amplifiers were presented in recent years [4.98–4.100].

Figure 4.13 [4.98] show one of the experimental setups. The gain-switched 1.55 μm distributed feedback laser diode (DFB-LD) generated a linear red-shift frequency-chirping optical pulse train at a 3.6418 GHz repetition rate. The optical pulse train passed through an isolator, and was amplified by a 15 m long Er^{3+}-doped polarization-maintaining fiber pumped with a 1.48 μm CW Fabry–Perot laser diode (FP-LD). The amplified optical pulses were launched into a 3.7 km-long polarization-maintaining optical fiber (PMF) to compensate for their chirping by the 18.7 ps/km/nm normal dispersion (at 1.55 μm) of the PMF [4.101]. At the end of the PMF, a compressed optical pulse train was obtained.

The compressed optical pulse train was modulated by a Ti:LiNbO$_3$ intensity modulator (insertion loss = 6.4 dB), and amplified by a 150 m long Er^{3+}-doped single mode fiber pumped by a 200 mW CW FP-LD at 1.48 μm wavelength. Amplified optical signal pulses of 700 mW peak power were obtained, when the optical pulses were modulated by a pseudo-random bit stream (PRBS $2^{15} - 1$) of a ½ mark ratio.

A 23 km long dispersion-shifted single mode optical fiber (DSF) was used as the transmission line. The amplified optical signal pulses were coupled into the DSF through a dichroic mirror DM3. The DSF had an anomalous dispersion of

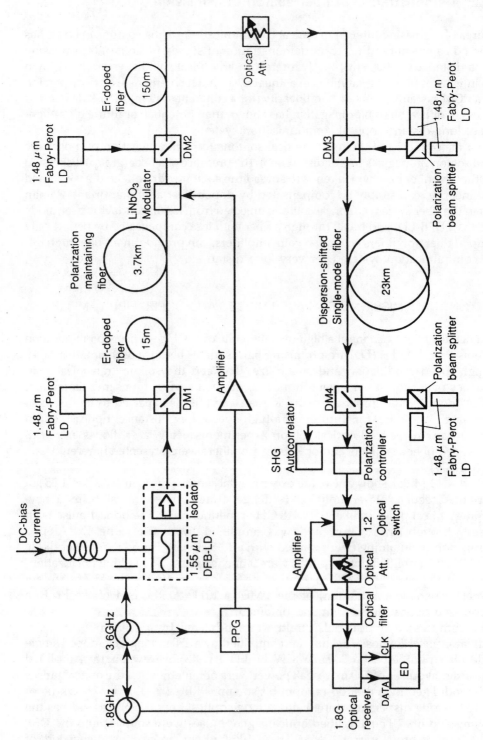

Fig. 4.13 3.6 Gbit/s optical soliton transmission. (From [4.98] Iwatsuki *et al.* (1990) *IEEE Photonic Technol. Lett.* **PTL-2**, 355–357. Reproduced by permission. ©1990 IEEE.)

4.0 ps/km/nm, a loss of 0.25 dB/km, and a mode field diameter of 6.0 μm at a 1.55 μm wavelength. The loss of the DSF was compensated for by both forward and backward Raman amplification [4.88]. To make the Raman gain insensitive to signal pulse polarization, two 1.48 μm CW FD-LDs, orthogonally polarized to each other, were coupled by a polarization beam splitter, and used as pump sources. The total loss of the DSF with Raman amplification was less than 1.0 dB, when forward and backward pump powers were 150 and 70 mW, respectively. The transmitted optical signal pulse was driven by a 1.8209 GHz sinusoidal signal. The state of polarization was adjusted by a polarization controller (PC) before the optical switch injection. An optical bandpass filter with a 1 nm bandwidth was employed to reduce beat noises between amplified spontaneous emissions. The demultiplexed 1.8209 Gbit/s optical signal pulses were received by a F-1.6 G optical receiver to measure the BER. The optical pulse widths after transmitting through the DSF and DM4 were measured by a SHG autocorrelator.

Assuming the optical signal pulse modulated by the PRBS pattern before transmission of the DSF has a Gaussian pulse shape, the optical pulse width (FWHM) is estimated as 17.8 ps. A reduction on pulse broadening due to optical soliton effect, as the input peak power is raised, can be observed. After 23 km transmission, the SHG autocorrelation trace of optical signal pulses shows soliton pulse width is about 12 ps when the input peak power is increased to 177 mW. As an advanced example, a 3.2–5 Gbit/s error-free soliton transmissions over 100 km using Er booster amplifiers and repeaters are reported [4.100]. Here, the new pre-emphasis technique is used for repeating the soliton, in which $N = 1.2$–1.4 solitons are excited in each soliton fiber. The experimental setup for soliton transmission is shown in Fig. 4.14 [4.100], in which the peak power of the output soliton pulse from EDFA (Er doped fiber amplifier) 2 exceeds the peak power of the $N = 1$ soliton.

The laser source for soliton generation is a DFB-LD, of which the average output power is about 2–3 mW. With gain-switching technique, a 27 ps pulse train with a GHz repetition rate can be easily generated, but it is far from being in a transform-limited condition. An FPR (Fabry–Perot resonator) is used as a narrow-band spectral filter to obtain transform-limited pulses although this decreased the average transmitted power to 1/10 of the input power. The bandwidth of the FPR is 0.18 nm, the cavity spacing is about 200 μm, and the finesse is about 30. Then the pulse is coupled into a LiNbO$_3$ Mach–Zehnder type light intensity modulator to switch the soliton train. The soliton transmission fiber is four 25 km dispersion shifted fibers with zero dispersions at 1.490–1.500 μm and -2.3 to -2.4 ps/km/nm dispersion at 1.545 μm. The fiber loss is 0.24 dB/km and the mode field diameter is 5.6–6.0 μm. Thus, the $N = 1$ soliton peak power for the 20 ps pulse is as small as 6–10 mW.

SOLITON TRANSMISSION OVER VERY LONG FIBER WITH LOSS COMPENSATED BY RAMAN GAIN

The possibility that soliton transmission over very long fiber with loss periodically compensated by Raman gain was predicted in [4.102]. Further theoretical studies [4.95] predicted that the rate–length product for such transmission could be as high as $\approx 30\,000$ GHz km for a single channel, and many times that figure with wavelength

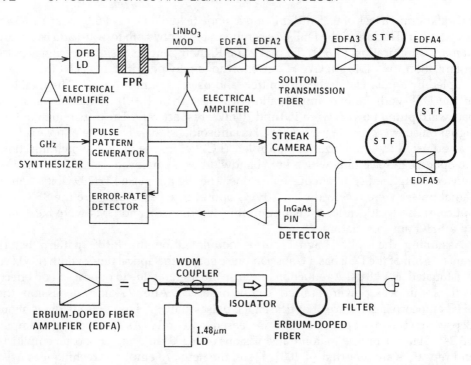

Fig. 4.14 Soliton transmission using erbium amplifiers. (From [4.100] Nakarawa *et al.* (1990) *IEEE Photonic Technol. Lett.* **PTL-2**, 216–219. Reproduced by permission. ©1990 IEEE.)

multiplexing. In order to understand the influence of energy changes on soliton propagation, some concepts [4.84, 4.103] have to be reviewed.

The area of a pulse S is defined as the time integral of the absolute value of the pulse amplitude. Therefore S scales as $A\tau$, where A is the peak amplitude of the pulse and τ is the pulse width. Since the pulse energy, E, scales as $A^2\tau$, τ scales as E^{-1}, provided that S is preserved during propagation. Both perturbation theory and direct numerical solution of the non-linear Schrödinger equation [4.95] have shown that S is preserved (here the adiabaticity and the preservation of S are essentially one and the same) provided that the loss (gain) rate is small enough. This is conveniently expressed by the condition $\alpha z_0 \ll 1$, where α is the average loss (gain) coefficient and z_0 is the soliton period [4.111]. For a typical case [4.104] in which the soliton energy is enhanced fourfold over a distance of 1000 km, the effective value of α is 0.0014/km. With $z_0 \approx 55$ km (for $\tau=50$ ps and the known fiber parameters) the above criterion is easily satisfied, with $\alpha z_0 \approx 0.08$.

A further condition that must be met relates to the recovery of the pulse area over a single amplification period, L (the length of the fiber loop). Numerical computations have shown that large deviations in S result in the neighbourhood of a resonance, occurring at $z_0 \approx L/8$, between the perturbation period (L and the soliton phase. To avoid that resonance, the long z_0 limit must be remained as $z_0 \gg L/8$ in order for the pulse area to be well recovered at the end of each amplification period. Soliton compression over many round trips should therefore remain distortionless and scale as E_{-1} until the compression itself makes the soliton period (proportional to τ^2)

become comparable with $L/8$. Thus, this second condition limits the maximum adiabatic compression that we can achieve.

The apparatus is shown schematically in Fig. 4.15. A 41.7 km length of low-loss (0.22 dB/km at 1600 nm) single mode fiber with group delay dispersion $D \approx 17$ ps/nm/km at 1600 nm is closed on itself with an all-fiber version of a Mach–Zehnder interferometer. The interferometer allows pumped light at 1497 nm (≈ 300 mW CW from a KC1:T1 colour-centre laser) to be efficiently coupled into the loop, while at the same time allow the signal light (50 ps, minimum bandwidth pulses from a mode-locked 100 MHz pulse repetition rate, NaCl colour-centre laser operating at 1600 nm) to recirculate around the loop. The signal pulses are coupled into the fiber loop through a fixed 5% fiber fusion coupler rather than by means or the all-fiber Mach–Zehnder interferometer (FMZ). Thus one can avoid variation in signal coupling from the environmentally sensitive FMZ, such that an absolute value of the signal power within the loop can be precisely and stably determined and set to that of the fundamental soliton [4.104].

The difference in pump and signal frequencies, ≈ 430 cm^{-1}, corresponds to the peak of the Raman gain band in quartz glass. The 50 ps pulse width makes the soliton period $z_0 \approx 55$ km, so that we easily meet the criterion $z_0 \gg L/8$ for stable soliton transmission [4.95] (L is the amplification period, here 41.7 km). Two fiber polarization controllers are set here for adjusting the polarization of the signals entering the (slightly polarizing) FMZ orthogonal to that of the previous round trip. In this way, as averaged over every two round trips, no one polarization is favoured. Thus the Raman gain, which depends on the relative polarization of pump and signal, remains more uniform over many round trips. Furthermore, no polarization component of the Raman spontaneous emission can have greater gain than the signals themselves.

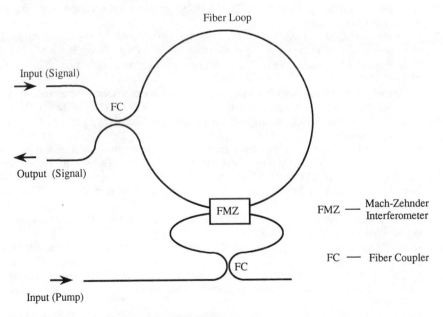

Fig. 4.15 Soliton loss compensated by Raman gain.

Much research remains to be done before a soliton-based communications system can be realized in practice. A truly single-frequency laser, a kind of high-efficiency amplifier, will make soliton transmission a strong candidate in large distance communications.

REFERENCES

[4.1]   P. Kaiser, J. Midwinter and S. Shimada (1987) "Status and future trends in terrestrial optical fiber systems in North America, Europe, and Japan" *IEEE Commun. Mag.* **25**(10), 8–22.

[4.2]   S. Shimada *et al.* (1986) "Gigabit/s optical fiber transmission systems—Today and tomorrow" *ICC'86* paper 48.3.

[4.3]   D. N. Payne and W. A. Gambling (1975) "Zero material dispersion in optical fibers" *Electron. Lett.* **11**, 176–178.

[4.4]   F. P. Kapron (1977) "Maximum information capacity of fiber optic waveguides" *Electron. Lett.* **13**, 96–97.

[4.5]   M. Miyagi and S. Nishida (1979) "Pulse spreading in a single-mode fiber due to third-order dispersion" *Appl. Opt.* **18**, 678–682.

[4.6]   D. Marcuse (1980) "Pulse distortion in single-mode fibers" *Appl. Opt.* **19**, 1653–1660.

[4.7]   T S. Kinsel (1970) "Wide-band optical communication systems: Part 1—time division multiplexing" *Proc. IEEE* **58**, 1666–1683.

[4.8]   R. S. Tucker *et al.* (1987) "Optical time-division multiplexing and demultiplexing in a multigigabit/second fiber transmission system" *Electron. Lett.* **23**, 208–209.

[4.9]   R. S. Tucker *et al.* (1987) "16 Gbit/s fiber transmission experiment using optical time-division multiplexing" *Electron. Lett.* **23**, 1270–1271.

[4.10] A. Hasegawa and F. Tappert (1973) "Transmission of stationary nonlinear optical pulses in dispersive dielectric fibers. 1. Anomalous dispersion" *Appl. Phys. Lett.* **23**, 142–144.

[4.11] K. J. Blow and N. J. Doran (1982) "High-bit rate communication systems using nonlinear effects" *Opt. Commun.* **42**, 403.

[4.12] J. Satsuma and N. Yajima (1974) "Initial value problems of one dimensional self-modulation of nonlinear waves in dispersive media" *Prog. Theor. Phys.* **55**, 284–306.

[4.13] K. J. Blow and D. Wood (1986) "The evolution of solitons from nontransform limited pulses" *Opt. Commun.* **58**, 349–354.

[4.14] Y. Hayashi *et al.* (1987) "System configuration for a gigabit/s optical transmission system utilizing the 1.55 μm wavelength region" in *Proc. GLOBECOM'87* 22.4.1–22.4.5.

[4.15] A. H. Gnauck *et al.* (1986) "8-Gbit/s transmission over 30 km of optical fiber" *Electron. Lett.* **22**, 600–602.

[4.16] S. K. Korotky *et al.* (1987) "8-Gbit/s transmission experiment over 68 km of optical fiber using a Ti:LiNbO$_3$ external modulator" *J. Lightwave Technol.* **5**, 1505–1509.

[4.17] K. Hagimoto *et al.* (1987) "8 Gbit/s optical regenerator employing DFB-LD direct modulation scheme and dispersive fibertransmission experiments" *ECOC'87, Helsinki, Finland, Sept. 1987.*

[4.18] A. H. Gnauck and J. E. Bowers (1987) "16 Gbit/s direct modulation of an InGaAsP laser" *Electron. Lett.* **23**, 801–803.

[4.19] R. L. Jungerman *et al.* (1987) "Coded phase reversal LiNbO$_3$ modulator with bandwidth greater than 20 GHz at 1.3-μm wavelength" *Electron. Lett.* **23**, 172–174.

[4.20] J. C. Cartledge (1990) "Theoretical performance of multigigabit-per-second lightwave systems using injection-locked semiconductor lasers" *J. Lightwave Technol.* **8**, 1017–1022.

[4.21] M. Shirasaki *et al.* (1990) "20 Gbit/s no-chirp intensity modulation by DPSH-IM method and its fiber transmission through 330 ps/nm dispersion" *Electron. Lett.* **26**, 33–35.

[4.22] J. C. Cartledge and G. S. Burley (1990) "Chirping-Induced waveform distortion in 2.4 Gbit/s lightwave transmission systems" *J. Lightwave Technol.* **8**, 699–703.

[4.23] K. Wakita *et al.* (1990) "High-speed InGaAlAs/InAlAs multiple quantum well optical modulators" *J. Lightwave Technol.* **8**, 1027–1032.

[4.24] R. A. Linke (1985) "Modulation induced transient chirping in single frequency lasers" *IEEE J. Quantum Electron.* **QE-21**, 593–597.

[4.25] H. Kressel and J. K. Butler (1977) *Semiconductor Lasers and Heterojunction LEDs* (New York: Academic), chapter 17.

[4.26] P. J. Corvini and T. L. Koch (1987) "Computer simulation of high-bit-rate optical fiber transmission using single-frequency lasers" *J. Lightwave Technol.* **5**, 1591–1595.

[4.27] S. Yamamoto *et al.* (1987) "Analysis of chirp power penalty in 1.55-μm DFB-LD high-speed fiber transmission systems" *J. Lightwave Technol.* **5**, 1518–1524.

[4.28] T. Okiyama *et al.* (1988) "Evaluation of 4 Gbit/s optical fiber transmission distance with direct and external modulation" *J. Lightwave Technol.* **6**, 1686–1692.

[4.29] K. Hagimoto and K. Aida (1988) "Multigigabit-per-second optical baseband transmission system" *J. Lightwave Technol.* **6**, 1678–1685.

[4.30] D. A. Atlas *et al.* (1988) "Chromatic dispersion limitations due to semiconductor laser chirping in conventional and dispersion-shifted single-mode fiber systems" *Opt. Lett.* **13**, 1035–1037.

[4.31] J. C. Cartledge and G. S. Burley (1989) "The effect of laser chirping on lightwave system performance" *J. Lightwave Technol.* **7**, 568–573.

[4.32] R. A. Linke and A. H. Gnauck (1983) "High speed laser driving circuit a gigabit modulation of injection lasers" *Proc. SPIE* **425**(23).

[4.33] W. T. Tsang *et al.* (1983) "High-speed direct single-frequency modulation with large tuning rate and frequency excursion in cleaved-coupled cavity semiconductor lasers" *Appl. Phys. Lett.* **42**, 650.

[4.34] I. P. Kaminow *et al.* (1983) "High-speed 1.55 μm single-longitudinal mode ridge waveguide laser" *Electron. Lett.* **19**, 784–785.

[4.35] T. L. Koch *et al.* (1984) "1.55 μm InGaAsP vapor phase transported (VPT) buried heterostructure distributed feedback lasers" in *Proc. 9th IEEE Laser Conf., Rio de Janeiro, Brazil, Aug. 7–10 1984*, post deadline-paper.

[4.36] G. H. B. Thompson (1972) "A theory for filamentation in semiconductor lasers including the dependence of dielectric constant on injected carrier density" *Opto-Electron.* **4**, 257–310.

[4.37] R. S. Tucker and I. P. Kaminow (1984) "High-frequency characteristics of directly modulated InGaAsP ridge waveguide and buried heterostructure lasers" *J. Lightwave Tech.* **2**, 385–393.

[4.38] T. L. Koch and J. E. Bowers (1984) "On the nature of wavelength chirping in directly modulated semiconductor lasers" *Electron. Lett.* **20**, 1038–1039.

[4.39] K. Iwashita *et al.* (1982) "Chirp pulse transmission through a single mode fiber" *Electron. Lett.* **18**, 873.

[4.40] K. Kishino *et al.* (1982) "Wavelength variation of 1.6 μm wavelength buried heterostructure GaInAsP/InP lasers due to direct modulation" *J. Quantum Electron.* **QE-18**, 343.

[4.41] C. Lin *et al.* (1983) "Picosecond frequency chirping and dynamic line broadening in InGaAsP injection lasers under fast excitation" *Appl. Phys. Lett.* **42**, 141.

[4.42] N. A. Olsson *et al.* (1984) "Dynamic linewidth of amplitude modulated single-longitudinal-mode semiconductor lasers operation at 1.5 μm wavelength" *Electron. Lett.* **20**, 121.

[4.43] R. A. Linke (1984) "Transient chirping in single-frequency lasers: Lightwave systems consequences" *Electron. Lett.* **20**, 472–474.

[4.44] P. J. Chidgey *et al.* (1984) "1.2 Gbit/s optical fiber transmission over 113.7 km using a 1.528 μm DFB ridge-waveguide laser" *Electron. Lett.* **20**, 707–709.

[4.45] R. A. Linke *et al.* (1984) "120 km lightwave transmission experiment at 1 Gbit/s using a new long-wavelength avalanche photodetector" *Electron. Lett.* **20**, 498–499.

[4.46] R. A. Linde (1984) "Direct gigabit modulation of injection lasers structure dependent speed limitations" *J. Lightwave Technol.* **2**, 40–43.

[4.47] J. E. Midwinter (1977) "A study of intersymbol interference and transmission medium instability for an optical fiber system" *Opt. Quantum Electron.* **9**, 299–304.

[4.48] B. L. Kasper *et al.* (1984) "A 130 km transmission experiment at 2 Gbit/s using silicon-core fiber and a vapor phase transported DFB laser" *ECOC'84, Stuttgart, Sept., 1984* post deadline paper PD-6.

[4.49] R. Olshansky and D. Fye (1984) "Reduction of dynamic linewidth in single-frequency semiconductor lasers" *Electron. Lett.* **20**, 928–929.

[4.50] L. Bickers and L. D. Westbrook (1985) "Reduction of laser chirp in 1.5 μm DFB lasers by modulation pulse shaping" *Electron. Lett.* **21**, 103–104.

[4.51] G. P. Agrawal and C. H. Henty (1988) "Modulation performance of a semiconductor laser coupled to an external high-Q resonator" *IEEE J. Quantum Electron,* **QE-24**, 134–142.

[4.52] N. A. Olsson *et al.* (1988) "Performance characteristics of a 1.5 μm single frequency semiconductor laser with an external waveguide Bragg reflector" *IEEE J. Quantum Electron,* **QE-24**, 143–147.

[4.53] J. C. Cartedge (1990) "Improved transmission performance resulting from the reduced chirp of a semiconductor laser coupled to an external high-Q resonator" *J. Lightwave Technol.* **8**, 716–721.

[4.54] S. Kobayashi *et al.* (1980) "Single-mode operation of 500 Mbit/s modulated AlGaAs semiconductor laser by injection locking" *Electron. Lett.* **16**, 746–748.

[4.55] H. Nishimoto *et al.* (1983) "Injection-locked 1.5 μm InGaAsP/InP laser capable of 450 Mbit/s transmission over 106 km" *Electron. Lett.* **19**, 509–510.

[4.56] C. Lin and F. Mengel (1984) "Reduction of frequency chirping and dynamic linewidth in high-speed directly modulated semiconductor lasers by injection locking" *Electron. Lett.* **20**, 1073–1075.

[4.57] C. Lin *et al.* (1985) "Frequency chirp reduction in a 2.2 Gbit/s directly modulated InGaAsP semiconductor laser by CW injection" *Electron. Lett.* **21**, 80–81.

[4.58] M. Shirasaki *et al.* (1990) "20 Gbit/s no chirp intensity modulation by DPSH-IM method and its fiber transmission through 330 ps/nm dispersion" *Electron. Lett.* **26**, 33–35.

[4.59] C. J. Buczek *et al.* (1973) "Laser injection locking" *Proc. IEEE* **61**, 1411–1431.

[4.60] R. Lang *et al.* (1976) "Suppression of relaxation oscillation in the modulated output of semiconductor lasers" *IEEE J. Quantum Electron.* **QE-12**, 194–199.

[4.61] T. Andersen *et al.* (1982) "Generation of single-mode picosecond pulses by injection locking of an AlGaAs semiconductor laser" *Appl. Phys. Lett.* **41**, 14–16.

[4.62] S. Kobayashi and T. Kimura (1982) "Optical FM signal amplification by injection locked and resonant type semiconductor laser amplifiers" *IEEE J. Quantum Electron,* **QE-18**, 575–581.

[4.63] H. Toba *et al.* (1984) "Injection locking technique applied to a 170 km transmission experiment at 445.8 Mbit/s" *Electron. Lett.* **20**, 370–371.

[4.64] R. Adler "A study of locking phenomena in oscillators" *Proc. IRE* **34**, 351–357.

[4.65] S. Kobayashi and T. Kumura "Injection locking in AlGaAs semiconductor lasers" *IEEE J. Quantum Electron.* **QE-17**, 681–689.

[4.66] N. A. Olsson *et al.* (1985) "Chirp-free transmission over 82.5 km of single mode fibers at 2 Gbit/s with injection locked DFB semiconductor lasers" *J. Lightwave Technol.* **3**, 63–66.

[4.67] Y. Suematsu *et al.* (1983) "Dynamic single-mode semiconductor lasers with a distributed reflector" *J. Lightwave Technol.* **1**, 161–176.

[4.68] R. A. Linke "Transient chirping in single-frequency lasers: lightwave systems consequences" *Electron. Lett.* **20**, 472–474.

[4.69] K. Wakita *et al.* (1990) "High-speed InGaAlAs/InAlAs multiple quantum well optical modulators" *J. Lightwave Technol.* **8**, 1027–1031.

[4.70] S. K. Korotky *et al.* (1985) "4 Gbit/s transmission experiment over 117 km of optical fiber using a Ti:LiNbO₃ external modulator" *J. Lightwave Technol.* **3**, 1027–1031.

[4.71] P. S. Henry (1985) "Lightwave primer" *IEEE J. Quantum Electron.* **QE-21**, 1862–1879.

[4.72] H. Nishimoto *et al.* (1988) "New method of analyzing eye patterns and its application to high-speed optical transmission systems" *J. Lightwave Technol.* **6**, 678–685.

[4.73] A. H. Guauck *et al.* (1986) "8 Gbit/s transmission over 30 km of optical fiber" *Electron. Lett.* **22**, 600–602.

[4.74] D. Marcuse and C. Lin (1981) "Low dispersion single-mode fiber transmission—the question of practical versus theoretical maximum transmission bandwidth" *IEEE J. Quantum Electron.* **QE-17**, 869–877.

[4.75] A. Alping *et al.* (1982) "20 Gbit/s optical time multiplexing with TJS GaAlAs lasers" *Electron. Lett.* **18**, 422–424.

[4.76] R. S. Tucker (1988) "Optical time-division multiplexing for very high bit-rate transmission" *J. Lightwave Technol.* **6**, 1737–1749.

[4.77] T. S. Kinsel and R. T. Denton (1968) "Terminals for a high-speed optical pulse code modulation communication system: 11. Optical multiplexing and demultiplexing" *Proc. IEEE.* **56**, 146–154.

[4.78] R. C. Alferness (1982) "Waveguide electrooptic modulators" *IEEE Trans. Microwave Theory Tech.* **MTT-30**, 1121–1137.

[4.79] S. K. Korotky *et al.* (1985) "Fully connectorized high-speed Ti:LiNbO$_3$ switch/modulator for time-division multiplexing and data encoding" *J. Lightwave Technol.* **3**, 1–6.

[4.80] H. Haga *et al.* (1985) "An integrated 1×4 high-speed optical switch and its applications to a time demultiplexer" *J. Lightwave Technol.* **3**, 116–120.

[4.81] G. Eisenstein *et al.* (1987) "Optical time-division multiplexed transmission system experiment at 8 Gbit/s" *Electron. Lett.* **23**, 1115–1116.

[4.82] P. J. Chidgey and D. W. Smith (1987) "Sampled optical time-division multiplexing of asynchronous data" *Electron. Lett.* **23**, 1228–1229.

[4.83] B. L. Kasper *et al.* (1985) "SAGM avalanche photodiode optical receiver for 2 Gbit/s and 4 Gbit/s" *Electron. Lett.* **11**, 982–984.

[4.84] L. F. Mollenauer *et al.* (1980) "Experimental observation of picosecond pulse narrowing and solitons in optical fibers" *Phys. Rev. Lett.* **45**, 1095–1098.

[4.85] E. Shiojiri and Y. Fujii (1985) "Transmission capability of an optical fiber communication system using index nonlinearity" *Appl. Opt.* **24**, 358–360.

[4.86] B. Hermansson and D. Yevick (1983) "Numerical investigation of soliton interaction" *Electron. Lett.* **19**, 570–571.

[4.87] P. L. Chu and C. Desem (1985) "Mutual interaction between solitons of unequal amplitudes in optical fiber" *Electron. Lett.* **21**, 1133–1134.

[4.88] A. Hasegawa (1984) "Numerical study of optical soliton transmission amplified periodically by the stimulated Raman process" *Appl. Opt.* **23**, 3302–3309.

[4.89] L. F. Mollenauer and K. Smith "Ultralong-range soliton transmission" *OFC, Houston, TX,* paper WO1.

[4.90] M. Nakazawa *et al.* (1989) "Soliton amplification and transmission with Er^{3+}-doped fiber repeater pumped by a GaInAsP laser diode" *Electron. Lett.* **25**, 199–200.

[4.91] M. Nakazawa *et al.* (1989) "Efficient Er^{3+} doped optical fiber amplifier pumped by a 1.48 μm InGaAsP laser diode" *Appl. Phys. Lett.* **54**, 295–297.

[4.92] Y. Kodama and A. Hasegawa "Amplification and reshaping of optical solitons in glass fiber-11" *Opt. Lett.* **7**, 339–341.

[4.93] H. Kubota and M. Nakazawa (1990) "Long-distance optical soliton transmission with lumped amplifiers" *IEEE J. Quantum Electron.* **QE-26**, 692–700.

[4.94] D. Yevick and B. Hermansson (1983) "Soliton analysis with the propagation beam method" *Opt. Commun.* **47**, 101–103.

[4.95] L. F. Mollenauer *et al.* (1986) "Soliton propagation in long fibers with periodically compensated loss" *IEEE J. Quantum Electron.* **QE-22**, 157–173.

[4.96] A. Hasegawa ans Y. Kodama (1981) "Signal transmission by optical solitons in monomode fiber" *Proc. IEEE* **69**, 1145–1150.

[4.97] D. Anderson and M. Lisk (1984) "Modulational instability of coherent optical-fiber transmission signals" *Opt. Lett.* **9**, 463–470.

[4.98] K. Iwatsuki *et al.* (1990) "3.6 Gbit/s all laser-diode optical soliton transmission" *IEEE Photonic Technol. Lett.* **PTL-2**, 355–357.

[4.99] K. Iwatsuki *et al.* (1990) "2.8 Gbit/s optical soliton transmission employing all laser diodes" *Electron. Lett.* **26**, 1–2.

[4.100] M. Nakazawa *et al.* (1990) "3.2–5 Gbit/s 100 km error-free soliton transmissions with Erbium amplifiers and repeaters" *IEEE Photonic Technol. Lett.* **PTL-2**, 216–219.

[4.101] K. Ieatsudi *et al.* (1989) "2.8 Gbit/s error-free optical soliton transmission employing all laser diodes" *IOOC'89, Kobe, Japan* Paper 20PDA-1.

[4.102] A. Hasegawa (1983) "Amplification and reshaping of optical solitons in a glass fiber IV: Use of the stimulated Raman process" *Opt. Lett.* **8**, 650–655.

[4.103] L. F. Mollenauer (1985) "Soliton in optical fibers and the soliton laser" *Phil. Trans. R. Soc. London* Ser. A **315**, 437–450.

[4.104] K. Smith and L. F. Mollenauer (1989) "Experimental observation of adiabatic compression and expansion of soliton pulses over long fiber paths" *Opt. Lett.* **14**, 751–753.

5

OPTICAL FIBERS—
WAVEGUIDES AND DEVICES

5.1 INTRODUCTION

The optical fiber, regardless of type, is seen by light propagating along it as a dielectric waveguide. To understand how light propagates, we must follow the same route as for analysing any other electromagnetic waveguide and solve Maxwell's equations using the appropriate boundary conditions. However, before launching into that, let us make a few general observations about the optical fiber.

The typical structure is shown below in Fig. 5.1. It consists of a circular core region of radius a embedded in a circular cladding which for the purposes of our theoretical modelling we will assume to be infinite in extent although in practice this is obviously not the case. The core has a higher refractive index or optical dielectric constant to the cladding and can be assumed to be essentially lossless, with its attention normally dominated by the Rayleigh scattering fundamental to the random structure of the glass from which it is made.

Fibers used for information transmission purposes fall into two broad classes: single mode and multimode, depending primarily upon their core size. Typical values are given below:

$$\text{Cladding index } n_2 = 1.45$$

$$\text{Core index } n_1 = 1.4645$$

and hence fractional index difference $[n_1 - n_2]/n_2 = 0.01$ (range 0.003–0.015) (these refractive index values correspond to those for silica and doped silica). Note $n_1 > n_2$ to ensure total internal reflection in the core.

$$\text{Core diameter } (2a)\text{—single mode fiber—8 } \mu\text{m}$$
$$\text{—multimode—50 } \mu\text{m}$$

$$\text{Cladding diameter—125 } \mu\text{m}$$

$$\text{Numerical aperture (NA} = \sqrt{\{n_1{}^2 - n_2{}^2\}}) = 0.2055.$$

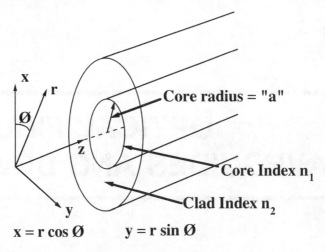

x = r cos Ø   y = r sin Ø

Fig. 5.1 The structure of the typical optical fiber.

Light travels (largely) within the core as a result of being totally internally reflected at the core cladding interface. Since the refractive index difference is small, this clearly implies grazing-incidence reflection with the light travelling at a small angle to the fiber axis. It is readily shown using Snell's law that for light entering the end face of the fibre at angle θ to the axis, then $\sin(\theta) = NA$ and hence for the example above $\theta = 12°$.

In the case of the single mode fiber, the core is so small that light can only travel by one path, the single mode, but for the bigger core fibers, there are many pathways and light can travel by each, corresponding to different ray paths within the core structure. We shall see later that a good measure of the 'size' of the fiber when viewed by an electromagnetic wave is the V parameter, defined as:

$$V = (2\pi a/\lambda) \sqrt{\{n_1^2 - n_2^2\}}$$

and we shall show later that, to a good approximation, the number of allowed modes is given by $M = V^2/2$. For our example above, with a $50\,\mu m$ core diameter and operating wavelength of $1\,\mu m$, we find that $M = 521$ and $V = 32.3$ so that over 500 different mode paths would exist. However, to understand their properties better, we need to turn to Maxwell's equations and find solutions for the field distributions.

5.2   THE ELECTROMAGNETIC THEORY LEADING TO LINEARLY POLARIZED (LP) MODES

We have already noted that we are concerned with a cylindrical dielectrical waveguide. It follows that we must seek solution to Maxwell's equations in cylindrical polar coordinates. The wave equation takes the general form:

$$\nabla^2 A = \mu\mu_0\epsilon\epsilon_0 \frac{d^2 A}{dt^2}.$$

In cylindrical polar coordinates, we then have the relation that

$$\nabla^2 = \frac{1}{r}\frac{d}{dr}\left(r\frac{d}{dr}\right) + \frac{1}{r^2}\frac{d^2}{d\phi^2} + \frac{d^2}{dz^2}$$

and we must meld these results and then look for solutions. We start off by assuming that a solution exists, and the r, ϕ, z and t functional dependences are separable so that it can be written in the form:

$$A = F_1(r)F_2(\phi)F_3(z)F_4(t).$$

The next step is then to postulate trial solutions for each of these functions. We note that we are interested in monochromatic wavelike solutions propagating along the fiber axis, so that it is natural to postulate that

$$F_3(z)F_4(t) = \exp\ i(\omega t - k_z z).$$

We also follow well established practice and postulate a ϕ dependence of the field of the form:

$$F_2(\phi) = \exp[\pm(il\phi)]$$

where l takes the values 0,1,2,3, etc.

Given these trial solutions, it is then a simple matter to substitute them to obtain an equation for the remaining function $F_1(r)$ as follows:

$$\left(\frac{d^2 F_1}{dr^2} + \frac{1}{r}\frac{dF_1}{dr}\right) \quad \text{(remaining from the } r \text{ variation)}$$

$$-\left(\frac{l^2}{r^2}\right)F_1 \quad \text{(from the } \phi \text{ variation)}$$

$$-(k_z^2)F_1 \quad \text{(from the } z \text{ variation)}$$

$$+(\omega^2\mu\mu_0\epsilon\epsilon_0)F_1 = 0 \quad \text{(from the } t \text{ variation)}$$

by substituting in turn for the different functions. We now gather this result together and make the following substitution:

Put

$$k_t^2 = \omega^2\mu\mu_0\epsilon\epsilon_0 - k_z^2 = -\gamma^2$$

so that our wave equation reduces to the following forms, according to the size of k_z and hence depending upon whether k_t or γ is real:

$$\frac{d^2F_1}{dr^2} + \frac{1}{r}\frac{dF_1}{dr} + \left(k_t^2 - \frac{l^2}{r^2}\right)F_1 = 0$$

or

$$\frac{d^2F_1}{dr^2} + \frac{1}{r}\frac{dF_1}{dr} - \left(\gamma^2 + \frac{l^2}{r^2}\right)F_1 = 0.$$

Readers whose mathematics is reasonably well oiled will recognize these equations as the Bessel and modified Bessel equations respectively to which the solutions are well known. The general form of solution to the first equation takes the form:

$$F_1 = A_1 J_l(k_t r) + A_2 Y_l(k_t r)$$

while the general solution for the second takes the form:

$$F_1 = B_1 W_l(\gamma r) + B_2 K_l(\gamma r).$$

In order to proceed further, we need to examine the physics of the guidance process and the boundary conditions associated with our problem. In order to do that more readily, let us simplify what appears to be a rather messy expression to something that is easier to work with. We first note some general relations taken from our basic electromagnetic theory. In the MKS system:

$$c = 1/\sqrt{(\epsilon_o \mu_o)}$$

and for a lossless dielectric (like optical glass at optical fiber wavelengths):

$$\epsilon = n^2 \text{ and } \mu = 1.$$

Hence we find that:

$$\omega^2 \epsilon \epsilon_o \mu \mu_o = \omega^2 \epsilon \mu / c = n^2 \omega^2 / c = n^2 k_o^2$$

where we have made the substitution $k_o = 2\pi/\lambda$, corresponding to the propagation constant for the wave in vacuum at wavelength λ. Given these substitutions, we see that if we use the parameter k_z to describe propagation in the core and the parameter γ to describe propagation in the cladding then we have the following identities:

$$k_x = \sqrt{(n_1^2 k_o^2 - k_z^2)} \qquad 0 < r < a$$

$$\gamma = \sqrt{(k_z^2 - n_2^2 k_o^2)} \qquad r > a.$$

If we now consider Snell's law for a 'ray' in the core medium propagating at an angle θ to the interface (see Fig. 5.2), then it is self-evident that $k_z = n_1 k_o \cos(\theta)$ and a little simple algebra will show that the condition for total internal reflection, namely that

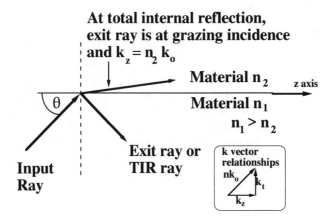

Fig. 5.2 Ray diagram for wave reflected (or totally internally reflected) at core–cladding boundary.

the exit ray travels at grazing incidence along the boundary in medium 2, implies that for TIR to occur, $k_z > n_2 k_o$. Hence it is clear that the substitutions we have made above lead to the conclusion that both k_x and γ are real quantities in the core and cladding respectively under the conditions that the wave is totally internally reflected at the core/cladding boundary. Hence we use the first equation, the ordinary Bessel equation, to describe the radial field variation in the core and the modified Bessel equation to describe the field variation in the cladding. We can now use our physical knowledge of the required solutions to progress.

The first equation describes the field variation with radius in the core region of the guide, from $0 < r < a$. Clearly, within this range, we must require the field to be finite at all points. If we look at the dependence of the Bessel function $Y_l(k_{tr})$ with r, then as r approaches 0, $Y_l(k_t r)$ tends to infinity. Hence we are forced to conclude in our general solution above that the parameter A_2 must be zero. For information, we show in Fig. 5.3 the general form of the first few J Bessel functions, showing that, away from argument zero, they rapidly begin to look very much like the sine or cosine

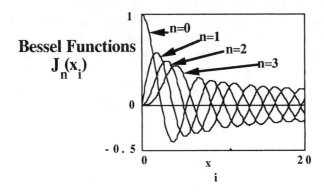

Fig. 5.3 Form of the J Bessel functions in the region of small argument.

functions albeit with slowly decaying amplitude. The Y functions are similar apart from the fact that, as x tends to zero, they head for minus infinity.

Likewise, if we examine the functional variation in the cladding, given by the general solution of the second equation, then we find that the modified Bessel function $B_1 W_l(\gamma r)$ approaches infinity as r goes to infinity. Both the W and K Bessel functions look very much the exponentials, although the K function falls to zero for large argument whilst the W function generally falls to zero at zero argument and diverges for large argument. Since the cladding region is defined as $r > a$, this clearly will not do if we are to have a guided wave field bound to the core and we must conclude that B_1 is also zero. Hence we conclude that the functional variation of the electric field with radius in the core and clad regions must be given by:

$$F_1 = A J_l(Ur/a) \qquad 0 < r < a$$

$$F_1 = B K_l(Wr/a) \qquad a < r < \infty$$

where we have dropped the subscripts on the scaling parameters A and B since they are no longer required and have introduced the widely used U, V and W parameters which are defined as related as follows:

$$U = a\sqrt{(n_1^2 k_o^2 - k_z^2)}$$

$$W = a\sqrt{(k_z^2 - n_2^2 k_o^2)}$$

$$V^2 = W^2 + U^2.$$

The final remaining step in solving Maxwell's equations to obtain the eigenvalue equation for the modes in the guide requires us to 'stitch together' the fields at the core–cladding interface so that they are continuous across the boundary.

The most complete way to proceed at this point is to derive expressions for all the vector field components in the core and cladding, match the appropriate ones at the core–cladding boundary and solve the resulting set of simultaneous equations. Doing this leads to a full description of the fields in the guide as well as the eigenvalue equation for the guided modes. However, the mathematics involved is rather messy and is given in great detail elsewhere (see for example [1]).

Starting along that route, we note that Maxwell's equations in cylindrical polar form are as follows:

$$\nabla \times E = \frac{-dB}{dt} = -\mu\mu_o \frac{dH}{dt}$$

$$\nabla \times H = \frac{dD}{dt} = \epsilon\epsilon_o \frac{dE}{dt}$$

where the del operator expands out as:

$$\nabla \times A = \begin{vmatrix} \dfrac{\mathbf{r}}{r} & \phi & \dfrac{\mathbf{k}}{r} \\[2mm] \dfrac{d}{dr} & \dfrac{d}{d\phi} & \dfrac{d}{dz} \\[2mm] A_r & rA_\phi & Az \end{vmatrix}$$

and where $\mathbf{r}$, ϕ and $\mathbf{k}$ are unit vectors directed in the r, ϕ and z directions. Now, given the form already established for the electromagnetic fields in the guide, we can write down a set of differential equations linking the 6 vector field components E_z, E_r and E_ϕ as well as H_z, H_r and H_ϕ as follows:

$$\frac{1}{r}\left[\frac{dE_z}{d\phi} + ik_z(rE_\phi) \right] = iw\mu\mu_o H_r$$

$$ik_z E_r + \frac{dE_z}{dr} = iw\mu\mu_o H_\phi$$

$$1/r \left[r\frac{dE_\phi}{dr} - \frac{dE_r}{d\phi} \right] = -iw\mu\mu_o H_z$$

$$\frac{1}{r}\left[\frac{dH_z}{d\phi} + ik_z rH_\phi \right] = iw\epsilon\epsilon_o E_r$$

$$ik_z H_r + \frac{dH_z}{dr} = -iw\epsilon\epsilon_o E_\phi$$

$$1/r \left[r\frac{dH_\phi}{dr} - \frac{dH_r}{d\phi} \right] = iw\epsilon\epsilon_o E_z.$$

By equating the tangential (z and ϕ) components of E and H, the required solution emerges after much algebra. On the way, one finds that because $[n_1 - n_2]/n_2 \ll 1$, the fields in the guide are effectively linearly polarized (along the x or y axis and essentially perpendicular to z). This result is not too surprising when one recognizes that because $[n_1 - n_2]/n_2 \ll 1$, the wave must propagate at a very small angle to the axis of the fiber if it is to be totally internally reflected. This leads to a considerable simplification in the mathematics and also to what are known as the linearly polarized mode solutions (LP modes). We can take advantage of this fact by adopting a somewhat more brute-force approach to finding a solution for the eigenvalue equation as follows.

We observe that an equally good boundary condition to apply (in place of equating E_z, E_r and E_ϕ as well as H_z, H_r and H_ϕ) is to observe that the normal component of D and also its first differential with respect to the boundary normal direction must be continuous. Hence we require, at $r = a$, that $D(\text{core}) = D(\text{clad})$ and $\partial D(\text{core})/\partial r = \partial D(\text{clad})/\partial r$. Knowing the r dependence of the electric fields, we can substitute for them in the above relations to obtain the LP mode eigenvalue equation directly. We first note the following relations from mathematics (see for example Abramowitz and Stegun [5.8]):

$$\left(\frac{1}{z}\frac{d}{dz}\right)^k [z^l J_l(z)] = z^{l-k} J_{l-k}(z)$$

$$\left(\frac{1}{z}\frac{d}{dz}\right)^k [z^{l-k} e^{il\pi} K_l(z)] = z^{l-k} e^{i(l-k)\pi} K_{l-k}(z).$$

Since we are only interested in the first differential, we can immediately form the following results:

$$\frac{d}{dz}[z^l J_l(z)] = z^l J_{l-1}(z)$$

$$= l z^{l-1} J_l(z) + z^l \frac{d}{dz}[J_l(z)]$$

and hence by collecting the terms above, we obtain:

$$\frac{d}{dl}[J_l(z)] = J_{l-1}(z) - \frac{l}{z} J_l(z)$$

and by using the same series of steps, we obtain for the differential of the K Bessel functions the result that:

$$\frac{d}{dz}[K_l(z)] = e^{-i\pi} K_{l-1}(z) - \frac{l}{z} K_l(z)$$

where we have used the well known result that $e^{-i\pi} = -1$. Now if we make the following substitutions in the expressions for J and K functions respectively:

$$z = Ur/a \qquad dz = (U/a)dr \qquad z = Wr/a \qquad dz = (W/a)dr$$

then we obtain the following relationships by equating the D and dD/dr at the core–cladding interface (i.e. setting $r = a$)

$$An_1^2 J_l(u) = Bn_2^2 K_l(u)$$

$$An_1^2 \left[J_{l-1}(u)\left[\frac{u}{a}\right] - \frac{l}{a} J_l(u) \right]$$

$$= Bn_2^2 \left[-K_{l-1}(w)\left[\frac{w}{a}\right] - \frac{l}{a} K_l(w) \right]$$

where as before, we have used A and B to represent the unknown field amplitudes in the core and cladding. Taking the ratio of the two equations, we eliminate A and B to obtain the desired eigenvalue equation for the LP modes as follows:

$$\frac{UJ_{l-1}(U)}{J_l(U)} = \frac{-WK_{l-1}(W)}{K_l(W)}.$$

This apparently intractable relationship is a functional relationship involving only one unknown, namely k_z, since the parameters U and W, apart from k_z, involve only known quantities that describe the guide itself, thus:

$$U = a\sqrt{(n_1^2 k_o^2 - k_z^2)}$$

$$W = a\sqrt{(k_z^2 - n_2^2 k_o^2)}.$$

Hence given the values for n_1, n_2, a and $k_o = 2\pi/\lambda$, in principle we can solve for k_z for any given mode and value of l. Normally, this requires a computer to carry out a numerical solution of the equation. However, there is one special case that we can solve exactly rather simply and that at least allows us to understand how the sequence in which the LP modes build up as the fibre V value is increased from zero. This is to solve for the case of mode cut-off which is the situation when $W=0$ and, in effect, there is a finite field at an infinite distance from the fibre core and hence the mode is leaking and no longer truly guided.

5.3 MODE CUT-OFF CONDITION

We are thus concerned with the situation when W tends to zero. Taking the right-hand side of the eigenvalue equation, if we examine the behaviour of the K Bessel functions for small W, then we find that the ratio:

$$\frac{K_{l-1}(W)}{K_l(W)}$$

remains finite so that the right-hand side of the eigenvalue equation approaches zero as W does. Hence we must seek solutions to the equation:

$$\frac{UJ_{l-1}(U)}{J_l(U)} = 0.$$

One solution is evidently the case that $U=0$. However, if both U and W are zero, then so also is V and there is no guide. However, the lowest order mode does propagate down to $V=0$ and has fields described by the J_o and K_o Bessel functions. For $U>0$, then we must look for zeros of the J_{-1} Bessel function. Since $J_{-1}(x) = -J_1(x)$ it is clear that this leads us to the zeros of $J_1(x)$ which occur at the (approximate) values of x given in Table 5.1.

Table 1. Zeros of the $J_l(x)$ Bessel function

Values of l	0	1	2	3	4	5	6
Values of x for $J_l(x)=0$	2.40	3.83	5.14	6.40	7.60	8.77	9.94
	5.52	7.01	8.40	9.70	11.00	12.30	13.60
	8.65	10.20	11.60	13.00	14.40	15.70	17.00
	11.80	13.30	14.80	16.20	17.60	19.00	20.30
	14.90	16.50	18.00	19.40	20.80	22.20	23.60
	etc.						

Thus for the $l=0$ set of modes, we find the lowest order, described as the LP$_{01}$ mode, starts to propagate as soon as a guide exists (e.g. $V>0$). Then, when the V value exceeds the first zero of J_1, ($V=3.83$) the next mode of that set cuts in, namely the LP$_{02}$ followed at $V=7.01$ by the LP$_{03}$ and so on. Another set of modes with radial field distributions described by the J_1 function and corresponding to solutions from setting $l=1$ are dictated by the zeros of J_0 so that when $V=2.40$, the first of this set cuts in, designated the LP$_{11}$ mode followed at $V=5.52$ by the next, the LP$_{12}$ and so on for the LP$_{1m}$ set.

A further set of solutions of the eigenvalue equation arise from setting $l=2$, the cut-offs of which are dictated by the zeros of the J_1 function, so that the LP$_{21}$ mode starts to propagate at $V=3.83$ followed at $V=7.01$ by the LP$_{22}$ mode. Notice that, apart from the LP$_{01}$ mode, whenever an LP$_{0m}$ mode cuts in, so also does an LP$_{2(m-1)}$ mode. Thereafter, we can set down the V values at which the sets corresponding to larger values of l cut-in by looking up the zeros for the $J_{l-1}(V)$, so for example, the LP$_{73}$ mode cuts in at $V=17.0$. Taking these ideas together, we can now construct a table that shows how the modes build up in an optical fiber as the V value increases. The result is shown in Table 5.2 which shows the rather complex way the different LP$_{1m}$ modes build up.

Table 5.2 Development of the LP mode sequence with V value

	Mode number			Total number of states		
V value	l value	m value	$(l+2m)$	Pol.	Orient.	total
0	0	1	2	2	1	2
2.4	1	1	3	2	2	4
3.83	0	2	4	2	1	2
3.83	2	1	4	2	2	4
5.14	3	1	5	2	2	4
5.52	1	2	5	2	2	4
6.4	4	1	6	2	2	4
7.01	0	3	6	2	1	2
7.01	2	2	6	2	2	4
7.6	5	1	7	2	2	4
8.4	3	2	7	2	2	4
8.65	1	3	7	2	2	4
8.77	6	1	8	2	2	4
9.7	4	2	8	2	2	4
9.94	7	1	9	2	2	4
	Total number of modes in $V=10$ fiber					54

In this table we have also included some other information. The column of values of $(l+2m)$ shows that this increases monotonically. The reason for this will emerge later but for the moment we will only note that it indicates that there are groups of modes sharing a single quantum number $(l+2m)$ that have something physical in common. The other columns entitled Pol. and Orient. refer to other issues of importance; firstly, polarization. The mode patterns that we have derived are plain polarized fields that can be launched into the fiber. However, being circularly symmetric, there are always two orthogonal polarization states available, horizontal and vertical (x or y in Cartesian coordinates), either or both of which can be excited by polarizing the exciting laser or source appropriately. The orientation factor is more subtle. We recall that the solution we proposed for the ϕ dependence of the electric field took the form $\exp[\pm il\phi]$ and since both solutions will in general be present, according to their relative phase we can expect to see real fields that vary as $\cos(l\phi)$ and/or $\sin(l\phi)$, i.e. two identical patterns around any diameter but rotated one from the other by an angle $\pi/(2l)$, (for $l>0$). In the case that $l=0$, then there is perfect rotational symmetry and there can only be one set of patterns. Hence all the LP_{0m} give rise to only one orientational symmetry but to two polarization states and hence two degenerate states each whereas the LP_{lm} for $l>0$ each give rise to two rotational states and two polarization states for each rotational state and hence four states in total. These effects are illustrated below in Fig. 5.4.

5.4 VISUALIZING THE HIGHER-ORDER MODE FIELDS

We have established from the mode cut-off condition how the mode sequence builds up and we established earlier the general form of the field distributions, namely for the variation around any diameter it would be given by:

$$F_2(\phi) = \exp[\pm(il\phi)]$$

and for the variation along a radius it would be given in the core region by:

$$F_1 = AJ_l(Ur/a) \qquad 0 < r < a$$

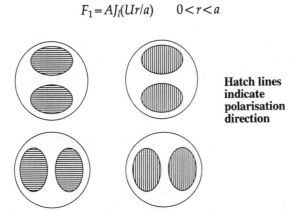

Hatch lines
indicate
polarisation
direction

Fig. 5.4 Allowed states of the LP_{11} mode pattern showing how polarization and orientation combine to give four possible states.

and in the cladding region by:

$$F_1 = BK_1(Wr/a) \qquad a < r < \infty .$$

We saw from the eigenvalue equation that the condition for a mode to be cut-off was that $J_{l-1}(V) = 0$. To obtain some more feel for the implication of this condition, we need to examine the properties of the J Bessel function. For large x, there is a well known relationship which states that:

$$J_l(x) = \sqrt{\frac{2}{\pi x}} \cos\left(x - \frac{l\pi}{2} - \frac{\pi}{4} \right).$$

Applying this to the mode-cut off condition, we see that it reduces directly to:

$$J_{l-1}(V) = \sqrt{\frac{2}{\pi V}} \cos\left(V - \frac{l\pi}{2} + \frac{\pi}{4} \right) = 0 \qquad \text{for cut-off}$$

and for this to be true, then we must have the following conditions satisfied:

$$\left(V - \frac{l\pi}{2} + \frac{\pi}{4} \right) = \left(m + \frac{1}{2} \right)\pi$$

or

$$V_{LO} = \left(\frac{l\pi}{2} + m\pi + \frac{\pi}{4} \right) + \frac{\pi}{2}\left(l + 2m + \frac{1}{2} \right).$$

If we substitute this back into the relationship for the mode field at cut-off and at $r = a$, i.e.

$$J_l(V) = \sqrt{\left[\frac{4}{\pi^2(l + 2m + \frac{1}{2})} \right]} \cos(m\pi)$$

we see that at the core cladding boundary, at cut-off, the field function is at one of the zero-slope regions of the cosine variation. Hence we can deduce that for the guided mode, the value of U must be such that it varies between this maximum in the cosine function and the next zero of the cosine, thus:

$$\frac{\pi}{2}\left[l + 2m + \frac{1}{2} \right] < U_{lm} < \frac{\pi}{2}\left[l + 2m + \frac{3}{2} \right] .$$

For $V < V_{CO}$, the mode does not propagate. At cut-off, $W = 0$ and $U = V_{CO}$. Then as V increases, U will increase a very small amount and the balance will be taken up by a rapidly increasing W. This is shown schematically in Fig. 5.5 for a low V value mode.

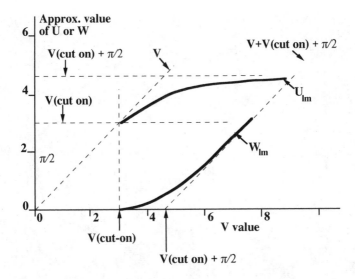

Fig. 5.5 Relationships between U, V, W as V increases through cut-on.

For low-order modes, this relationship will not be a good approximation. For example, for the LP_{01} mode, we know from the table of Bessel function zeros that the maximum of J_0 occurs at 0 and its first zero at 2.4, significantly more than $\pi/2 = 1.57$ but thereafter, the zero quite rapidly approach a spacing of π (e.g. $5.52 - 2.4 = 3.12$, $8.65 - 5.52 = 3.13$, $11.8 - 5.52 = 6.28$ etc) and the same applies to the other zeros as is apparent from the plot of the J functions. Hence, if we were to do a detailed computer solution of the eigenvalue equation for any given V value fiber, we would find allowed values of k_z for all those modes that were allowed to propagate (as given by our table above for the $V = 10$ fiber for example) and the actual values of k_z (or U_{lm}) would lie within the ranges indicated by our result immediately above. The result is that the field distributions observed as a function of radius would each follow the relevant $J_l(U_{lm}r/a)$ variation and would meet up at the core–cladding boundary ($r = a$) after the field had gone through the appropriate number of maxima. If we plot E^2 versus radius, between $r = 0$ and $r = a$ for any given LP_{lm} mode, there will be m field maxima. Hence the general form for the field pattern for an LP_{lm} mode when viewed by the eye (through a microscope) consists of a circularly symmetric pattern of bright spots having m spots along the radius and $2l$ spots around any diameter. A good example is shown in Fig. 5.6. which shows a fine example of an $LP_{17.16}$ mode pattern photographed for a fiber having a V value of about 78 and hence very close to cut-off. In the case of LP_{om} modes, the pattern shows no variation around a diameter so that the LP_{om} mode shows a central circular spot surrounded by $(m - 1)$ circular rings.

Of greatest interest to most modern systems users, however, is the fiber having a value of $V < 2.405$ and hence supporting only the LP_{01} mode whose appearance is almost indistinguishable from a circular spot with a Gaussian cross section, almost perfectly matching that from most gas lasers although with a much smaller diameter, typically in this case about 8–10 μm.

Fig. 5.6 $LP_{17,16}$ mode field pattern.

5.5   PHASE AND GROUP VELOCITY

Of greatest interest to the designer of a transmission system is probably the information governing how this pulse of signal energy will travel from transmitter to receiver. If we stop for a second to remind ourselves how this occurs in a uniform dielectric medium, then we can rapidly move on to analyse how it occurs in a fiber. In a bulk glass or lossless dielectric, light travels at a phase velocity given by:

$$V_p = c/n$$

so that the propagation constant of the wave in the medium is given by $k_z = k_o\, n$, as we have already seen. The phase velocity represents the velocity at which the electric field maximum or crest of any peak in the cosine or sine variation of a monochromatic infinite plane wave travels. Of more general interest is the velocity at which a wave packet or pulse travels and this is given by the group velocity. In the case of the lossless glass medium, this takes the form:

$$V_g = c/N$$

where the group index N is related to the phase index n by the relationship:

$$\frac{dk_z}{dk_o} = n + k_o \frac{dn}{dk_o} = N.$$

The velocity at which a pulse of energy will travel in an optical fiber is similarly given by this expression, where k_z is the propagation constant we found by solving the eigenvalue equation for a given mode but this means that writing down the general form of the group index for the fiber mode is rather messier.

We observe that by defining a parameter $B = 1 - (U/V)^2$, then by using the relationships already given for U, V and W, we can write down an expression for k_z as follows:

$$k_z = k_o \sqrt{[Bn_1^2 + (1-B)n_2^2]}.$$

Such a simple expression is suggestive of there being some simple physical interpretation of this parameter B. It can be shown (see [5.1]) that the parameter B represents the following:

$$B = \int_0^a rE(r)dr \bigg/ \int_0^\infty rE(r)dr$$

and thus represents the fraction of the total modal field within the core. A detailed computer solution of the variation of B with V value for the lower-order modes is shown in Fig. 5.7 which shows that as the mode cuts-off, the fraction of field in the core, for the $l=0$ and $l=1$ modes goes to zero and as $V \gg V_{CO}$, B tends to unity.

Examining the formula above for k_z and noting that we can define an effective phase index for the guided mode, by analogy to the equivalent behaviour in a bulk medium as $n_{eff} = (k_z/k_o)$ we see that the fiber behaves as a composite medium whose average property, so far as monochromatic wave propagation is concerned, it is given

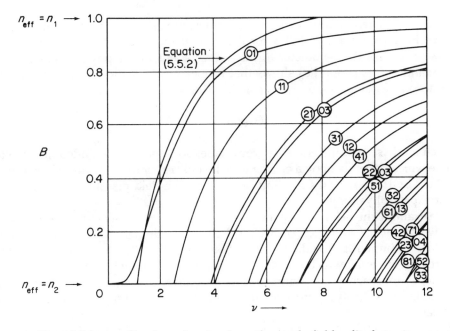

Fig. 5.7 Plot of B versus V value, showing how the mode field splits between core and cladding regions as cut-off is passed.

by an average of the core and clad material properties weighted by the split of the total modal field between them.

From this result, we can now formally set out to deduce an expression for the group velocity characteristics of the medium as follows. Formally, we simply need to differentiate the expression for k_z with respect to k_o to obtain an expression for the effective group index, N_{eff}, although doing this in practice is a little laborious. The calculation proceeds as follows:

$$\frac{dk_z}{dk_o} = [Bn_1^2 + (1-B)n_2^2]^{1/2} + (k_o/2)[Bn_1^2 + (1-B)n_2^2]^{1/2}G$$

where the parameter G represents the following expression:

$$G = \left\{ \frac{dB}{dk_o}[n_1^2 - n_2^2] + 2B \left[n_1\frac{dn_1}{dk_o} - n_2\frac{dn_2}{dK_o} \right] + 2n_2\frac{dn_2}{dK_o} \right\}.$$

This result does not look too promising but by dint of further algebra and a substitution of the following parameter:

$$\frac{1}{2} \left[B + \frac{d(VB)}{dV} \right] = \Gamma$$

we find that whole expression simplifies to the remarkably tractable form:

$$n_{eff}N_{eff} = \Gamma n_1 N_1 + (1-\Gamma)n_2 N_2.$$

We see immediately that there is strong formal similarity to the expression already derived for effective phase index for the guided mode and it is evident that the parameter Γ, like the parameter B, must have some strong physical significance. Examination of its properties shows that it does just this and formally represents the fraction of guided mode power within the fiber core. Thus, we have:

$$\Gamma = \int_0^a rE^2(r)dr / \int_0^\infty rE_2(r)dr$$

for each mode so that the effective group index for a pulse in a guided mode takes a value which depends upon a blending of the indices for the core and clad materials weighted by the power distribution rather than the field distribution, just as one might expect physically.

If we examine the behaviour of the parameter Γ for a representative set of modes, then we discover (see Fig. 5.8) that for the LP_{01} mode, as V approaches zero, so also does Γ, implying that the group index approaches that for a pulse travelling in pure clad material, which is of course exactly what is happening, whilst for large V value, Γ approaches unity implying all the power is contained within the core and the group velocity approaches that for a pulse travelling exclusively on core material.

Similar behaviour applies to all the guided modes as V tracks from $V = V_{CO}$ to $V \gg V_{CO}$ although the reader who is interested in the properties of the modes in

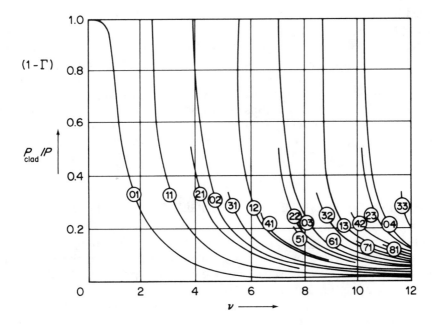

Fig. 5.8 Variation of fractional guided-mode power, Γ, in the core as V varies.

large V value fibers should note that, for reasons we will not discuss here, the minimum value of Γ approaches $[1-1/l]$ as V approaches V_{CO} rather than zero for large l modes (see for example [5.1]). However, we can say with confidence that within a given large core multimode fiber, there will be modes with values of Γ ranging from near zero to near unity and hence there will be values of group index ranging from near to N_2 to near to N_1. This has major implications for pulse propagation.

To identify these, we should first note that, for optical glasses in their region of transparency, N and n are very nearly equal. Since the transit time through a distance L is given by $T=LN/c$, we see that the minimum and maximum transit times for the modes in a fiber fabricated from materials having group indices N_1 and N_2 will be approximately:

$$T(\text{max}) = LN_1/c \quad \text{and} \quad T(\text{min}) = lN_2/c.$$

If we define $\Delta T = T(\text{max}) - T(\text{min}) = $ pulse spreading, then for a typical value of $N_1 - N_2 = 0.015$, we see that over 1 km, $\Delta T = 50$ ns, an unacceptable pulse spreading for most wideband applications. Accordingly, we find that uniform core multimode fibers find very limited applications and then only for very short distance links.

The solution favoured to this problem during the 1970s and deployed in the early production fiber systems was the graded-index multimode fiber. Here, the refractive index of the core was graded to form a roughly parabolic distribution of refractive index versus radius according to a formula: $n(r)=\sqrt{[n_1^2\{1-2\Delta(r/a)^\alpha\}]}$ where the parameter α is a constant with a value close to 2 and $\Delta=(n_1-n_2)/n_2$ with the cladding index taking the regular value of n_2 as normal. Fibers of this type, known as graded-index fibers, have the property that light travelling by different guided modes travel at approximately the same group velocity, with the result that the pulse spreading calculated above for a uniform core multimode fiber of 50 ns/km is reduced to a value in the range 0.1–1 ns/km. Evidently this is much more acceptable but for many purposes it is not good enough. It also enormously complicates the problems of manufacturer since the core profile must be controlled with very great precision to achieve less than a few ns/km pulse spreading. Accordingly, single mode fibers have become the dominant type for high performance systems use. For the interested reader, the properties of graded index fiber are discussed in great detail in other books, see for example [5.2] or [5.1].

5.6   THEORY OF DISPERSION IN SINGLE MODE FIBERS

Evidently in a single mode fiber, it will not be possible for light to travel by different mode paths characterized by different group velocities so that the multipath or multimode dispersion effects discussed above will not apply. Note also that the fiber with $V < 2.405$, although called single mode, actually propagates two degenerate modes having orthogonal polarizations. However, for dispersion purposes, we will regard these as one. To establish what dispersion effects remain in the single mode fiber, we must examine one more effect. We recognize that a pulse of light consists of a spread of optical frequencies or wavelengths and, in a dispersive medium, they cannot all travel with identical group velocity. Hence, the spread in wavelengths in our transmitter will mean that there will be some pulse spreading even in our single mode fiber. We can calculate this by evaluating the pulse spreading arising from a spread of wavelengths $\Delta\lambda$ by evaluating the following:

$$\Delta T = \{L/c\}\{\partial N_{eff}/\partial\lambda\}\Delta\lambda.$$

When we evaluate this expression and collect the terms we find that they can be collected into two different groups thus:

$$\Delta T = \{M_{cmd} + M_{wd}\}\Delta\lambda$$

where the two new terms take the following approximate form:

$$M_{cmd} \cong \frac{\lambda}{c}\left[\Gamma\frac{d^2n_1}{d\lambda^2} + (1-\Gamma)\frac{d^2n_2}{d\lambda^2}\right]$$

(we have approximated $n_1/n_{eff} = n_2/n_{eff} = 1$) and:

$$M_{wd} = [N_1 - N_2]\frac{d\Gamma}{d\lambda}.$$

The former term is known as the composite material dispersion while the latter terms is known as the waveguide dispersion. (Note that we have used slightly different definitions to those used in some other texts (such as [5.1]) since these lead to essentially similar conclusions with less mathematics.)

The material dispersion is easily identified. Consider a wave propagating in a solid block of material having phase and group indices n and N respectively. Then we already know that $N = n - \lambda \, dn/d\lambda$ so that the pulse spreading due to wavelength spread in the source calculated above would be:

$$\Delta T = \{L/c\}\{\partial N_{eff}/\partial\lambda\}\Delta\lambda = \{L\lambda/c\}\{\partial^2 n/\partial\lambda^2\}\Delta\lambda$$

and immediately we see the presence of the term $(\lambda/c)\{\partial^2 n/\partial\lambda^2\}$ matching that of the composite material dispersion expression. Hence we recognize the CMD term as describing the pulse spreading due to the change in index with wavelength of the materials from which the fiber is composed, suitably weighted by the distribution of guided mode power between the core and clad materials. However, we should also recognize the critical importance of the term $\{\partial^2 n/\partial\lambda^2\}$ in fiber design and operation. If we examine the properties of silica-based glasses and plot the general form of the refractive index versus wavelength, then we find that, as a result of the absorption in the UV (at wavelengths < 200 nm) and the IR (at about $9\,\mu m$), the refractive index rises towards the UV and drops towards the IR, with a point of inflection in the region of 1300 nm. Plotting the first and second differentials of this curve with wavelength, we then obtain the result shown in Fig. 5.9 which highlights the important fact that, for silica glasses, $\partial^2 n/\partial\lambda^2 = 0$ at a wavelength of about

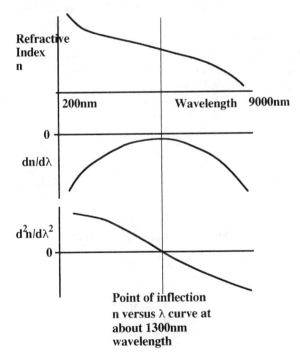

Fig. 5.9 Plot of refractive index versus wavelength for silica glass, showing also the first and second differential with wavelength.

1300 nm. Hence, with other things being equal, we would expect this effect to go to zero at about this wavelength.

The waveguide dispersion term arises from a different effect, with the clue being contained in the term $d\Gamma/d\lambda$. Remembering that Γ describes the fraction of power in the core, we see that an expression depending upon $d\Gamma/d\lambda$ evidently arises from the change in power distribution between core and cladding with wavelength. Not surprisingly, we see that its effect is modulated by the scale factor

$$\left[\frac{n_1 N_1}{n_{\text{eff}}} - \frac{n_2 N_2}{n_{\text{eff}}} \right]$$

which is a measure of the different behaviour of the core and clad materials.

5.7 DESIGN OF SINGLE MODE FIBER DISPERSION CHARACTERISTICS

Many detailed fiber designs have been described in the published literature to satisfy particular features. However, here we will concentrate simply on two basic designs, the normal single mode fiber and the 'dispersion shifted' design.

If we take the standard telecommunications grade fiber for use in the low attenuation window in the region of 1300 nm wavelength, with a relatively small index difference of and reasonable size core, say $8\,\mu$ diameter, then we find that in general, the waveguide dispersion is small and the overall dispersion is dominated by the composite material dispersion. A typical set of curves is shown in Fig. 5.10. The resultant dispersion, formed by adding the CMD and WD terms, goes through zero close to 1300 nm but the actual wavelength of zero dispersion has been shifted to a slightly longer wavelength than that for pure silica alone. The significance of the resulting numbers is readily appreciated.

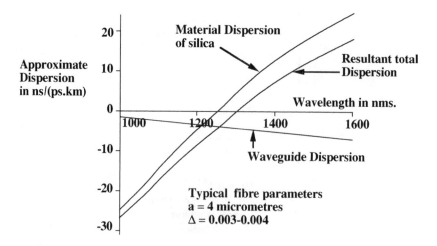

Fig. 5.10 Plot of typical values of dispersion for a standard telecommunications type single mode fiber designed for use at about 1300 nm wavelength.

A typical simple semiconductor laser, operating with pulsed current drive, emits on several longitudinal modes separated by about 1 nm and spread over 3–5 nm. Hence if we take the spectral spread to be about 3 nm, and observe that in a reasonable spread of wavelength either side of the zero dispersion wavelength the total dispersion is in the range $-4 < M_{td} < 4$ ps/(nm km) then we see that the pulse spreading will be in the region of 12 ps/km or less. Note that the sign of the M_{td} expression affects which wavelengths arrive first, long or short, but not the actual pulse spreading in time. The result is that we could seriously contemplate using such a laser and fiber combination transmission at 1 Gbit/s over a distance approaching 80 km and still see less than 1 ns pulse broadening. In practice, fiber attenuation at 1300 nm wavelength would not allow transmission over such a distance but it contrasts dramatically with the results we derived earlier for multimode or graded-index fibers.

We note also from the figure that, at a longer wavelength of 1500–1550 nm, the same fiber exhibits a M_{td} of about 18–20 ps/(nm km) so that with a laser of similar linewidth, a pulse spreading of the order of 60 ps/km would be observed. This is less good news since, as we shall see later, it is in this wavelength window that the minimum fiber attenuation occurs and hence it is at this wavelength that one would like to operate one's long distance transmission systems. This problem can be solved in two ways, by redesign of the fiber and/or the laser.

Let us examine the laser solution first. We commented that a typical simple semiconductor emits light over a spread of a few nm wavelength. This is the result of it oscillating on several longitudinal modes. However, if we consider the electrical implications of this, we see that a spread of 3 nm centred on 1500 nm is equivalent to a spectral spread of 400 GHz centred on 2×10^{14} Hz, an enormous bandwidth and clearly unrelated to the data bandwidth being impressed through the laser drive. If we take the alternative approach and assume that we had a single frequency laser before modulation and that we then modulate with 1 GHz spectral spread of signal, we might expect to see an optical spectral spread of 1 GHz on either side of the central emission line. 2 GHz centred on $2 \cdot 10^{14}$ Hz corresponds to an emission linewidth of 0.015 nm. If we now repeat our dispersion calculation using 20 ps/(nm km) and 100 km distance we see that the expected pulse spreading is just 30 ps, negligible for 1 GBit/s transmission. The sole requirement for this solution, then, is a semiconductor laser that can operate on single frequency, single-longitudinal mode even when modulated through its drive current. Such lasers now exist and are restricted to the single optical (centre) frequency by means of distributed feedback gratings impressed along the gain region which heavily select the chosen wavelength and suppress others. However, such lasers have proved difficult to make and in parallel with their development, an alternative solution was proposed and proven based upon redesign of the fiber.

We note that the effect of the small WD term was to shift the zero dispersion wavelength of pure silica to a slightly longer wavelength in the composite fiber structure. Evidently, if we could increase the WD term, we could increase the shift and perhaps move all the way to 1550 nm, the minimum attenuation wavelength (see the later section on fiber attenuation). If we examine the expression for the WD, then to maximize its value we must seek to maximize each component of it. The term

$$\left[\frac{n_1 N_1}{n_{\text{eff}}} - \frac{n_2 N_2}{n_{\text{eff}}} \right]$$

is purely a function of the materials from which the fiber is constructed. Maximizing it involves using core and clad glasses as dissimilar as possible, with as large a difference in their phase and group indices as possible. The other term, $d\Gamma/d\lambda$, is less obviously maximized. However, an examination of the plots of Γ versus V give a strong clue, when one notes that, for fixed mechanical dimensions, V varies as $1/\lambda$. For large V, Γ rapidly saturates at approximately unity while for very small V, it stabilizes near zero. The approach to stabilization is close to the cut-on value for the next mode (LP_{11}) so that one may deduce that the maximum value for $d\Gamma/d\lambda$ is somewhere in the range $V=1$ to $V=2$. Other studies suggest a value of $V=1.7$. Hence, we have the outlines for a fiber designed to exhibit maximum waveguide dispersion, a V value close to 1.7 and a large index difference between core and cladding. Taken together, these also clearly imply a very small core diameter. In Fig. 5.11 we show the dispersion curve for such a dispersion shifted fiber in which the zero of total dispersion has been successfully shifted to the longer wavelength of 1550 nm using this recipe.

Theory and practice agree well and such fibers can be produced. However, some additional problems do arise, a notable one being the increase in attenuation caused by the higher Rayleigh scattering associated with the higher index difference glass pair, adding a few tenths of a dB/km to the minimum attainable attenuation, figures which may not sound much but when integrated over a 100 km section make the difference between success and total system failure. The preferred engineering solution is still the subject of debate and further advances may well be seen.

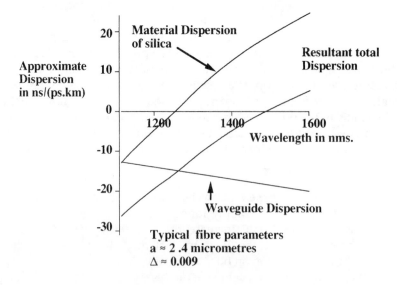

Fig. 5.11 Plot of dispersion value for a dispersion shifted fiber.

5.8 FIBER ATTENUATION

So far we have concentrated exclusively on the electromagnetic guidance and propagation properties of the fiber and have assumed that it is lossless. For the purposes of Maxwell's theory, this is perfectly valid but for propagation of signals over many tens of kilometres, it is not.

5.8.1 Rayleigh scattering

We have already commented that most fibers are formed from silica, the glassy form of quartz which is composed of SiO_2. Because it is in glassy form, its structure is not regularly and perfectly ordered like that of the crystal and this leads to density fluctuations, regions of greater or lesser density than the average for the solid material. Since these occur over distances small compared to one wavelength, they give rise to Rayleigh scattering. The property of this scattering is that the scattered power varies as $(1/\lambda)^4$ and, at a wavelength of one micrometer, typically takes values between 0.6 and 1 dB/km. Extrapolating these to other wavelengths of interest leads to the table below:

Rayleigh scattering in dB/km				
Wavelength (nm)	900	1000	1300	1500
Upper limit	1.52	1.0	0.35	0.17
Lower limit	0.91	0.6	0.21	0.10

Since the longer wavelength leads to lower attenuation by Rayleigh scattering, it is natural to think of moving to still longer wavelengths. However, other effects prevent that and also dictate the significance of the particular wavelengths chosen above, as we shall see shortly.

Note that when a single component glass is doped to increase its refractive index as might be required for forming the core of a fiber (e.g. silica with germania (GeO_2)), it is normal for the Rayleigh scatter loss to increase for two reasons. Firstly, the dopant chosen to increase the index will have a larger mass and polarizability (which is why it increases the index) and this would be expected to increase the scattering from the inevitable fluctuations in density of its components within the glass but a second reason is that composition fluctuations can now occur, with greater concentrations of one component in one place and more of another somewhere else. Since the two components have different polarizabilities, this produces a proportionately bigger effect.

Experimentally, then, we find that increasing the dopant concentration leads to quite rapid increases in Rayleigh scatter loss, as shown in Fig. 5.12.

5.8.2 Ultra-violet 'band-edge' and infra-red 'lattice' absorption

Apart from the Rayleigh scattering arising from the SiO_2 of the glass, there is also strong absorption in both the infra-red and ultra-violet that are fundamentally linked

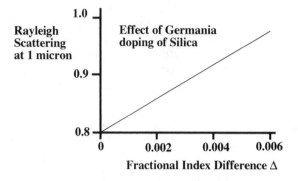

Fig. 5.12 Plot of variation of Rayleigh scatter with germania dopant concentration.

to the silica material itself. The Si—O bond connects the silicon and oxygen atoms elastically so that they can vibrate towards and apart from each other. The characteristic frequency is determined by the mass of the oxygen (lighter) and the strength of the 'elastic' and results in a characteristic vibrational frequency of about 3.3×10^{13} which corresponds to an infra-red wavelength in electromagnetic radiation of about $9 \mu m$. Since the Si—O bond has a strong dipole associated with it, it absorbs strongly and the result is that silica glass is jet black even in very thin samples at a wavelength of about $9 \mu m$. Since we are interested in samples 100 km thick (!), we must worry about the tails of this absorption peak many many orders of magnitude below the peak. Study shows that the tail of this absorption is found falling rapidly towards shorter wavelengths in the region of 1600–1700 nm.

In Fig. 5.13 we show both the Rayleigh scatter and the measured infra-red absorption for a silica-based fiber. Together they lead to a fundamental minimum of attenuation for this glass in the region of 1550 nm and having a value in the range 0.15–0.2 dB/km. If we note that a typical laser transmitter might emit 1 mW mean

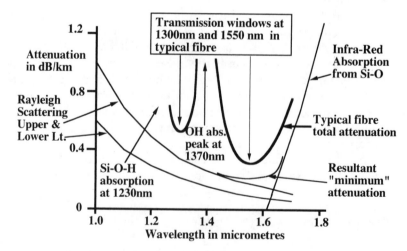

Fig. 5.13 Plot of fiber attenuation with wavelength in the region from 1000 to 2000 nm of greatest interest for fiber transmission.

power and a good optical receiver operating at low error rate and at 1 Gbit/s might require −30 to −35 dBm mean optical power, we see that a fiber attenuation of, say, 0.18 dB/km allows a range of 167 km before an attenuation of 30 dB has been reached, leaving 5 dB for margin. The clear implication of these figures is that very long unbroken fiber sections should be possible carrying data at very high data rates.

Just as silica absorbs in the infra-red, so it also exhibits a 'band-edge' type electronic absorption in the ultra-violet, cutting in at about 200 nm wavelength. This also exhibits a tail towards longer wavelengths, towards the desired fiber transmission window. However, experiment shows that, compared to the above effects, it can be ignored in the wavelength region 800–1600 nm of systems interest.

5.8.3 O—H absorption

The sole remaining absorption mechanism of major concern arises from contamination in the glass by water, or more precisely, by the O—H ion which becomes bonded into the silica glass matrix as Si—O—H. Like oxygen, it is bonded elastically but being much lighter than oxygen, the hydrogen can vibrate much more rapidly. The basic O—H vibration frequency is about 1.1×10^{14} Hz, corresponding to an infra-red wavelength of about 2.8 μm. By holding the level of O—H contamination very low (i.e. by making the glass very dry), the effect of this fundamental oscillation in the window of interest could be made negligible. However, unfortunately, the bonding of hydrogen and silicon to oxygen is not perfectly elastic so that the oscillation is not perfectly harmonic but there is some distortion of the motion, leading to the generation of harmonics and combinations of the two fundamental. In particular, we see the second and third harmonic of the O—H vibration at about 1.37 μm and about 0.95 μm and then combinations of these with the Si—O vibration at about 1.23 μm and 880 μm. These are also apparent in Fig. 5.13 where we plot the total attenuation of a real fiber. The result is low attenuation windows for systems use in the region of 1550, 1300 and 900 nm as tabulated above when we calculated the Rayleigh scattering loss.

The window at 900 nm first attracted systems interest because GaAs based semiconductor lasers were known and appeared to be the ideal sources for such systems as well as to be able to operate at about that wavelength. Later, when the advantages of the 1300 nm window became clear (zero material dispersion and lower attenuation), much interest was focused on the fabrication of lasers specifically for use there and this led to the development of lasers using GaInAsP/InP materials. Later still, this materials system was extended to allow 1550 nm operation once the ultimate advantage of that window became apparent. Today, the system designer thus has all three windows available for use and must make a judgement as to which represents the optimum solution for his application.

5.9 SPECIAL FIBER DESIGNS FOR POLARIZATION MAINTAINING FIBERS

We have already commented that the 'single mode fiber' actually supports two degenerate modes of orthogonal optical polarization. In such fibers, even if plane polarized light is launched, it very rapidly couples to the other polarization state and some elliptical and unstable state is formed. For many purposes, particularly information transmission, this does not matter at all. However, for connections between some components such as semiconductor lasers and integrated-optic planar electro-optic modulators, it can be very serious since the modulator depends upon maintaining the polarization state for its normal operation (excepting special polarization insensitive designs). We also note that coherent optical detection systems are inherently polarization sensitive unless special precautions are taken to avoid it. For all such systems, it is very useful to have a fiber which will propagate and maintain a single polarization state either as the only mode of propagation or at least as a very distinct mode of propagation from the other polarization state so that coupling is very weak between them.

The key requirement for such a fiber is to split the degeneracy between the two orthogonally polarized modes. This requires that the circular symmetry of the fiber structure be broken decisively and a variety of techniques have been proposed to achieve this. For example, one technique has been to deposit, at the time of preform fabrication, two different materials to act as claddings, using a low thermal expansion material for the sectors $\phi=0$–$90°$ and 180–$270°$ and a relatively high thermal expansion material for the sectors $\phi=90$–$180°$ and 270–$360°$ with the normal core structure and a thin layer of normal cladding. The resulting structure has the appearance of a 'bow-tie' [5.3] in cross section (see Fig. 5.14(a)) and leads to very

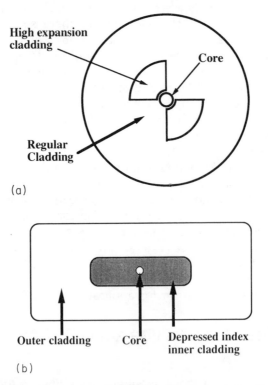

(a)

(b)

Fig. 5.14 Schematic cross section of two polarization maintaining fiber designs, the 'bow tie' design at (a) and the flattened ATT-Bell Labs design at (b).

strong strain gradient uniaxially across the core material. Acting through the photo-elastic effect, this mechanical strain ensures that the two orthogonal polarizations of light see very different refractive indices and hence propagate with different propagation constants.

Another technique has been to take a normally circularly symmetric fiber preform and to hot-roll it into a markedly rectangular structure, once again massively breaking the circular symmetry of the whole structure [5.4]. On pulling such a preform, the fiber retains a rectangular cross section (see Fig. 5.14(*b*)) which has the additional advantage that not only are the polarization states split but they are aligned to the axes of the rectangle, making their alignment with other planar structures relatively simple.

5.10 FIBER COUPLERS

We have seen from our analysis of the fields of optical waveguides that the electromagnetic field extends beyond the immediate core material into the surrounding cladding. This field is known as an evanescent field since its propagation constant in one direction (the radial direction) is imaginary and this results in an exponential (*K* Bessel function) decay rather than the sinusoidal/cosinusoidal or *J* Bessel function variation. However, evaluation of the decay constant for these evanescent fields shows that they penetrate a distance that is typically of the order of 1 wavelength into the cladding. The result of this is that, if another guide (core region) is brought to within a comparable distance, electromagnetic energy will tunnel from one core to the other. Hence if the cladding of two fibers is reduced and their cores brought together, we could expect to have a two-input port, two-output port element entirely composed of fiber in which light launched into either input port might appear at either or both of the output ports, depending upon the strength of the coupling. In practice, such devices can be fabricated with relative ease although producing them with well controlled characteristics is more difficult.

The typical two-port fiber coupler is formed by taking two standard single mode fibers, lightly twisting them together and tensioning them and then heating the twisted region whilst maintaining a controlled tension. When the glass softens, the regions fuse together and elongate, at the same time shrinking down and forming low taper regions on either size of the fusion region. The result is that the guided mode spreads out further into the cladding as the core tapers and eventually couples strongly to the adjoining core field. When the fibers then taper up again, the fields separate and launch power back into both the guided modes of the two fibers (see Fig. 5.15).

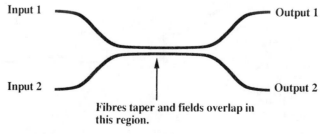

Fig. 5.15 Schematic layout of fused fiber coupler.

To control the tapering and coupling process, the operation is normally performed on a computer controlled jig and light at the appropriate wavelength is launched into one fiber and monitored at both output ports. These output signals are then used to control the tapering process so that it stops at exactly the right point to achieve the desired splitting ratio, e.g. 3 dB 50 : 50, 10 dB 90 : 10, etc.

We also know from the general theory of directional coupler type structures and coupled modes that the direction of energy flow between coupled wave or harmonic oscillators depends upon their relative phase. Hence, we can expect the energy flow between one fiber and another to be sensitively dependent on the phase of the waves in each and a relative phase shift of π radians can be expected to alter the direction of energy flow. As a result, in a coupler having a long coupling region, the energy from one input may cycle back and forth many times and small changes in excitation wavelength can produce a change in the number of times the energy crosses back and forth. The result is that, in long couplers, one finds the coupling between the either input port and a particular output port varies approximately sinusoidally with wavelength, with the important result that a device can be constructed entirely of fiber in which two lasers of different wavelength are injected, one through each input port and the outputs of both appear at a single output port, an efficient way of wavelength multiplexing two sources or detector channels to a single transmission channel.

In other applications, such a passive optical networks, there is often a strong interest in splitting the power from a single transmitter between many different receivers to operate a broadcast mode network. This can be done by forming a binary tree of 3 dB couplers, perhaps fusion welded together, as shown schematically in Fig. 5.16 or it can be done by extending the idea of the two port coupler to larger numbers of fibers. Several designs have been proposed.

One such extended fiber coupler uses 7 fibres formed into a hexagonal close-packed array, one in the centre surrounded by 6 identical fibers and drawn down as appropriate [5.5]. By symmetry, if the central fiber is excited, its power must coupled

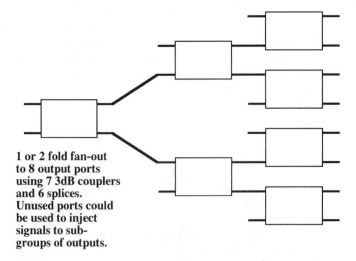

**1 or 2 fold fan-out
to 8 output ports
using 7 3dB couplers
and 6 splices.
Unused ports could
be used to inject
signals to sub-
groups of outputs.**

Fig. 5.16 A binary tree of seven 'two-port' fiber couplers forming a 1 to 8 fan-out.

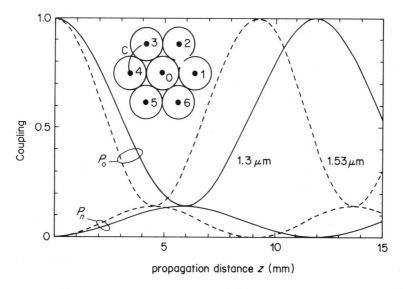

Fig. 5.17 Properties of hexagonal close packed 7-fiber fused coupler.

equally to the surrounding 6 and conditions can be found in which the power has been shared equally among them, thus forming a 1×6 coupler. Building trees of such higher-order couplers much more rapidly generates a large fan-out. The coupler cross section is shown in Fig. 5.17 together with the variation of power with distance through the coupling region for two different wavelengths.

Because of the higher-order structure of this device, the coupling rules are rather more complex and it has been shown that they take the following form. Because of the physical symmetry of the device, it is natural to consider injecting power through the central fiber and then to follow its distribution among all seven fibers. Using the notation of Fig. 5.17 above, we find that:

$$P_0(z,\lambda) = \{1 - (6/7)\sin^2[\sqrt{(7)}C(\lambda)z]\}$$

$$P_n(z,\lambda) = (1/7)\sin^2[\sqrt{(7)}C\lambda)z] \qquad \text{for } n = 1 \text{ to } 6.$$

The parameter $C(\lambda)$ is the reciprocal coupling length which is obviously wavelength dependent. A typical device length to obtain reasonably flat wavelength response might be about 5 mm.

5.11 OPTICAL FIBER AMPLIFIERS AND LASERS

Another class of devices of very rapidly growing importance based on specially doped fibers have led to a class of optically pumped amplifiers in fiber form [5.6]. The basic concept is not new, since it has been known for several decades that rare-earth ions such as erbium and neodymium, when doped into the right glass or crystal matrix, could be used to build optical amplifiers or lasers with optical pumping.

Level 3 has long lifetime, many µs to ms

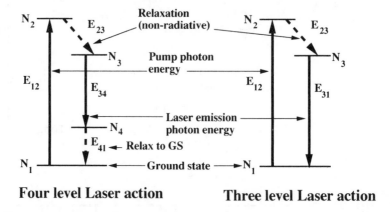

Four level Laser action   **Three level Laser action**

Fig. 5.18 Energy level diagrams (schematic) for 3 and 4 level optically pumped lasers.

The principle of optically pumped solid state lasers is illustrated in the energy level diagram shown in Fig. 5.18. The rare-earth ion would normally exist with its electrons in the ground state, level 1. Pumping the ions with light having a photon energy $h\nu$ equal to the energy gap E_{12} leads to the pump photons being absorbed and electrons being excited to energy level 2. To obtain laser action, these electrons must then relax back to a metastable level 3 from which the transition to the lower levels (1 or 4) is almost forbidden, resulting in the electron being trapped in level 3 for a considerable period of time, typically many tens or hundreds of microseconds. Ultimately, relaxation occurs with the emission of a photon having energy E_{31} or E_{34} depending upon whether the ion is of the 3-level or 4-level type.

Once the electron is stored in level 3, then gain can start to occur through stimulated emission. If a photon comes along having an energy $h\nu = E_{31}$ or E_{34} as appropriate, then it can stimulate the emission of another, a process that occurs coherently, with the additional photon being emitted in phase with the stimulating one, leading to a coherent build up of the wave. The gain is proportional to the difference in the number of electrons stored in the two levels, e.g.

$$G = K(N_3 - N_{(1 \text{ or } 4)})$$

with the obvious result that, unless $N_3 > N_{(1 \text{ or } 4)}$, the gain will be negative. In the case of the 3-level laser, this implies that more than half of the ions must be excited before any gain is observed. In the case of the 4-level system, since level 4 would normally be nearly empty, a small fraction of excited ions will start to produce gain. As a result, 4-level laser ions are usually preferred to 3-level ones. Examples are Nd, which provide 4-level laser action, and erbium and chromium which are 3 level.

Most of the ions of interest have more than one absorption or pump wavelength and some have more than one emission wavelength. Table 5.3 lists (using spectroscopic notation for the energy levels involved) the ones of primary interest for use with optical fiber systems.

Table 5.3 Fiber lasers in phosphate glasses

	Transition	Λ (nm)	Absorption or emission
Neodymium	$^4I_{9/2}$ to $^2H_{9/2}$	800	A
	$^4I_{9/2}$ to $^4F_{9/2}$	800	A
	$^4I_{9/2}$ to $^4F_{3/2}$	900	A
	$^4F_{3/2}$ to $^4I_{9/2}$	900	E
	$^4F_{3/2}$ to $^4I_{11/2}$	1060	E
	$^4F_{3/2}$ to $^4I_{13/2}$	1350	E
Erbium	$^4I_{15/2}$ to $^4I_{9/2}$	800	A
	$^4I_{15/2}$ to $^4I_{11/2}$	980	A
	$^4I_{15/2}$ to $^4I_{13/2}$	1550	A
	$^4I_{13/2}$ to $^4I_{15/2}$	1550	E

Other wavelengths at which laser action has been observed in fiber lasers

Ion	Λ (nm)	Glass
Holmium	1380	Fluorozirconate
Holmium	2080	Fluorozirconate
Thulium	2300	Fluorozirconate
Erbium	2702	Fluorozirconate

Note that the precise wavelengths at which absorption or emission occur are influenced by the matrix in which the ion is placed. Hence, there can be small changes in the above values between different composition glasses or within different crystalline matrices, such as YAG. For example, by doping neodymium into a phosphate glass with varying levels of phosphorus concentration, the emission wavelength at '1060 nm' can be shifted from about 1055 nm to 1090 nm. Practical examples of such 4-level lasers are the semiconductor laser pumped Nd: YAG laser at about 1.06 μm and the Nd glass lasers and amplifiers with flash lamp pumping used to generate very high power optical pulses. These can also be made to operate at about 1300 nm wavelength. In the case of the fiber devices, the dopant ion is placed in the core glass of the fiber and is optically pumped by light from a semiconductor laser coupled into the fiber via a fiber wavelength-multiplexer or directional-coupler of the type just described above, as shown in Fig. 5.19.

The signal to be amplified is injected through the other port of the directional coupler. For an Er doped fiber, very substantial gain can be obtained in the fiber system wavelength of 1530–1560 nm using pump lasers centred at about 810, 980 or 1480–1490 nm. The amplifiers saturate when the output power at the signal wavelength being amplified starts to approach that of the pump but can exhibit small signal gains in the 15–40 dB range, making them extremely attractive for long-haul transmission systems and probably also for passive optical distribution networks which should become capable of massive signal fan-out values of many thousands.

The wavelength dependence of the gain in the fiber amplifier depends on both the dopant ion and the glass matrix. The most studied is the erbium ion because of its obvious relevance to 1500 nm fiber systems. In Fig. 5.20 we show the gain spectra for erbium in two different glass systems [5.7].

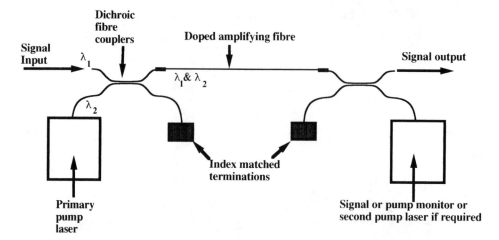

Fig. 5.19 Schematic layout of optically pumped fiber amplifier.

Much interest obviously centres on finding glass systems that will broaden and flatten the spectrum so that the amplifiers can be used over a broader spectral bandwidth in WDM systems. At present it appears that a spectral spread of about 40 nm is accessible, in electrical terms about 5000 GHz. Also of great interest for WDM systems is the finding that, because of the long lifetime of the metastable upper state, crosstalk between WDM channels is maintained at very low levels provided the amplifier is not run into saturation. The initial indications are thus that erbium doped fiber amplifiers are poised to play a very major role in the future development of optical fiber systems.

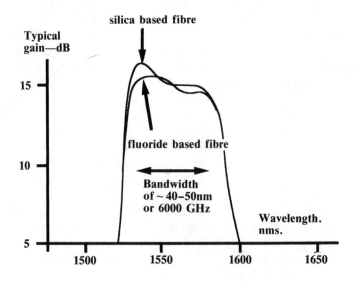

Fig. 5.20 Gain spectrum for erbium ions in two different types of glass matrix.

REFERENCES

[5.1] M. J. Adams (1981) *An Introduction to Optical Waveguides* (Chichester: Wiley).

[5.2] J. E. Midwinter (1979) *Optical Fibers for Transmission* (New York: Wiley).

[5.3] R. D. Birch, D. N. Payne and M. P. Varnham (1982) *Electron. Lett.* **18**, 1030.

[5.4] R. H. Stolen, W. Pleibel and J. R. Simpson (1984) *IEEE J. Lightwave Technol.* **LT-2**, 639.

[5.5] D. B. Mortimore and J. W. Arkwright (1989) *Electron. Lett.* **25**, 606.

[5.6] P. Urquhart (1988) *Proc. IEE Pt. J, Optoelectron.* **135**, 385.

[5.7] D. W. Hall and M. J. Weber (1991) "Glass Lasers" *Handbook of Laser Science & Technology, Supplement 1: Lasers* (Boca Raton, FL: CRC Press).

[5.8] M. Abramowitz and I. A. Stegun (1965) *Handbook of Mathematical Functions* (New York: Dover)

SEMICONDUCTOR LASERS AND QUANTUM WELL DEVICES

6.1 INTRODUCTION

Optical transmission facilities are expanding rapidly in all segments of telecommunications. The advantages of optical fiber, semiconductor lasers and different kinds of optoelectronic components for communication applications are well documented and some of these devices have already been installed in many practical systems. For developing the next generation communication systems with large bi-rate–distance product trasmission or advanced transport networks (such as ISDN or SONET), some kinds of advanced lightwave technology (such as coherent optical transmission, dense frequency division multiplexing, high-speed TDM) have already been developed in laboratory research and field experiments. In research and development (R&D) laboratories, current efforts are focusing on narrow linewidth and single-longitudinal mode devices for coherent systems, tunable and single-frequency devices for FDM, high-speed devices for TDM, integrated optical waveguide switches and modulators, quantum well lasers and modulators, doped fiber amplifiers and lasers, and other kinds of novel technologies. Recently there has been increasing interest in direct-detection DWDM systems more than in coherent optical fiber transmission systems. The most stringent requirements on semiconductor lasers with narrow linewidth, single-longitudinal mode and frequency-tuning characteristics are indispensable for these systems as mentioned in Chapter 2. In multichannel wavelength/frequency division multiplexed systems, the tuning range of the tunable laser used limits the maximum number of channels. Therefore, it is important to realize a frequency tunable and frequency stable narrow linewidth semiconductor laser having a wide tuning range for these systems as mentioned in Chapter 3. Because of the chromatic dispersion in fibers, high-speed and single-frequency no-chirp semiconductor lasers and high-speed detectors are also needed in long-haul transmission at speeds higher than a few gigabits per second as mentioned in Chapter 4.

6.2 NARROW-LINEWIDTH AND SINGLE-LONGITUDINAL MODE LASER

The special purity of the optical transmitter and local oscillator is often the first issue confronting semiconductor lasers for coherent applications. The measure used to quantify the required level of spectral purity is usually the ability to achieve a 10^{-9} bit-error rate (BER) in a digital transmission application for each modulation format and receiver architecture. 'Spectral purity' is intentionally vague and encompasses a number of quantities. Since most of the same issues confront both the transmitter and the local oscillator, we refer to both as 'the laser'.

To achieve low error rate the optical power should be concentrated in a single resonator mode. Mode partition events are inherently expected when the net gain minus loss of one particular mode does not exceed that of the others by a sufficient margin. This is true for an unmodulated continuous wave laser, but generally poses more of a problem for modulated lasers.

The laser linewidth stems from the spontaneous emission into the lasing mode. The random phase of the spontaneous emission results in a phase noise in the cavity field and, in turn, a frequency noise in the laser output. The line shape is predominantly Lorentzian, with a typical 3 db width of 100 MHz at 1 mW output for a conventional single-frequency laser.

Recently the transmission distance and capacity of optical fiber communications have been expanding rapidly through research and development of coherent optical communication systems [6.4]. Distributed Bragg reflector (DBR) lasers are expected to be good candidates for light sources for these systems because of their stable single-longitudinal mode (SLM) oscillation. In addition, a very narrow spectral linewidth of less than 1 MHz is required for the laser diodes in some kinds of coherent optical communication systems [6.5].

The requirement of linewidth varies widely depending on the modulation format and bit rate of the system. Detailed analysis can be found in [6.6]. Briefly, for an FSK heterodyne system, the required linewidth is 9% of bit rate, assuming a 1 dB receiver sensitivity penalty. For DPSK systems, the requirement is 0.16% of bit rate. Homodyne systems, which promise the best sensitivity, require a laser linewidth of only 0.03% of bit rate. Thus, the FSK format is the most practical choice for use of the conventional solitary single-frequency lasers. Most other systems need special lasers with external cavities to obtain extremely narrow linewidths.

The semiconductor laser with narrow spectra linewidth is also very useful for expanding its transmission distance and capacity in order to avoid the chromatic dispersion of single-mode fiber in a 1.5 μm wavelength region in direct modulation systems.

In a conventional Fabry–Perot laser (with two parallel cleaved mirrors), the modal losses are primarily due to the mirror loss, which is independent of frequency and equal for all modes. Because of the wide spectral gain in semiconductors, net gain differences between various longitudinal modes are very small ($<1 cm^{-1}$). As a result, many modes can reach lasing threshold, leading to multimode oscillations, as depicted in Fig. 6.1(*a*) [6.1].

To increase the gain difference, a frequency-dependent loss is introduced, which consists of corrugated structure built inside the laser cavity. If the period of the corrugation is equal to $\lambda_B/2n$ (λ_B is the Bragg wavelength, and n is the refraction index of the corrugation medium), only the mode near the Bragg wavelength λ_B is reflected constructively (Bragg reflection), and this mode will build up to lase while other modes having higher losses are suppressed from oscillation, as shown in Fig. 6.1(*b*). The concept of a DFB laser using a corrugated grating was demonstrated by Kogelnik and Shank in 1971 [6.2], but was not realized until 1974 [6.3]. In the last few years, such devices have become commercially available.

In theory, in a DFB laser with both end facets anti-reflection (AR) coated, two modes located symmetrically on either side of the Bragg wavelength would experience the same lowest threshold gain in the idealized symmetrical structure and would lase simultaneously. In practice, however, the randomness in the cleaving process results in different end phases, removing the degeneracy of the modal gain and giving rise to single mode operation. Facet asymmetry can be further enhanced by putting a high-reflection coating on one facet and a low-reflection coating on the other (the hi–lo structure) to improve the single-frequency yield.

6.2.1 The linewidth of semiconductor lasers [6.19]

Interest in the linewidth was present from the very beginning of laser physics. In their first paper proposing the laser, Schawlow and Townes [6.7] derived a formula for the linewidth, predicted that the lineshape would be Lorentzian and that the linewidth would narrow inversely with laser power, so that the linewidth–power product $\Delta\lambda P_0$ is constant. This formula is only valid below threshold. Lax [6.8] pointed out that above threshold, the amplitude fluctuations of the laser are

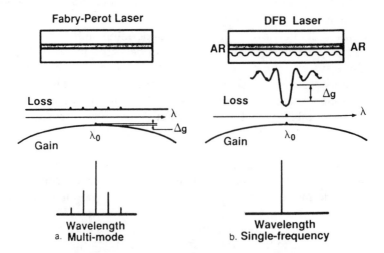

Fig. 6.1 Schematic of F-P and DFB lasers. (From [6.1] Lee and Zah (1989) *IEEE Commun. Mag.* **27**(10), 42–52. Reproduced by permission. ©1989 IEEE.)

stabilized and this is accompanied by a ×2 reduction in ΔP_0. The detailed change in the linewidth through the threshold region was calculated by Hempstead and Lax [6.9]. The extreme narrowness of the linewidth of gas lasers made measurement of the intrinsic linewidth a formidable problem. However, Gerhardt et al. [6.10] finally succeeded by using a 500 m folded interferometer and by operating their He–Ne laser at microwatt power levels. The interest in linewidth was renewed in the last few years by a burst of activity stimulated by semiconductor lasers. The first careful linewidth studies of AlGaAs laser were made by [6.11]. They observed the Lorentzian shape and the line narrowing inversely with power as expected, but surprisingly, they found that the linewidth was about 50 times greater than that predicted by the modified Schawlow–Townes formula. This linewidth enhancement was explained as due to the change in the cavity resonance frequency with gain [6.12]. This results in a correction of $1+\alpha^2$ to the modified Schawlow–Townes formula, where the linewidth enhancement factor $\alpha = \Delta n'/\Delta n''$ is the ratio of the changes in the real and imaginary parts of the refractive index with change in carrier number. InAlGaAs and InGaAsP laser α is about 4–7 [6.12–6.17]. In a semiconductor laser, the laser line occurs at the foot of a steep absorption edge and this causes α to be large [6.13]. An additional factor of about ×2 in linewidth results because, in semiconductor lasers, the population associated with laser transition is not fully inverted [6.11, 6.12]. In addition to the increased linewidth, semiconductor lasers depart from the expectations of the classical laser theory in another respect. It was found in [6.18] that the lineshape is not a perfect Lorentzian, but has satellite peaks far out in the wings that are separated from the main peak by multiples of the relaxation oscillation frequency of the laser.

In general, the broadening of semiconductor laser linewidth is due to the fluctuations in the phase of optical fields. These fluctuations are caused by the instantaneous phase change and the additional phase shift through the change in carrier density because of the coupling between optical phase and intensity in a semiconductor. This broadening of linewidth due to the additional phase shift is explained by using the linewidth enhancement factor α. The linewidth of a single-mode semiconductor laser is expressed as [6.12]

$$\Delta \nu = \frac{K}{4\pi S}(1+\alpha^2) \tag{6.1}$$

where S is the photon density in a laser cavity, K is an average spontaneous emission rate for unit volume, and α is a linewidth enhancement factor, respectively.

When the semiconductor laser operates in a single-mode condition, the rate equations are given by [6.20]

$$\frac{dS}{dt} = g(N-N_g)S - \frac{S}{\tau_p} + \frac{C'}{\tau_s}N \tag{6.2}$$

$$\frac{dN}{dt} = \frac{1}{eV_a} - g(N-N_g)S - \frac{N}{\tau_s} \tag{6.3}$$

where g is the gain coefficient, N_g is the carrier density for positive gain, τ_p is the photon lifetime, τ_s is the carrier lifetime, V_a is the volume of the active region, and

C' is the spontaneous emission factor [6.21, 6.22]. From the rate equations, S and K are obtained for a steady state. S is given by

$$S = \frac{1}{eV_a} \tau_p(I - I_{th}).$$

(6.4)

K is given approximately by

$$K \simeq C \frac{N_{th}}{\tau_s}$$

(6.5)

where N_{th} is the carrier density at threshold.

By using (6.1)–(6.5), the linewidth is given by

$$\Delta \nu_{FP} = \frac{C'}{4\pi} \frac{(1/\tau_p)}{(I/I_{th} - 1)}(1 + \alpha^2).$$

(6.6)

For the analysis of the linewidth of a DBR laser, the external cavity model proposed by [6.20] was adopted. Figure 6.2 [6.19] shows the schematic diagram of a DBR laser and its equivalent cavity.

In [6.23], the reflectivity of the DBR region $\hat{r}$ is given by

$$\hat{r} = r \, \exp(-j\phi)$$

$$= \left\{ \frac{\gamma \, r_0 \, \exp(-j\Phi) \, \cosh(\gamma \, L_B) - [(\alpha_g + j\Delta\beta) \, r_0 \, \exp(-j\Phi) + j\kappa] \, \sinh(\gamma \, L_B)}{\gamma \, \cosh(\gamma \, L_B) + [(\alpha_g + j\Delta\beta) + j\kappa \, \exp(-j\Phi)] \, \sinh(\gamma \, L_B)} \right\} \exp(j\varphi)$$

(6.7)

$$\Phi = 2\beta_0 \, L_B + \varphi$$

(6.8)

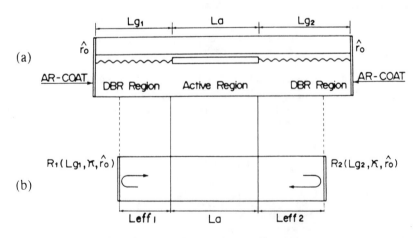

Fig. 6.2 Model for LD linewidth analysis. (From [6.19] Takahashi *et al.* (1989) *IEEE J. Quantum Electron.* **QE-25**, 1280–1287. Reproduced by permission. ©1989 IEEE.)

$$\gamma^2 = (\alpha_g + j\Delta\beta)^2 + \varkappa^2 \tag{6.9}$$

$$\Delta\beta = \beta - \beta_0 \tag{6.10}$$

where $\varkappa$ is the grating coupling coefficient, $\Delta\beta$ is the deviation of the propagation constant β from the propagation constant at Bragg condition β_0, L_B is the length of the Bragg region, α_g is the loss coefficient of the waveguide, and $\hat{r}_0$ is the complex reflectivity of a facet which is given by $r_0 \exp(-j\varphi)$.

In (6.7), the reflectivity r and the phase ϕ of the complex reflectivity of the DBR region depend on the light frequency ω. When the lasing wavelength is nearly equal to the Bragg wavelength ($\beta \approx \beta_0$), the phase change of the DBR region in the round trip is assumed by $d\phi = L_{eff} \Delta\beta$ in [6.20]. The DBR reflectors can be treated as the equivalent external reflectors with the ω-independent effective length L_{eff} and ω-dependent reflectivity $R(=r^2)$ as shown in Fig. 6.2. The relation between the effective length L_{eff} and the DBR length L_B is obtained by the coupled mode theory as follows [6.20]:

$$L_{eff} = \left\{ \frac{1}{2} \frac{\alpha_g L_B \left[\tanh(\gamma_0 L_B)/(\gamma_0 L_B) - 1/\cosh(\gamma_0 L_B) \right] + \tanh^2(\gamma_0 L_B)}{\alpha_g \tanh^2(\gamma_0 L_B) + \gamma_0 \tanh(\gamma_0 L_B)} \right\} \tag{6.11}$$

$$\gamma_0^2 = \alpha_g^2 + \varkappa^2. \tag{6.12}$$

The effective length L_{eff} and the coupling coefficient $\varkappa$ were estimated by measuring the mode spacing according to [6.20]

$$\Delta\lambda = \frac{\lambda^2}{2 n_{eff}(L_a + L_{eff\ 1} + L_{eff\ 2})} \tag{6.13}$$

where L_a is the length of the active region, $L_{eff\ 1}$ is the front effective length, $L_{eff\ 2}$ is the rear effective length, $n_{eff} = n_{eq} - \lambda(\partial n_{eq}/\partial\lambda)$, and n_{eq} is the equivalent refractive index. The $\varkappa$ was estimated to have a value from 40 to 50 cm^{-1} for the DBR lasers in experiments [6.19]. The value of $\varkappa = 50$ cm^{-1} was assumed in the calculation [6.19].

For the DBR laser, the electric field F is distributed not only in the active region, but in the DBR region in the longitudinal direction. The spontaneous emission factor C' and the photon lifetime τ_p in (6.2) must be rewritten as $C_{DBR} = \xi_1 C'$ and $(1/\tau_p)_{DBR} = \xi_1(1/\tau_p)$ where ξ_1 is the longitudinal confinement factor in the active region defined by [6.20]

$$\xi_1 = \int_{-L_a/2}^{L_a/2} |F(z)|^2 \, dz$$

$$= \frac{L_a}{L_a + L_{eff\ 1} + L_{eff\ 2}}. \tag{6.14}$$

The photon lifetime τ_p is given by

$$\tau_p = \left[\frac{c}{n_{eq}} \left(\alpha_{int} + \frac{1}{L_a} \ln \frac{1}{r_1 r_2} \right) \right]^{-1} \tag{6.15}$$

where c is the light velocity, r_1 is the front reflectivity of the DBR region, r_2 is the rear reflectivity of the DBR region, α_{int} is the internal loss coefficient, and we assumed 25 cm^{-1} in the calculation.

Then the linewidth of a DBR laser is given by

$$\Delta \nu_{DBR} = \xi_1^2 \, \Delta \nu_{FP}$$

$$= \frac{C'}{4\pi} \frac{(1/\tau_p)}{(I/I_{th}-1)} \xi_1^2 \, (1+\alpha^2). \tag{6.16}$$

The bias current I and the output power P from the front facet are approximately related by

$$P \approx \frac{h\nu}{e} \eta_d \, (I - I_{th}) \tag{6.17}$$

where η_d is the external differential quantum efficiency from the front facet defined by

$$\eta_d = \eta_i \frac{1}{2} \frac{(1-r_1^2) \, r_2}{(1-r_1^2) \, r_2 + (1-\frac{2}{2}) \, r_1} \frac{\ln(1/r_1 \, r_2)}{\alpha_{int} \, L_a + \ln(1/r_1 \, r_2)} \tag{6.18}$$

and η_i is the internal quantum efficiency.

From (6.16) and (6.17), the linewidth is obtained as follows:

$$\Delta \nu_{DBR} = \frac{C'}{4\pi P} \frac{h\nu}{e} I_{th} \frac{1}{\tau_p} \eta_d \, \xi_1^2 \, (1+\alpha^2). \tag{6.19}$$

In (6.6) and (6.16), the difference of linewidth between an FP laser and a DBR laser is the term ξ_1^2. It is possible to obtain a narrower linewidth for a DBR laser than for an FP laser at the same bias level. The term ξ_1^2 is taken into consideration in calculation.

6.2.2 λ/4-shifted DFB laser diodes [6.24]

DFB and DBR lasers should show the lowest threshold and the largest wavelength selectivity when they operate at the Bragg wavelength, since the Bragg reflection occurs most effectively at the Bragg wavelength. In a DBR laser, carefully designed waveguides are required for that purpose [6.25]. On the other hand, a DFB laser inherently has a two-mode property [6.26]. A phase shift of the corrugation corresponding to a quarter of the lightwave (λ/4) breaks the two-mode degeneracy [6.26] and readily leads to the single mode operation at the Bragg wavelength [6.27–6.30].

The best structure for a stable single-frequency laser is the quarter-wave-shifted DFB laser. In this structure, a phase shift is incorporated in the corrugation at the centre of the laser cavity with both end facets AR coated. When the phase is π/2, the lowest threshold gain is obtained for the main mode exactly at the Bragg wavelength, and the gain difference between the central mode and the nearest side mode has the largest value.

Because of the large gain difference between the main mode and the side mode, performance of λ/4-shifted DFB lasers is superior to that of conventional DFB lasers in terms of dynamic single mode stability and negligible mode partition noise [6.31] at multi Gbit/s modulation speeds. Also, a narrow linewidth (3 MHz) has been obtained under CW operation [6.32].

THEORETICAL ANALYSIS OF λ/4-SHIFTED DFB LD [6.24]

A theoretical model of the Ω-shifted DFB laser in the 1.5 μm range comprising an InGaAsP/InP semiconductor heterostructure is shown in Fig. 6.3 [6.24]. Three quaternary layers, the waveguide layer (Q_1), the active layer (Q_2), and the buffer (anti-meltback) layer (Q_3), are sandwiched by InP. The layer thicknesses are denoted by h, d, and t, respectively and the energy bandgap wavelengths are assumed to be 1.3, 1.55, and 1.3 μm, respectively. The first-order corrugation grating with the period Λ is formed at the hetero-interface between the InP and the waveguide layer. The phase of the corrugation is shifted by 2Ω at the centre of the device with length L.

The equivalent refractive index variation $n(z)$ is written in the form of

$$n(z) = \begin{cases} n_0 + \Delta n \ \cos\left(\dfrac{2\pi}{\Lambda}z + \Omega\right) & z < 0 \\[3mm] n_0 + \Delta n \ \cos\left(\dfrac{2\pi}{\Lambda}z - \Omega\right) & z \geqslant 0 \end{cases} \tag{6.20}$$

where n_0 is the mean value and Δn is the amplitude of the variation. $\hat{p}_l$ and $\hat{p}_r$ are the field reflectivities at the left ($z = -L/2$) and right ($z = L/2$) ends, respectively. The output power and the external differential quantum efficiency per facet, for instance, from the right facet, are denoted by P_r and η_{dr}, respectively. The right-going wave $R(z) \exp(-j\beta z)$ and the left-going wave $S(z) \exp(j\beta z)$ couple with each other due to the corrugation, in which β is the propagation constant, and the slowly varying amplitudes $R(z)$ and $S(z)$ obey the following coupled-wave equations [6.25]:

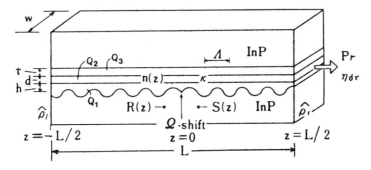

Fig. 6.3 Model of Ω-shifted DFB lasers. (From [6.24] Abiba *et al.* (1987) *J. Lightwave Technol.* **LT-5**, 1564–1573. Reproduced by permission. ©1987 IEEE.)

$$-\frac{dR}{dz}+(\alpha-j\delta\beta)R=j\varkappa\ \exp(\mp j\Omega)S\qquad z\gtreqless 0 \qquad (6.21)$$

$$\frac{dS}{dz}+(\alpha-j\delta\beta)S=j\varkappa\ \exp(\pm j\Omega)R\qquad z\gtreqless 0 \qquad (6.22)$$

where $\varkappa$ is the coupling coefficient between the two waves determined from the corrugated waveguide parameters, α the gain induced by the current, and $\delta\beta$ the deviation of the propagation constant β at λ from β_B at the Bragg wavelength λ_B, which is

$$\lambda_B=2\pi n_0/\beta_B=2\Lambda n_0 \qquad (6.23)$$

in the first-order corrugation. The sign of Ω in (6.21) and (6.22) are inverted in the left and right halfs. The general solutions of (6.21) and (6.22) are in the form of

$$R(z)=r_1\ \exp(\gamma z)+r_2\ \exp(-\gamma z) \qquad (6.24)$$

$$S(z)=s_1\ \exp(\gamma z)+s_2\ \exp(-\gamma z) \qquad (6.25)$$

$$\gamma^2=(\alpha-j\delta\beta)^2+\varkappa^2. \qquad (6.26)$$

The boundary conditions at the Ω-shifted position and at both ends determine the threshold and the field distributions and the general cases were discussed in [6.30].

Now we assume non-reflective ends. $\hat{\rho}_l=\hat{\rho}_r=0$, since the largest wavelength selectivity is expected [6.28]. Then in [6.24], the following transmission gain T_{gain} was defined by the ratio of the output from the right end to the input at the left end:

$$T_{gain}=\left|\frac{4e^{-j\Omega}\gamma^2}{(\Gamma e^{-\gamma L/2}-\hat{\Gamma}e^{\gamma L/2})^2+\varkappa^2 e^{-2j\Omega}(e^{-\gamma L/2}-e^{\gamma L/2})^2}\right|^2$$

$$\Gamma=\gamma+\alpha-j\delta\beta\qquad\hat{\Gamma}=-\gamma+\alpha+j\delta\beta. \qquad (6.27)$$

The threshold condition is derived by setting the denominator of (6.27) to be zero.

Figure 6.4 [6.24] shows the threshold gain α_{th} and the propagation constant in the form of $\delta\beta$, both normalized by the length, L, of the resonant modes. The parameter is the phase shift Ω and $\varkappa L=2$ is assumed $\Omega=0$ corresponds to a uniform corrugation and the two modes with the identical lowest threshold gain are present [6.26]. With increasing Ω, the resonant modes move toward smaller $\delta\beta L$ and the threshold of the λ_0 mode decreases, while that of the next λ_1 mode increases.

The lowest threshold occurs exactly at the Bragg wavelength λ_B when $\Omega=\pi/2$, and the threshold gain difference defined by

$$\Delta\alpha_{th}=\alpha_{th}(\lambda_1)-\alpha_{th}(\lambda_0) \qquad (6.28)$$

becomes the largest.

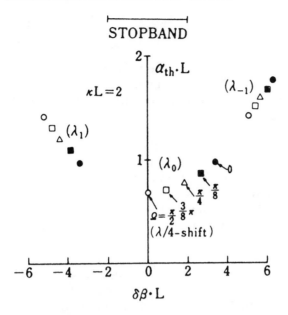

Fig. 6.4 Threshold gain α_{th} of Ω-shifted DFB LD. (From [6.24] Abiba *et al.* (1987) *J. Lightwave Technol.* **LT-5**, 1564–1573. Reproduced by permission. ©1987 IEEE.)

The threshold gain difference $\Delta\alpha_{th}L$ of (6.28) and the external differential quantum efficiency η_{dr} are given by

$$\eta_{dr} = \eta_i\gamma_c\frac{\alpha_{th}}{2\alpha_{th}+\alpha_{in}} \qquad (6.29)$$

as a function of the coupling strength $\varkappa L$ [6.34, 6.35]. In (6.29), η_i is the internal quantum efficiency, γ_c the ratio of the current flowing into the active region to the total current [6.33], and α_{in} the internal optical losses, which is given by

$$\alpha_{in} = \xi\alpha_{ac} + (1-\xi)\alpha_{ex}. \qquad (6.30)$$

In (6.30), ξ is the optical confinement factor defined by the ratio of the optical power in the active layer to the total power, and α_{ac} and α_{ex} are the optical losses in the active and cladding layers, respectively. The stability of the main mode in direct modulation was discussed in [6.36] and the threshold gain difference $\Delta\alpha_{th}$ of $10\,\mathrm{cm}^{-1}$, which corresponds to $20\,\mathrm{cm}^{-1}$ of mirror loss difference, is large enough to maintain the main-to-side-mode ratio over 30 dB. If the laser length is $350\,\mu m$, $\Delta\alpha_{th}L$ should be more than 0.35. The result in [6.24] indicates that the $\lambda/4$-shifted DFB structure has a fairly large threshold gain difference when $\varkappa L>1$. The output efficiency decreases with increasing $\varkappa L$. If we design the device with $\eta_{dr}>0.25$, the coupling strength $\varkappa L$ should be less than 2.

The threshold current is another important factor in the design. The expression of the threshold current density J_{th} was adopted in [6.24, 6.35, 6.37]:

$$J_{th} = \frac{eB_{eff}}{A_0^2} d \left(\frac{2\alpha_{th}}{\xi} + \alpha_0 + \alpha_{ac} + \frac{1-\xi}{\xi}\alpha_{ex} \right)^2 \qquad (6.31)$$

where e is the electronic charge, B_{eff} the recombination coefficient, A_0 and α_0 the slope and the effective loss obtained from the peak gain versus injected carrier density curve. The threshold current I_{th} is obtained from J_{th} multiplied by the stripe width W and the length L:

$$I_{th} = WLJ_{th}/\gamma_c. \qquad (6.32)$$

The calculated J_{th} and I_{th} (for $W = 1\,\mu m$ and $\gamma_c = 1$) are shown in Fig. 6.5 [6.24] as a function of the length L with a parameter $\varkappa$. The $\lambda/4$-shifted corrugated waveguide assumed in the calculation is depicted in the inset. If the length becomes shorter than $300\,\mu m$, abrupt increases in J_{th} are seen, though I_{th} has a minimum for $\varkappa > 90\,cm^{-1}$. In order to avoid such increases of J_{th}, a coupling strength $\varkappa L$ larger than 2 is necessary.

The optimum coupling strength $\varkappa L$ of about 2, for example, $\varkappa = 60\,cm^{-1}$ and $L = 340\,\mu m$, is the result from the above discussions on the threshold gain difference, the output efficiency, and the threshold current density; then an I_{th} of about 10 mA is expected, but we may choose smaller $\varkappa L$ when the output efficiency is a key point of the design. Moreover it was reported [6.38] that a larger coupling strength might lead to a poor single mode property due to a non-uniform power distribution along the cavity length.

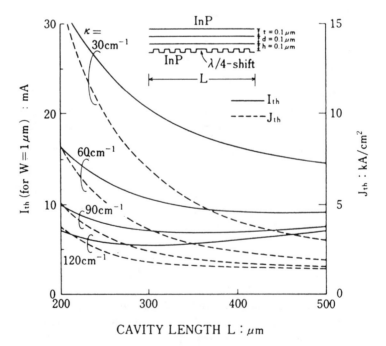

CAVITY LENGTH L : μm

Fig. 6.5 Threshold current I_{th} of Ω-shifted DFB LD. (From [6.24] Abiba *et al.* (1987) *J. Lightwave Technol.* **LT-5**, 1564–1573. Reproduced by permission. ©1987 IEEE.)

In the above calculations the electric field parallel to a layer and a fundamental transverse mode are assumed and the spatial mode is a so-called TE_0 mode. As for the stability of the fundamental transverse mode [6.39], the layer thicknesses and the stripe width should be controlled so as to make the higher order modes cut off and the same condition as in Fabry–Perot (FP) type diode lasers applies to the present case. The polarization characteristic in DFB lasers, however, is different from that in the FP lasers [6.40]. In FP lasers the TE mode has a lower threshold since the facet reflectivity of the TE mode is larger than that of the TM mode [6.41]. The $\lambda/4$-shifted DFB lasers here have non-reflective ends, and the coupling coefficient and the optical confinement factor play a role in polarization behaviour.

The computed coupling coefficient [6.42, 6.26] and the threshold gain α_{th} for TE_0 and TM_0 modes are functions of the corrugation depth a. A rectangular shaped corrugation is assumed. $\varkappa$ for TE_0 is larger than that for TM_0 by about 10% and the threshold gain difference $2\{\alpha_{th}(TM)-\alpha_{th}(TE)\}$, corresponding to the mirror loss difference in FP lasers, is about 5–6 cm^{-1} for the DFB region length $L=350\,\mu m$. The estimated threshold gain difference due to different $\varkappa$ values is not large enough to stabilize the TE mode.

FABRICATION OF $\lambda/4$-SHIFTED DFB LD

The $\lambda/4$-shifted corrugation grating was fabricated by a negative and positive photoresist method [6.30, 6.44]. A positive photoresist (MP1350) was formed in the right half, an SiN thin film was deposited by an electron cyclotron resonance (ECR) plasma deposition technique, and then the negative photoresist (RU1100N) was spin-coated on the entire surface. A conventional two-beam holographic exposure was carried out using a He–Cd laser. In the developing process of the positive and the negative photoresists, the holographic interference pattern is inverted in the two regions. Therefore we get π phase shift in the corrugation at the centre and the first-order corrugation with the $\lambda/4$-shift is realized. Electron-beam lithography, a phase-mask method, and a double exposure method are also interesting alternative ways to produce the $\lambda/4$-shifted corrugation.

A two-step liquid phase epitaxy was employed to make a double-channel planar buried heterostructure (DC-PBH) laser [6.45]. Figure 6.6 [6.24] is a schematic illustration with a cross sectional SEM view in the vicinity of the $\lambda/4$-shifted position. In order to realize, non-reflective ends of the DFB region were also buried and a so-called double-window structure was made. A residual reflectivity at the DFB region end is estimated to about 0.03% [6.46]. The DFB region length l_a+l_b was 350 μm and $l_a=l_b$. The SEM photograph shows uniform layers without any defect or step around the $\lambda/4$-shifted position.

6.2.3 MQW-DFB laser with narrow spectral linewidth [6.50]

Narrowing the spectral linewidth of a laser diode is a key technique for optical communication systems. Long-cavity distributed feedback (DFB) lasers having a multiple quantum well (MQW) structure in the active layer [6.47, 6.48] are very

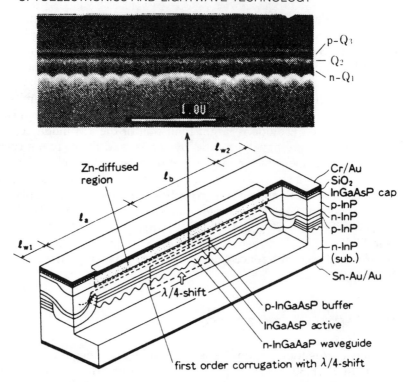

Fig. 6.6   Schematic structure of λ/4-shifted DFB LD. (From [6.24] Abiba *et al.* (1987) *J. Lightwave Technol.* **LT-5**, 1504–1573. Reproduced by permission. ©1987 IEEE.)

promising for achieving a sub-MHz spectral linewidth. However, in such DFB lasers, it is difficult to obtain stable single mode operation at high output power. This is because the spatial hole burning effect [6.49] limits the minimum spectral linewidth in these lasers.

Recently, a corrugation-pitch-modulated (CPM) [6.50, 6.51] structure is shown to stabilize the single mode and produce a narrower spectral linewidth. This structure is effective in maintaining the single-longitudinal mode at a high output power by suppressing the spatial hole burning effect. Therefore, a long-cavity CPM-MQW-DFB laser produces a spectral linewidth of 170 kHz, at 25 mW output power. In addition, the small linewidth–power product 400 kHz mW is attained.

A schematic drawing of the long-cavity CPM-MQW-DFB laser is presented in Fig. 6.7 [6.50]. An InGaAs/InGaAsP MQW structure was introduced in the active region to reduce the linewidth enhancement factor α and internal loss α_0. The InGaAs well layer thickness is 6 nm, and the InGaAsP ($\lambda_g = 1.15\,\mu$m) barrier layer is 10 nm. There are four MQW active layer wells sandwiched between the InGaAsP ($\lambda_g = 1.15\,\mu$m) cladding layers. The cavity length is 1200 μm, and the coupling constant is 35 cm^{-1}. Both of the structure's facets were coated with sputtered SiN$_x$ to reduce the reflectivity to below 1.0%. An effective phase shift of λ/4 was realized by introducing the CPM structure. In the 360 μm long phase-arranging region, the corrugation pitch was set so as to be only 0.08 nm longer than that of the other regions to obtain the λ/4 phase shift.

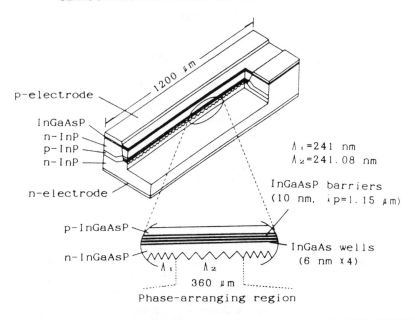

Fig. 6.7 Schematic drawing of MQW-DFB LD. (From [6.50] Okai *et al.* (1990) *IEEE Photonic Technol. Lett.* **PTL-2**, 529–530. Reproduced by permission. ©1990 IEEE.)

The CPM grating was fabricated by the photomask self-interference method [6.52] as follows. The CPM grating photomask was made by computer controlled mechanical ruling and successive replica processes. Then, the grating pattern on the photomask was photolithographically transferred to the surface of the InP substrate using a He–Cd laser as the light source.

The phase-arranging region's length is an important parameter for attaining a narrow spectral linewidth in the CPM structure. Although the light intensity profile becomes flatter along the laser cavity as the phase-arranging region becomes longer, the threshold gain difference between the main mode and its biggest submode $\Delta\alpha_{th}L$ becomes smaller. The phase-arranging region is optimized to be $360\,\mu$m long in the present device in order to produce a stable single mode. Thus, $\Delta\alpha_{th}L$ is more than 0.3, even if the residual reflectivity of both facets is 1%.

The spectral linewidth was measured with the self-heterodyne method [6.53], using a 150 MHz acoustic-optic modulator and a 5 km single mode fiber. Laser light was introduced to the optical fiber using a rod lens, to reduce the reflection into the laser, a 60 dB optical isolator was positioned just after the rod lens. The temperature of the laser was controlled so as to be $25\pm0.01°$C.

The minimum spectral linewidth is 170 kHz, which was achieved at 25 mW output power. Note that no rebroading is observed at high output powers. By employing the longer cavity (1200 μm instead of 400 μm), the linewidth–power product is reduced to 400 kHz mW, or down to 1/4.5 of the product obtained with the 400 μm cavity.

6.3 WAVELENGTH-TUNABLE LASER DIODES

The ability to tune the laser wavelength or frequency is important for a variety of applications in optical communications, such as wavelength or frequency division multiplexing (WDM or FDM), frequency modulation (FM) of lasers, local oscillator tuning in coherent communication schemes, and optical switching in local area networks. Large wavelength-frequency shifts (tens or hundreds of ångströms, which corresponds to hundreds or thousands of gigahertz) are desirable for WDM/FDM. On the other hand, frequency modulation requires small, but fast, frequency shifts, as well as a flat FM response. Continuous wavelength tuning is necessary for FM, and while it is also desirable for WDM/FDM applications, some combination of discrete mode hopping and continuous tuning is also acceptable.

Large gain bandwidth in semiconductors causes multilongitudinal mode operation in ordinary Fabry–Perot cavity lasers [6.54]. Thus a useful single mode tunable semiconductor laser requires a wavelength-selective element inside the cavity as in Fig. 6.8. Such wavelength selectivity can be provided by an internal grating (DFB and DBR lasers [6.55]) or a multiple-mirror cavity (e.g., C^3 lasers [6.56]). Distributed feedback lasers (DFB) provide gain in the grating region, while distributed Bragg reflector lasers (DBR) use gratings as mirrors and gave a separate gain region. Single mode, as well as tunable, lasers have been demonstrated with the above laser structures [6.56–6.59]. Continuous wavelength tuning is accomplished by changing the round-trip phase condition in the laser cavity. Shift of the mode wavelength changes the position of the mode relative to the gain window of the wavelength selective element, eventually causing a mode hop. Thus, to achieve wide continuous tuning of a single-mode laser, one has to provide for independent tuning of the round-trip phase and the wavelength selective element.

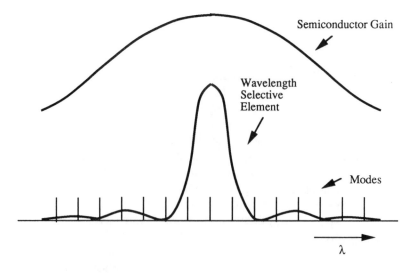

Fig. 6.8 Wavelength selection in laser diodes.

6.3.1 An ideal tunable laser diode [6.60]

An ideal continuously tunable single-longitudinal mode laser is one in which the output can be tuned smoothly over the entire gain bandwidth without any significant reduction in spurious mode suppression or change in power.

In order for this to be possible, three characteristics of the laser must be adjusted simultaneously in the proper proportion. These characteristics are: (1) the wavelength of the resonant mode λ_m, (2) the centre wavelength of the mode selection filter λ_f, which provides the single-longitudinal mode (SLM) feature, and (3) the pumping (gain) level which provides for constant power out. The parameter that controls the latter feature is most easily adjusted by an external power levelling feedback loop, as is the common practice. The remaining problem is to find a laser configuration and a means of control that will tune λ_m and λ_f over the bandwidth of the gain while keeping $\lambda_m = \lambda_f$ for maximum spurious mode suppression ratio S_R.

Figure 6.9 [6.60] gives three laser configurations that, in principle, can provide continuous, constant amplitude tuning over the full bandwidth of the gain. The first is simply a DFB laser with temperature control. The second two are extended cavity configurations with a phase shifter and a tunable narrow-band mirror in the passive waveguide region.

For the case of Fig. 6.9(a), the temperature-controlled DFB laser, a very short discussion should suffice, since this configuration has been widely demonstrated

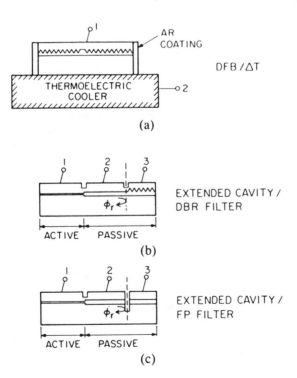

(a)

(b)

(c)

Fig. 6.9 Ideal continuously tunable single-frequency lasers. (From [6.60] Coldren and Corzine (1987) *IEEE J. Quantum Electron.* **QE-23**, 903–908. Reproduced by permission. ©1987 IEEE.)

[6.62]. The obvious problem with this solution is that the tuning rate is slow. Even with minimal thermal mass between the laser and the thermoelectric cooler, the tuning rate would be limited to a few tens of hertz. For proper operation of this device it is desirable to use the phase-shifted DFB geometry [6.63] (which provides a mode centred in the grating reflection band) for maximum S_r and antireflection facet coatings so that the feedback is due solely to the gratings. This latter condition is necessary to ensure that the tuning rate of the selected mode solution λ_m is identical to that of the grating stopband centre as the effective index n_{eff} of the entire waveguide region is changed. Of course, it is this natural alignment of the mode solution and the cavity loss minimum that guarantees the desired operation in the DFB structure. Figure 6.10(a) and (b) [6.60] shows the general form of the grating reflection characteristics and the modal solutions. For this case as for all others the round-trip cavity phase ϕ_{rt} must be an integer number of 2π for a given mode

$$\phi_{rt} = 2\pi m. \tag{6.33}$$

With the mode centred in the grating stopband $\lambda_m = \lambda_f$, the reflection phase varies linearly with λ [6.64]; therefore

$$2(2\pi n_{eff} l_{eff})\lambda_m = 2\pi m \tag{6.34}$$

where l_{eff} is the effective cavity length and m is the mode number. Thus, if the index changes by Δn_{eff},

$$\Delta\lambda_m/\lambda_m = \Delta n_{eff}/n_{eff} \tag{6.35}$$

as for any single cavity laser, and for the DFB case, $n_{eff} = n_f$.

The specification of required tuning and power levelling signals is now obvious. A power levelling feedback loop controls I_1, and I_2 provides a temperature change ΔT, which tunes λ_m and λ_f according to (6.35). For linear tuning, the change in index due to the change in junction temperature and carrier density caused by the predictable changes in I_1 to level power as the gain bandwidth is traversed must be taken into account. The gain maximum will be tuned at about 5 Å/°C in the same direction, so the gain will sweep past the more slowly moving lasing mode. If 200 Å of usable gain bandwidth was assumed, then about 50 Å of continuous tuning should be possible by changing the junction temperature by 50°C before this occurs. Figure 6.9(a) indicates the relative movements of the three characteristics of concern due to changes in I_1 and ΔT.

For a completely electronic device the three-section laser configurations are considered in Fig. 6.9(b) and (c). ('Completely electronic' does not imply that thermal effects might not be importance, especially at low tuning rates. Rather, this term is used to indicate that thermal effects are not essential.) Existing two-section devices cannot provide alignment of λ_f and λ_m while maintaining constant power out. However, some range of constant amplitude tuning is possible in two-section devices before the misalignment becomes sufficiently great to cause a mode jump.

For the three-section extended cavity cases, we refer to Fig. 6.10(a) and (d) for the form of the reflection filters, Fig. 6.10(c) for the mode locations, and Fig. 6.10(e)

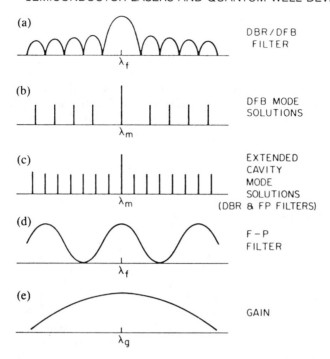

Fig. 6.10 Mode selection filters in LD. (From [6.60] Coldren and Corzine (1987) *IEEE J. Quantum Electron.* **QE-23**, 903–908. Reproduced by permission. ©1987 IEEE.)

for the gain. In the Fabry–Perot case (Fig. 6.9(c)), the length of the third cavity must be sufficiently small that a second mode is not reinforced within the gain bandwidth. Also, it should not be excessively short, or very little difference in reflection will occur for the mode adjacent to the desired one. A spacing of successive maxima of about twice the desired tuning range should be sufficient. For 100 Å of tuning at 1.55 μm, this gives $l_3 \approx 20$ μm.

As always, the mode wavelength is determined by the solution of (6.33), where in this case

$$\phi_{rt} = 2\phi_1 + 2\phi_2 + \phi_r \tag{6.36}$$

and $\phi_1 = 2\pi n_1 l_1 / \lambda_m$, $\phi_2 = 2\pi n_2 l_2 / \lambda_m$, and ϕ_r is the reflection phase of the grating [6.64] or compound mirror [6.65] (Fabry–Perot) at λ_m. Equation (6.36) assumes no impedance discontinuity between sections. In order to specify the control signals at terminals 1, 2 and 3, the required index changes are determined that must occur in each of the respective sections to keep the mode in the centre of the reflective mode selection filter band ($\lambda_m = \lambda_f$), while also maintaining constant amplitude. Under the boundary condition that $\lambda_m = \lambda_f$, a reference plane is selected at the interface between sections 2 and 3 such that $\phi_r \equiv 0$. Then, for either the grating or the Fabry–Perot reflector

$$\Delta\lambda_m / \lambda_m = \Delta\lambda_f / \lambda_f = \Delta n_3 / n_3 \tag{6.37}$$

and this requires from (6.36)

$$\Delta\phi_2 = -\Delta\phi_1. \tag{6.38}$$

In order to satisfy (6.38), we can solve for the required index shift in region 2, assuming a mode wavelength shift as given by (6.37) and allowing for an index shift in region 1 due to this wavelength shift as well as gain adjustments for power levelling:

$$\Delta n_2/n_2 = \Delta\lambda_m/\lambda_m (1 + n_1 l_1/n_2 l_2) - \Delta n_1/n_1 (n_1 l_1/n_2 l_2). \tag{6.39}$$

Equation (6.39) gives us the necessary conditions on n_2 to maintain alignment of λ_m and λ_f for the best SLM while tuning.

6.3.2 External cavity tunable laser diodes

Wavelength tuning can be achieved by providing a frequency tunable loss element inside the laser cavity. Elements such as a diffraction grating [6.66], an electro-optic (EO) filter [6.67], and an acousto-optic (AO) filter [6.68] have been used. The most common method is to use a diffraction grating in an external cavity.

The tuning range of a grating-loaded external cavity laser is limited, in principle, only by the width of the gain spectrum of the semiconductor material employed as the gain medium in the laser. For InGaAsP materials, a maximum tuning range of 55 nm has been achieved [6.69]. While this large tuning range is desirable, tuning must be done mechanically. Electronically tunable external cavity lasers using either an EO filter or an AO filter are more attractive. However, neither method can provide continuous wavelength tuning over a useful range.

The most conventional method of constructing a tunable laser is to use a semiconductor chip as the gain medium in an external cavity with a diffraction grating serving as both a mirror and a narrow-band filter. One facet of the laser diode is AR-coated, and the light from this end is collimated by a lens while the other cleaved facet and the diffraction grating from the external cavity. The lasing frequency is tuned by rotating the grating. Fine tuning can be achieved by axial displacement of the grating, or by adding an adjustable phase plate. In principle, a tuning range over the entire width of the gain spectrum is possible. In practice, however, the maximum range obtained was 55 nm [6.66], centred at 1.5 μm wavelength, limited by the gain available in the semiconductor chip to overcome the total loss of the assembly. Thus, the light-coupling efficiency of the lens imposed a major limitation. Because of the long cavity length, such a grating tuned external cavity laser has exhibited a very narrow linewidth of 10 kHz. An extremely wide tuning range of 105 nm centred about 0.8 μm wavelength has been obtained using an optimized GaAs/AlGaAs single QW laser [6.70].

An acousto-optically tunable laser is another configuration of an external cavity laser. Inside the cavity are an acousto-optically tunable filter (AOTF), an AO modulator, and a semiconductor chip [6.68]. The AOTF has the property that an incident linearly polarized light is diffracted into the orthogonal polarization by the acoustic beam. For a given acoustic frequency, only a small range of the optical frequencies, which satisfy the law of conservation of momentum, is diffracted. The optical frequency and the acoustic frequency are related by

$$f_o = (c/v)(1/\Delta n)f_a \qquad (6.40)$$

where c/v is the ratio of the velocity of light in the vacuum to the acoustic velocity in the crystal, and Δn is the birefringence of the crystal. The ratio of the optical frequency to the acoustic frequency for TeO_2 is of the order of 3×10^6. Thus, the peak wavelength of the filter with a 3 nm passband can be tuned over the entire wavelength range from $1\,\mu m$ to $1.6\,\mu m$ by varying the acoustic frequency from 50 MHz to 90 MHz. In this configuration polarization of the incident beam is restored after being reflected by the end mirror and passing through the crystal back to the semiconductor chip. However, the diffracted beam is frequency shifted by f_a. Therefore, the acousto-optic modulator is needed to correct the frequency shift of the optical beam. The maximum tuning range of the laser achieved was 83 nm centred at $1.3\,\mu m$ wavelength with the acoustic frequency varying from 69.494 MHz to 74.570 MHz and a driving power of 3W. The tuning speed is $3\,\mu s$.

An electro-optically tunable laser is also a kind of external cavity laser in which the tunable bandpass filter employs a wavelength tunable TE-to-TM polarization converter and a passive polarization filter fabricated from single mode $Ti:LiNbO_3$ strip waveguide [6.67]. Again, a $1.55\,\mu m$ wavelength semiconductor laser chip with one facet AR-coated is oriented in such a way that it provides maximum gain for the TM polarized light in the $LiNbO_3$ waveguide. A maximum tuning range of 7 nm has been obtained by a $\pm 50\,V$ bias applied to the TE-to-TM converter.

6.3.3 Wavelength-tunable DFB/DBR lasers

Through the change in the effective index of refraction, the wavelength of solitary lasers such as DFB lasers can be tuned continuously by temperature and injected carriers. The rate of wavelength tuning by temperature is about $0.1\,nm/°C$. The tuning speed is limited by the thermal impedance. Thus, it is useful only for a small tuning range (about ± 1 nm) and low-speed (a few milliseconds) applications.

The carrier-induced refractive index change is approximated by:

$$\Delta n_{eff} = -\Gamma(\Delta n/\Delta N)N \qquad (6.41)$$

where $\Gamma = 0.3$–0.5 is the optical mode confinement factor, $(\Delta n/\Delta N) = 10^{-20}\,cm^{-3}$ at $1.55\,\mu m$ is the index change per carrier density, and N is the injected carrier density. The wavelength tuning range can be estimated by:

$$\Delta\lambda/\lambda = \Delta n_{eff}/n_{eff} \qquad (6.42)$$

and is limited by the amount of effective index change achievable without excessively heating the laser, and excessive non-radiative recombination at high carrier densities. In practice, the maximum index change is about 1%, resulting in an estimated best tuning range of 10–15 nm. The high injected carrier density (10^{18} cm^{-3}) in the semiconductor laser reduces the effective index of refraction in the corrugation region (the Bragg region), thereby decreasing the lasing wavelength. In a single-electrode DFB laser operated above threshold, however, most injected carriers recombine to produce photons, resulting in a very small increase in carrier density, which in turn leads to a small change in lasing wavelength. The range of wavelength tuning can be improved by using two electrodes in a DFB laser, with a large current applied to one electrode and a small current to the other. A schematic diagram is shown in Fig. 6.11. The operational principle of wavelength tuning can be qualitatively understood as follows. In the asymmetric structure of the DFB laser, the optical field is higher in the region near the output port where the facet is not reflecting (AR-coated for example), and the wavelength is primarily determined by the effective index of refraction in this region. With this section I_2 pumped at current densities at or slightly below the normalized threshold current density (under uniform pumping) just to overcome the absorption loss, it serves as a Bragg reflector. In addition, because of the low pumping level, the injected carriers do not contribute significantly to photon generation, resulting in a large change of refractive index, thereby a substantial wavelength tuning. The gain is provided by the other section I_1 pumped substantially above threshold.

The wavelength tuning characteristics [6.71] are shown in Fig. 6.12(*a*) [6.1]. The tuning range of 3.3 nm has been obtained at 1 mW output power. The tuning range reduced to 2 nm at a power output of 5 mW. The linewidth was 15 MHz (Fig. 6.12(*b*)).

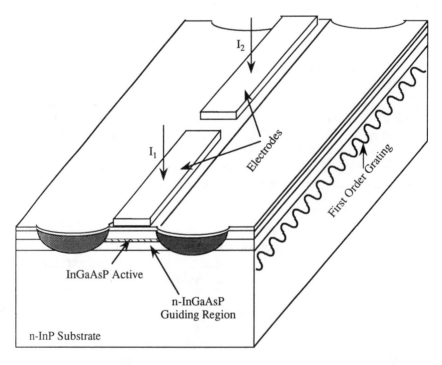

Fig. 6.11 Two-sectioned DFB LD.

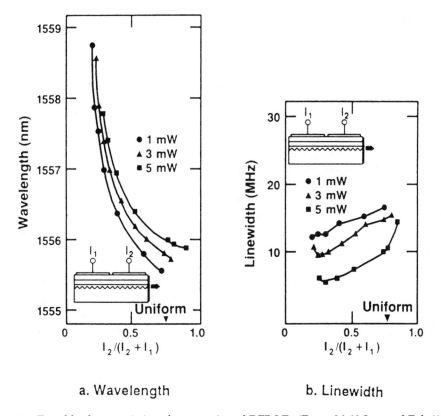

a. Wavelength   b. Linewidth

Fig. 6.12 Tunable characteristics of two-sectioned DFB LD. (From [6.1] Lee and Zah (1989) *IEEE Commun. Mag.* **27**(10), 42–52. Reproduced by permission. ©1989 IEEE.)

PRINCIPLES OF TUNING IN TWO-SEGMENT DFB LASERS [6.54]

The structure of a two-segment DFB laser is illustrated schematically in Fig. 6.13. The laser has a uniform grating throughout the device; a λ/4 phase shift region can also be introduced to break the degeneracy of the two main DFB modes [6.72, 6.73]. Two independent contacts control carrier injection into the two segments. To analyse the laser, the device is divided mentally into two parts and consider complex amplitude reflections Γ_1 and Γ_2 from two grating segments shown in Fig. 6.13:

$$\Gamma_i = |\Gamma_i| \exp(j\phi_i) \qquad i = 1, 2. \qquad (6.43)$$

The reflections introduce gain and change the phase of the optical signal. Assume that saturated gain in an individual section is not sufficient to cause lasing of that section because of its own distributed feedback and gain. The lasing condition can then be written as

$$\Gamma_1 \Gamma_2 = 1. \qquad (6.44)$$

The round-trip phase condition in this equation

$$\phi_1 + \phi_2 = 2m\pi \qquad m = \text{integer} \tag{6.45}$$

defines the modes of the structure. Considering absolute magnitudes

$$|\Gamma_1|\ |\Gamma_2| = 1 \tag{6.46}$$

the combined threshold gains of the two segments could be obtained. The above equations will be used to study wavelength shifts of the lasing mode under different bias conditions on the two sections.

For simplicity, in this analysis there are no reflections from the laser facets is assumed. In this case the amplitude reflection coefficient Γ for reflection from a section of waveguide grating of length L can be obtained from simple coupled mode theory [6.72, 6.73]:

$$\Gamma = \frac{j\varkappa\ \sinh(\mu L)}{\mu\cosh(\mu L) - j\delta\sinh(\mu L)} \tag{6.47}$$

where $\varkappa$ is the grating coupling coefficient and δ is the complex detuning parameter:

$$\begin{aligned}
\delta &= \frac{(\omega - \omega_B)}{v_g} - j\frac{g}{2} \\
&= \left(\frac{2\pi n_g}{\lambda^2}\right)(\lambda - \lambda_B) - j\frac{g}{2}
\end{aligned} \tag{6.48}$$

where ω is the angular frequency and λ is the free space wavelength. The Bragg frequency ω_B is given by the Bragg condition

$$\beta(\omega_B) = \pi/\Lambda \tag{6.49}$$

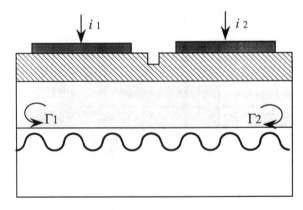

Fig. 6.13 Schematic structure of two-sectioned DFB lasers.

where β is the propagation constant of the waveguide and Λ is the first-order grating period. The power gain coefficient in the waveguide is g, v_g is the group velocity, n_g is the group index, and

$$\mu^2 = \varkappa^2 - \delta^2. \tag{6.50}$$

The $\exp(+j\beta z - j\omega t)$ sign convention is for the propagation. If desired, the effect of facet reflection can be included by modifying expression (6.50) for the grating reflection coefficient (see [6.72, equation 8.18]).

The laser behaviour, such as wavelength tuning, mode hopping, and power output, is determined by the amplitude and phase of the reflection coefficients Γ_i under different bias conditions. Parameter values typical for semiconductor lasers are used ($\varkappa = 50 \text{ cm}^{-1}$, $L = 125\,\mu\text{m}$, $n_g = 3.5$, operating wavelength near $\lambda_0 = 1.3\,\mu\text{m}$); wavelength is given explicitly rather than in normalized form to give a better feeling for the wavelength scale involved. Note that at the Bragg condition (zero detuning) the reflection phase is $\pi/2$, which gives a round-trip phase of π, in contradiction with the mode requirement. Thus, for a uniformly excited laser, there is no mode at this point, and the two degenerate modes with reflection phases 0 and π have symmetrical detuning about the Bragg point. To put a mode at the Bragg wavelength, one has to add (or subtract) an extra phase of π to the round-trip phase; this is achieved in the $\lambda/4$ shifted DFB lasers [6.72, 6.73].

A waveguide grating is tuned by varying the refractive index of the waveguide. A refractive index change of Δn gives, to lowest order, a propagation constant change of

$$\Delta\beta = \Delta n \, \varkappa_0 \tag{6.51}$$

where $\varkappa_0 = \omega/c = 2\pi/\lambda$ and c is the velocity of light in free space. This, in turn, shifts the Bragg frequency ω_B by

$$\Delta\omega_B = -v_g\Delta\beta. \tag{6.52a}$$

The corresponding Bragg wavelength shift is

$$\Delta\lambda_B = (\lambda^2/2\pi n_g)\Delta\beta = (\Delta n/n_g)\lambda. \tag{6.52b}$$

Figure 6.14 [6.54] illustrates schematically the shift of the reflection amplitude and phase as a result of the refractive index change.

In semiconductor lasers the index change is usually achieved by varying the carrier density in the material. The carrier density change causes change in both the refractive index of the material and the gain coefficient. The two changes are linked by the Kramers–Krönig relations; the relation between the two change can also be written phenomenologically as

$$\Delta\beta/\Delta g = -\alpha/2. \tag{6.53}$$

Here α is the linewidth enhancement factor [6.74] defined by

$$\alpha \equiv \frac{\Delta n_r/\Delta N}{\Delta n_i/\Delta N} \tag{6.54}$$

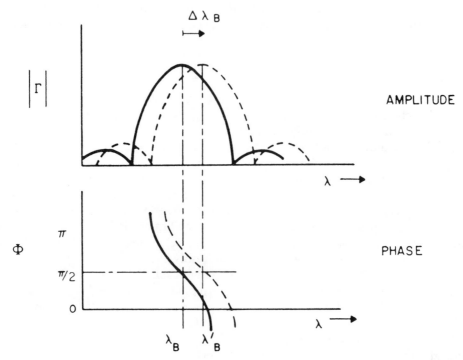

Fig. 6.14 Tuning of waveguide Bragg gratings. (From [6.54] Kuznetsov (1988) *IEEE J. Quantum Electron.* **QE-24**, 1837–1844. Reproduced by permission. ©1988 IEEE.)

where Δn_r and Δn_i are the changes in the real and imaginary parts of the complex refractive index and ΔN is the change in carrier density. Thus a large value of α implies large real index changes (and hence strong tuning) accompanied by small changes of gain. The major difference, as far as wavelength tuning is concerned, between DBR and DFB tunable semiconductor lasers is the magnitude of α in the grating region. In the DBR lasers the operating wavelength is below the bandgap of the material in the grating region; under these conditions the material is transparent and the value of α is very large [6.75, 6.76]. An infinite value of α is often implicitly assumed in the grating and phase control regions. At the same time in the gain region the bandgap is smaller and the α factor is much smaller. This inhomogeneity of α is the basic reason for the substantial tunability of the DBR lasers [6.77]. On the other hand, for the DFB laser the operating wavelength is close to the bandgap, the value of α is relatively small (≈ 3–5) and essentially the same in both sections. With modal gain clamped above threshold, the index and the gain are strongly coupled in both sections; this considerably complicates the tuning mechanism, as will be shown below.

TUNING IN MULTI-SECTION DBR LASER DIODES [6.1, 6.89]

A multi-section wavelength-tunable distributed Bragg reflector (DBR) laser has a wider tuning range than some types of tunable semiconductor lasers [6.77–6.79], so it is a particularly suitable light source for coherent optical fiber transmission systems and multichannel FDM systems, in which the tuning range of the tunable laser used for the local oscillator limits the maximum number of channels [6.80–6.83]. Another important performance required for the tunable laser is a narrow linewidth. For example, the linewidth of below 3.2 MHz is required for 1 Gbit/s CPFSK systems [6.84].

Improvement of the wavelength tuning range has been made by separating the Bragg region in the passive waveguide (a large bandgap material) from the active region (a small bandgap material) inside the laser cavity. The injected carrier density in the Bragg region can be high because carriers do not contribute to photon generation. The corrugated region merely serves as a tunable distributed Bragg reflector (DBR). The wavelength is electronically tuned by the current injected into the DBR region [6.85–6.88].

The DBR exhibits a high reflection within a certain wavelength band (the stop band), which is nominally between 2 and 4 nm wide. The mode that is nearest to the centre of the band and simultaneously satisfies the 2π round-trip phase condition will lase. Thus, by introducing a phase region in the waveguide, independently controlled also by the injected current, the lasing wavelength can be tuned around each Bragg wavelength. Figure 6.15 shows the schematic of a three-electrode wavelength-tunable DBR laser. With a proper design and independent adjustment of the three currents in the active, Bragg, and phase regions, quasi-continuous tuning ranges from 8 to 10 nm have been achieved.

For continuous wavelength tuning, one control current was divided in a prescribed proportion into the Bragg and phase regions. Continuous tuning ranges from 2 to 4 nm have been reported with this configuration. The maximum tuning range is limited by the amount of current that will heat the device. The increase in the junction temperature caused an increase in the lasing wavelength, which

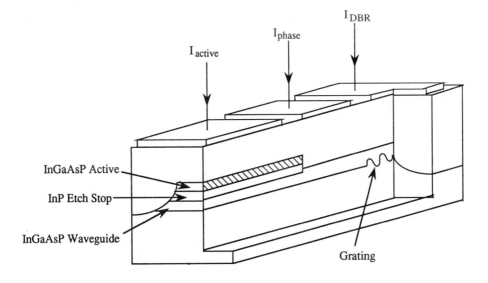

Fig. 6.15 Three-sectioned DBR LD.

compensates for the decrease in the lasing wavelength by the injection current, resulting in a reduction of the effective wavelength tuning. Such a thermal effect was also observed in a wavelength division packet switching experiment when the packet residence time exceeded a few μs.

The oscillation condition of the three-section tunable DBR laser and formulae for the threshold current, the lasing wavelength, the differential efficiency, the linewidth under wavelength tuning was discussed in [6.89]. Figure 6.16 shows a schematic of a tunable DBR laser. This laser has three sections: an active section for light amplification, a phase control (PC) section for phase matching of the lasing light reflected from DBR and a cleaved facet, and a DBR section for the selection of the longitudinal mode wavelength. The PC and DBR sections are made of passive material for the lasing wavelength. These three sections are assumed to be electrically isolated from each other.

The oscillation condition is written as

$$r_1 r_{DBR} C_{out} \exp\{(g_{th} - \alpha_a) - 2j\beta_a\} l_a \exp(-\alpha_p - 2j\beta_p) l_p = 1 \tag{6.55}$$

where r_1 is the field reflectivity of the cleaved facet, r_{DBR} is the complex field reflectivity of the DBR section, C_{out} is power coupling efficiency between the active section and the external passive waveguide, and g_{th} is the threshold modal power gain of the active section. α_a and α_p are the power absorption coefficients, β_a and β_p are the propagation constants, and l_a and l_p are the length of the active section and PC section.

From (6.55), the power condition and the phase condition for the threshold are obtained as

$$g_{th} = \alpha_a + \alpha_p l_p / l_a - 1/l_a \ln(r_1 |r_{DBR}| C_{out}) \tag{6.56}$$

$$\pi q = \beta_a l_a + \beta_p l_p - \text{Arg}(r_{DBR}) \qquad q \text{ integer} \tag{6.57}$$

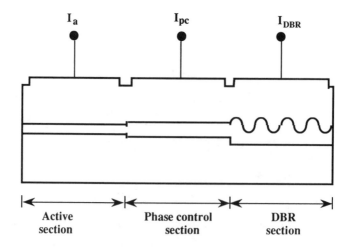

Fig. 6.16 Schematic of three-sectioned DFB model.

where r_{DBR} equals $|r_{DBR}|\exp\{jArg(r_{DBR})\}$ and r_{DBR} is a function of the equivalent refractive index $n_{d,eq}$ and the absorption coefficient α_d of the DBR section. β_a and β_p are given by

$$\beta_a = 2\pi n_{a,eq}/\lambda \qquad \beta_p = 2\pi n_{p,eq}/\lambda \tag{6.58}$$

where λ is the lasing wavelength and $n_{a,eq}$ and $n_{p,eq}$ are the equivalent refractive index of the active and PC sections, respectively. From (6.56), (6.57) and (6.58), the threshold gain lasing wavelength can be obtained.

When current is injected into the passive section, the carrier density increases and the refractive index decreases due to the free carrier plasma effect and band filling effect, and also the absorption increases simultaneously due to the same effects. The refractive index is also dependent on the lasing wavelength. Therefore, $n_{p,eq}$, $n_{d,eq}$, α_p and α_d are given by

$$n_{p,eq} = n_{0,eq}(\lambda) + \xi_p(\partial n/\partial N)N_p \tag{6.59}$$

$$n_{d,eq} = n_{0,eq}(\lambda) + \xi_d(\partial n/\partial N)N_d \tag{6.60}$$

$$\alpha_p = \alpha_0 + \xi_p(\partial\alpha/\partial N)N_p \tag{6.61}$$

$$\alpha_d = \alpha_0 + \xi_d(\partial\alpha/\partial N)N_d \tag{6.62}$$

where N_p and N_d are the injected carrier density at the PC and DBR sections, respectively, $\partial n/\partial N$ and $\partial\alpha/\partial N$ are the material parameters and $n_{0,eq}(\lambda)$ and α_0 are the equivalent refractive index and the absorption of the passive sections when there is no carrier injection. ξ_p and ξ_d are the confinement factors.

The carrier injection into the passive sections also causes the change of threshold gain and the refractive index of the active section. The threshold gain and refractive index of the active section are

$$g_{th} = A_0(N_{th} - N_0) \tag{6.63}$$

$$n_{a,eq} = n_{a0,eq}(\lambda) - \lambda/4\pi \; \alpha A_0(N_{th} - N_{th0}) \tag{6.64}$$

where A_0 is the modal gain coefficient, N_0 is the carrier density at which the gain becomes positive, $n_{a0,eq}(\lambda)$ and N_{th0} are, respectively, the equivalent refractive index of the active section and the threshold carrier density in the absence of carrier injection into the passive sections, and α is the linewidth enhancement factor.

To compare to the experimental results, the relationship between the carrier density and current can be written as

$$I - I_{leak} = eV(AN + BN^2 + CN^3) \tag{6.65}$$

where e is the electron charge, V is the volume of the active layer or that of the passive waveguide, A, B, and C are constants, and I_{leak} is the leakage current due to incomplete carrier confinement.

The lasing wavelength and threshold current under the application of the PC and DBR currents are obtained from (6.56)–(6.65). From (6.65), N_p and N_d can be estimated and from (6.59)–(6.62) $n_{p,eq}$, $n_{d,eq}$, α_p, α_d can be calculated. By solving (6.56)–(6.59), (6.63) and (6.64) the lasing wavelength (at which the threshold gain is minimum) can be determined. Once the wavelength and threshold were obtained, the output power and linewidth can easily be calculated.

The differential efficiencies of the tunable DBR laser are

$$\eta_{d,1} = \eta_i(g_{th}-\alpha_a)/g_{th}\ (1-r_1^2)r'/\{(r_1+r')(1-r_1r')\} \tag{6.66}$$

$$\eta_{d,2} = \eta_i(g_{th}-\alpha_a)/g_{th}\ r_1 C'_{out}|t_{DBR}|^2/\{(r_1+r')(1-r_1r')\} \tag{6.67}$$

where

$$C'_{out} = C_{out}\ \exp(-\alpha_p l_p) \tag{6.68}$$

$$r' = |r_{DBR}|C'_{out}. \tag{6.69}$$

$\eta_{d,1}$ and $\eta_{d,2}$ are the differential efficiency of the light output from the cleaved facet and DBR, respectively, η_i is the internal quantum efficiency, and $|t_{DBR}|^2$ is the power transmissivity of DBR. r_{DBR} and t_{DBR} are given as [6.90]

$$r_{DBR} = \frac{\varkappa/\Gamma\ (\rho-\Gamma/\varkappa)\ \exp(\gamma l_d)+\Gamma/\varkappa\ (\varkappa/\Gamma+\rho)\ \exp(-\gamma l_d)}{(\Gamma/\varkappa-\rho)\ \exp(\gamma l_d)+(\varkappa/\Gamma+\rho)\ \exp(-\gamma l_d)}e^{j\Omega}/j \tag{6.70}$$

$$t_{DBR} = \frac{\varkappa/\Gamma+\Gamma/\varkappa}{(\Gamma/\varkappa-\rho)\ \exp(\gamma l_d)+(\varkappa/\Gamma+\rho)\ \exp(-\gamma l_d)} \tag{6.71}$$

where

$$\rho = jr_2\ \exp\{-j(2\beta_0 l_d+\Omega)\} \tag{6.72}$$

$$\gamma^2 = \varkappa^2-(\beta_d-\beta_0+j\alpha_d)^2 \tag{6.73}$$

$$\Gamma = -\gamma-j(\beta_d-\beta_0+j\alpha_d) \tag{6.74}$$

$$\beta_d = 2\pi/\lambda\ n_{d,eq} \tag{6.75}$$

$$\beta_0 = 2\pi/\lambda_B\ n_{0,eq} \tag{6.76}$$

and λ_B is the Bragg wavelength in the absence of the carrier injection, $\varkappa$ is the coupling coefficient, r_2 is the facet reflectivity at the end of the DBR section, l_d is the length of the DBR section, and Ω is the corrugation phase at the facet.

$\eta_{d,1}$ is equal to that of the laser which has two mirrors with reflectivities of r_1 and r'. r' is the equivalent reflectivity seen from the active section to the external waveguide. From (6.66) and (6.67), the output power is

$$P_1 = hc/\lambda e\ \eta_{d,1}(I_a-I_{th})_0 \tag{6.77}$$

$$P_2 = hc/\lambda e\ \eta_{d,2}(I_a-I_{th})_0 \tag{6.78}$$

where P_1 and P_2 are the output power from the cleaved facet and DBR respectively, h is Planck's constant, c is the velocity of light, and $(I_a - I_{th})_0$ is the effective injection current after subtraction of leakage current. Large r', C_{out} and small $\alpha_p l_p$ are required to realize a tunable DBR laser with high output power.

The linewidth of the three-section tunable DBR laser is given by

$$\Delta\nu = \frac{R(1+\alpha^2)}{4\pi I_s} \left(\frac{L_{eff} + \alpha\lambda^2/4\pi \ d\{|\ln \ (r_{DBR})|\}/d\lambda}{L_{a,eff}} \right)^{-2} \quad (6.79)$$

where

$$R = n_{sp} c g_{th}/n_{a,eff} \quad (6.80)$$

$$I_s = (I_a - I_{th})_0 n_{a,eff}/g_{th} c e \quad (6.81)$$

$$L_{a,eff} = l_a n_{a,eff} \qquad n_{a,eff} = n_{a,eq} - \lambda\xi_a(\partial n/\partial\lambda) \quad (6.82)$$

$$L_{eff} = l_a n_{a,eff} + l_p n_{p,eff} - \lambda^2/4\pi \ \partial\{Arg(r_{DBR})\}/\partial\lambda \qquad n_{p,eff} = n_{p,eq} - \lambda\xi_p(\partial n/\partial\lambda). \quad (6.83)$$

where R is the spontaneous emission rate, n_{sp} is the spontaneous emission factor, I_s is the total photon number in the active section, and $L_{a,eff}$ and L_{eff} are the effective active section length and effective total cavity length, respectively. n_{sp} is regarded as constant. The last term of (6.83) is the equivalent length of the DBR section. This equation can easily be derived from [6.91, equation (37)]. The dependence of the gain coefficient on the wavelength does not have to be considered. Equation (6.79) indicates that the tunable laser with a short length of the active section and long length of the external passive sections gives narrow linewidth. However, long external passive sections also cause an increase of threshold gain because of the absorption loss of the PC and DBR sections. When the PC and DBR current are applied to the waveguide, the linewidth broadens due to the increase of threshold gain caused by the increased absorption loss in the waveguide.

Figure 6.17 [6.89] shows the measured change of the lasing wavelength versus the PC current and also the result of the theory. The DBR current was also changed at the same time. When the currents increased, the lasing wavelength moved to the shorter wavelength side. A complete wavelength change was about 1.8 nm. The linewidth was 5 MHz when no PC and DBR currents were applied. Experimentally, the linewidth increased by five times when the wavelength was tuned, but in theory, it increased only by 2.3 times. A pronounced broadening of the linewidth was observed when the bias current was started applying to the PC section.

6.4 HIGH-SPEED DFB/DBR LASER DIODES

High-speed transmitters with single-longitudinal mode laser diodes and broad-band low noise receivers are necessary to realize high capacity and long-span optical fiber transmission systems.

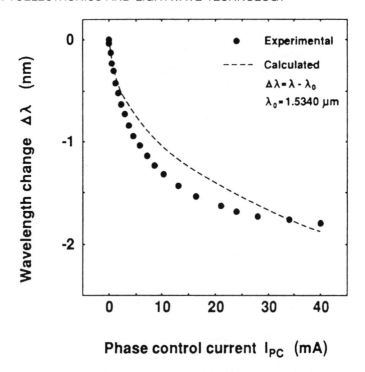

Fig. 6.17 Lasing wavelength tuning of three-sectioned DBR LD. (From [6.89] Kotaki and Ishikawa (1989) *IEEE J. Quantum Electron.* **QE-25**, 1340–1345. Reproduced by permission. ©1989 IEEE.)

There are some important factors to be considered to realize high-speed long-span transmission system: high frequency response of laser diodes, APDs, and electrical circuits; spectral characteristics of the LD outputs, when directly modulated with high-speed pulse streams; optical fiber dispersion and transmission loss characteristics. In order to realize higher bit-rate systems, very high-speed LDs have been developed by reducing residual capacitances of LD chips and by increasing the relaxation oscillation frequency. More than 15 GHz cutoff frequencies (at room temperatures) have been reported for Fabry–Perot LDs [6.92, 6.93]. Gain bandwidth (GB) products of InGaAs APDs were also improved by optimizing the structure of the APD avalanche regions. High GB product APDs, with more than 50 GHz [6.94, 6.95] have been utilized in Gbit/s transmission experiments [6.96, 6.97]. In order to realize longer span transmissions, several experiments were performed in the 1.5 μm wavelength region (the lowest transmission loss region), while it was reported that the wavelength chirping of the DFB LD output strongly restricted the transmission length when 1.3 μm zero dispersion fibers were used [6.98].

The intrinsic small-signal modulation response of a laser diode is characterized by the relaxation frequency [6.99]

$$f_R \frac{1}{2\pi}\sqrt{(AP_o/\tau_p)} \tag{6.84}$$

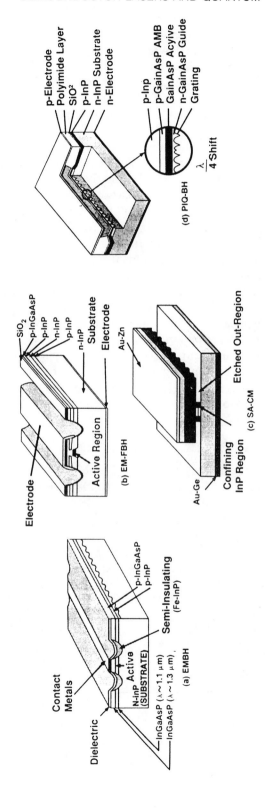

Fig. 6.18 High-speed DFB/DBR laser structures. (From [6.1] Lee and Zah (1989) *IEEE Commun. Mag.* **27**(10), 42–52. Reproduced by permission. ©1989 IEEE.)

where A is the differential optical gain, P_0 the average photon density in the gain region and τ_p the photon lifetime in the laser cavity. The direct modulation bandwidth $f_c(-3\mathrm{dB})$ may be markedly lower than f_r in the high-frequency range due to parasitic elements in the mounted laser chip and its connection to the driver circuit. High-speed response, single-longitudinal mode operation, and small wavelength chirping characteristics are important for Gbit/s range transmitter light sources. DFB-LD showed very attractive abilities on both high-speed response and narrow spectral linewidth.

Various high-speed DFB laser structures are shown in Fig. 6.18 [6.1]. For all devices shown above, the active region of the quaternary layer, approximately $0.15\,\mu\mathrm{m}$ thick by$1.5\,\mu\mathrm{m}$ wide, is sandwiched by higher-bandgap InP layers in the so-called buried heterostructure (BH). A low capacitance is achieved by using an etched mesa BH (EM-BH) or etched mesa flat BH (EM-FBH) stripe [6.100–6.102], a self-aligned constricted mesa (SACM) structure [6.103], or by applying semi-insulating Fe-doped InP or polyimide surrounding the active region in the lateral direction [6.32, 6.102, 6.104]. Lasing threshold currents are between 10 and 20 mA. Figure 6.19 [6.1] shows the best reported modulation bandwidths versus the root of power for high-speed DFB laser diodes. The slope is between 3.5 and 7 GHz/mW½. The best bandwidth is 17 GHz, obtained by optimizing the doping in the active layer and waveguide width, and detuning the wavelength [6.104]. Also shown in the same figure are preliminary results for InGaAsP/InP MQW lasers. Because of the large differential gain, MQW lasers are expected to exhibit higher bandwidths. Although a relaxation frequency of 30 GHz has been demonstrated inGaAs/AlGaAs MQW lasers [6.105], the value has been much smaller for InGaAsP/InP MQW lasers. However, in Fig. 6.19 the slope for MQW lasers is larger than that of the DH lasers.

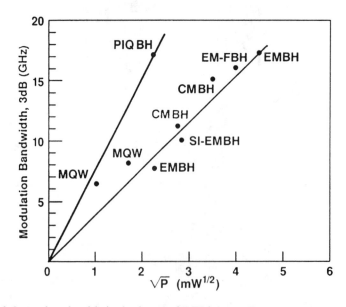

Fig. 6.19 Modulation bandwidth for high-speed DFB lasers. (From [6.1] Lee and Zah (1989) *IEEE Commun. Mag.* **27**(10), 42–52. Reproduced by permission. ©1989 IEEE.)

For an example, a cross-sectional view of a self-aligned constricted mesa (SA-CM) DFB laser is shown in Fig. 6.20 [6.106]. A self-aligning (SA) process [6.107] and a low-pressure MOCVD growth technique for 1.5 μm SA-CM DFB lasers were developed to control InP homojunction widths and the thickness of each layer. The lasing wavelength was 1.51 μm and the side-mode suppression ratio was greater than 40 dB. Preliminary results of aging tests at 50°C and 5 mW showed no serious degradations up to 3500 h of operation.

Direct high-frequency modulation of the SA-CM DFB lasers was investigated using the HP8510A network analyser and a mesa-type InGaAs p–i–n photodiode with 13 GHz bandwidth, which was measured by the optical heterodyne method [6.108]. A 3 dB bandwidth of 13 GHz was obtained at an output power of 12 mW. The bandwidth value was limited not by parasitics but by the response of the photodetector.

Pulse response of the SA-CM DFB lasers was measured using a pulse pattern generator and sampling oscilloscope. The wide opening of an NRZ eye pattern at 5 Gbit/s was measured directly from the photodetector output.

6.5 QUANTUM WELL DEVICES

Considerable attention has been paid to quantum well lasers, since they have superior characteristics, such as extremely low threshold current, less temperature dependence, narrow gain spectrum, etc, compared with conventional double heterostructure lasers.

6.5.1 Potential and density of states in a quantum well [6.115]

When the thickness L_z of a semimetal or semiconductor layer, for example, the active layer of a double heterojunction laser, is reduced to the order of a carrier de Broglie wavelength ($\lambda = h/p \approx L_z$), effects not typical of the bulk material, known as quantum size effects (QSE), occur [6.109–6.112]. That these effects could occur in a semiconductor, where the carrier mass can be small and $\lambda \approx L_z$, was appreciated over 35 years ago when the problem of determining the nature of carrier transport in the (thin)

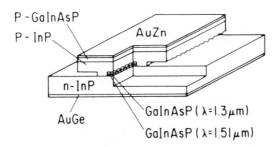

Fig. 6.20 Self-aligned constricted mesa DFB laser. (From [6.106] Hirayama *et al.* (1989) *IEEE. J. Quantum. Electron.* **QE-25**, 1320–1323. Reproduced by permission. ©1989 IEEE.)

channel formed by the inversion layer on a semiconductor surface was studied [6.113]. (This is now a well known problem in field-effect transistor research [6.114].) A simple case, which is generic to the problem of light emission in quantum well heterostructures, is that of a thin narrow-gap III–V semiconductor layer (a direct-gap active layer) sandwiched between two wider gap confining semiconductor layers, for example, a thin (epitaxial) layer of GaAs sandwiched between wider gap $Al_xGa_{1-x}As$ (Fig. 6.21) [6.115] or a thin layer of $In_{1-x}Ga_xP_{1-z}As_z$ sandwiched between InP [6.116]. When well thickness $L_z < \lambda$ (the carrier de Broglie wavelength), the size quantization occurs and a results in a series of discrete energy levels given by the bound state energies of a finite square well. A potential well exists in both the conduction band and the valence band giving rise to a series of bound state E_n for the electrons, E_{hhn} for heavy holes, and E_{lhn} for light holes. For the case shown in Fig. 6.21, which can readily be adapted (as below) into various forms of photopumped or p–n quantum well heterostructure lasers, it is obvious that electrons or holes in the GaAs layer are restricted (confined) to a finite potential well, and the usual band-to-band recombination process is then modified in a fundamental manner.

The energy spectrum of carriers in a thin layer (Fig. 6.21, GaAs) falling within the QSE regime ($\lambda = h/p \approx L_z < 500$ Å) is typically determined by assuming that the Hamiltonian (single-particle approximation) can be separated into a component (z) normal to the layer and into the usual (unconfined) Bloch function components (x,y) in the plane of the layer [6.117]. The resulting carrier energy eigenvalues are of the form

$$E(n,k_x,k_y) = E_n + (\hbar^2/2m_{n,p}^*)(k_x^2 + k_y^2) \tag{6.85}$$

where E_n is the nth confined particle (particle-in-a box) energy eigenvalue of the z component of the Hamiltonian, $m_{n,p}^*$ is the electron or hole effective mass, $\hbar$ is

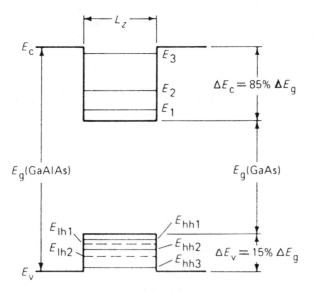

Fig. 6.21 Well potential of a QW heterostructure. (From [6.115] Holonyak *et al.* (1980) *IEEE. J. Quantum Electron.* **QE-16**, 170–185. Reproduced by permission. ©1980 IEEE.)

Planck's constant $(h/2\pi)$, and k_x and k_y are the usual (unconfined quasi-continuous x and y components of the crystal 'momentum.' The value of E_n in (6.85) is designated in Fig. 6.21 by E_1, E_2, E_3 for electrons, by E_{hh1}, E_{hh2}, E_{hh3} for heavy holes, and by E_{lh1}, E_{lh2} for light holes. As shown in Fig. 6.21, below the energy E_1 there are no allowed electron states. Then, as shown in Fig. 6.22 [6.115], a sub-band with a constant density of states (per unit area) [6.118].

$$g(E)\mathrm{d}E = \{m_n^*/(\pi\hbar^2)\}\mathrm{d}E \tag{6.86}$$

begins at E_1, followed by another sub-band (step) with a constant (upward, $E\uparrow$) block of states appearing at E_2, etc. This behaviour applies also to heavy holes and to light holes ($E\downarrow$) as is clear in Fig. 6.22. In Fig. 6.22, the half-parabolas that originate from the conduction band edge E_c and valence band edge E_v correspond to the densities of states of a bulk sample. The step-like densities of states are characteristic of a two-dimensional or quantum well structure. Interband recombination transitions ($\Delta n = 0$ selection rule) occur from a bound state in the conduction band E_n to a bound state in the valence band E_{hhn} or E_{lhn}. The energy of the transition (or recombination radiation) is given by

$$\hbar\omega = E_g(\mathrm{GaAs}) + E_n + E_{hhn} \ (\text{or } E_g + E_n + E_{lhn}).$$

Also, Fig. 6.22 shows that $n = 1$ (or 2, 3, . . .) electrons can recombine with $n = 1$ (or 2, 3, . . .; $\Delta n = 0$) heavy holes (n, $e\rightarrow$hh transition) or with $n = 1$ (or 2, 3, . . .; $\Delta n = 0$) light holes (n', $e\rightarrow$lh transition). The selection rule $\Delta n = 0$ has been assumed for electron–hole recombination but might be upset by flaws in an actual structure.

Figure 6.22 illustrates a very important and advantageous feature of quantum well heterostructures: recombination can proceed from a block of electrons, all in principle at nearly a fixed energy (say at E_1), with a similar block of holes also all at nearly a fixed energy (say at E_{hh1}). In a bulk sample, the recombining carriers are distributed in energy over parabolically varying densities of states, which are small at the band edges, and thus in principle the electrons and the holes cannot all be located at fixed energies nor then recombined in a narrow linewidth (or fixed wavelength).

Figure 6.23 [6.115] illustrates still another important feature of quantum well heterostructures: carriers injected in a sample at higher energy scatter downward in energy (thermalize) to ultimately a lesser density of states. In Fig. 6.23, carriers injected in a bulk sample (parabolic density of states) at higher energy scatter downward in energy to a lesser density of states, which becomes constraining. Within each sub-band of a quasi-two-dimensional structure, however, the step-like density of states is constant and downward electron scattering is not constrained, making possible phonon-assisted recombination and laser operation. Photon emission (LO phonon) I_{em} in this process is proportional in a bulk sample to the final density of states $\sqrt{(E - \hbar\omega_{LO})}$ which is to be compared with phonon absorption I_{abs} proceeding as $\sqrt{(E + \hbar\omega_{LO})}$ [6.119], giving

$$I_{abs} \propto \sqrt{E + \hbar\omega_{LO}} > \sqrt{E - \hbar\omega_{LO}} \propto I_{em}. \tag{6.87}$$

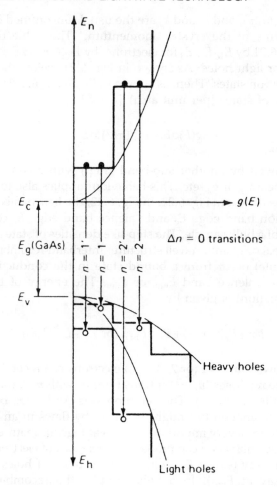

Fig. 6.22 Density of states for a QW heterostructure. (From [6.115] Holonyak *et al.* (1980) *IEEE J. Quantum Electron.* **QE-16**, 170–185. Reproduced by permission. ©1980 IEEE.)

Therefore, in a bulk sample, carrier thermalization by phonon generation has some tendency to be limited by the decreasing density of states (particularly at the band edge), whereas in a quasi-two-dimensional quantum well system, within a constant-density of-states region, no such limitation occurs, not to mention that, in the latter case, the basic electron–phonon interaction is enhanced and possibly even stimulated phonon emission occurs [6.131]. Thus, the carrier thermalization for the quantum well case [6.23] tends to be more efficient and, in fact, can transfer an electron (before recombination) to well below the confined-particle states, for example, below E_1. This is an important effect that can, in fact, lead to laser operation at energies $\hbar\omega < E_g$ instead of, as expected without phonon participation, at $\hbar\omega > E_g$. Thus, contrary to the behaviour of a bulk III–V sample, for which phonon participation in recombination is weak, in a quantum well hetero-structure, phonon involvement in electron–hole recombination is a major effect.

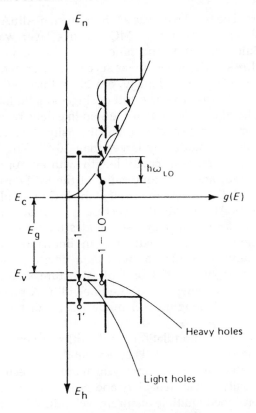

Fig. 6.23 Downward scattering in QW and in a bulk semiconductor. (From [6.115] Holonyak *et al.* (1980) *IEEE J. Quantum Electron.* **QE-16**, 170–185. Reproduced by permission. ©1980 IEEE.)

6.5.2 Quantum well lasers

Quantum well lasers have yielded several advantages over bulk lasers in the same wavelength range, such as low internal loss, high quantum efficiency, low threshold current operation, small linewidth enhancement factors, and large bandwidth.

RIDGE-TYPE STRIPE QUANTUM WELL LASERS

A narrow spectral linewidth of MBE-grown GaInAs/AlInAs MQW laser is discussed here as an example of ridge-type stripe MQW lasers.

   A schematic structure of a ridge-type GaInAs/AlInAs MQW laser is illustrated in Fig. 6.24 [6.120]. The laser wafers were grown by conventional metal-source molecular beam epitaxy (MBE) on Sn-doped (100)-InP substrates. The wafers were grown in a three-chamber RIBER-2300 system, and high purity metal were used as sources. Typical substrate temperature for the growth was about 560°C as measured by an IR pyrometer. The growth rate was 0.8–1 μm/h under an arsenic beam pressure of about 2×10^{-5} Torr. The first layer was a 0.8 μm thick n-AlInAs buffer layer doped with Si to $n = 1 \times 10^{18}$ cm^{-3}. An n-type MQW active layer was composed of ten GaInAs wells (thickness $L_Z = 80$Å) and nine AlInAs barriers (thickness $L_B = 20$ Å). A 2.5 μm-thick p-AlInAs cladding layer was successively grown on the MQW layer, doped with Be to $p = 1 \times 10^{18}$ cm^{-3}. Finally, a 0.15 μm thick p-GaInAs contact layer doped to $p = 2 \times 10^{18}$ cm^{-3} was grown. A rather single-longitudinal mode dominated spectrum was considered to indicate that the step-like density of states in the MQW active layer gave a narrow gain spectrum. The intensity of a main mode was higher than that of neighbouring modes by larger than 10 dB. The linewidth seems to become narrow as the increase of cavity length with an approximate relation of $\Delta\nu \propto 1/L$. These facts indicate good uniformity of the MBE-grown wafers.

These obtained linewidths are less than half of the InGaAsP/InP DH lasers [6.121]. Therefore, the GaInAs/AlInAs MQW structure is considered to be a promising material system as an active layer of laser diodes for future coherent lightwave communication systems. Further reduction in linewidth will be possible by optimizing a well number [6.122] as well as a device structure, since single-longitudinal mode lasers with DFB or DBR structure should be preferable to multimode lasers from the viewpoint of the linewidth. A precise characterization of the linewidth for single-longitudinal mode laser will be necessary in future works.

For high-speed use, even modulation bandwidths of beyond 10 or 20 GHz for different types of AlGaAs or GaInAsP BH lasers, but these results have been obtained for relatively complicated laser structures, which are not desirable with respect to processing reproducibility, device reliability and production costs. Alternatively, the use of properly dimensioned multiple quantum well (MQW) structures for the active laser region may improve a variety of characteristics, and is also an efficient means to increase the differential gain A and consequently the relaxation frequency f_R. A simple ridge waveguide laser was fabricated from a separate confinement hetrostructure (SCH)-MQW structure grown by MOVPE. Very high modulation bandwidth (up to 14 GHz) and relaxation frequencies (up to 36 GHz) have been achieved [6.123], despite the lower optical and carrier confinement in the RW in comparison to the BH structure.

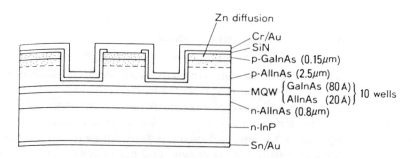

Fig. 6.24 Ridge-type stripe QW lasers. (From [6.120] Matsushima *et al.* (1989) *IEEE J. Quantum Electron.* **QE-25**, 1376–1380. Reproduced by permission. ©1989 IEEE.)

STRAINED-LAYER QUANTUM WELL LASERS

There have been several reports predicting that further improvements are possible by employing strained layer quantum wells [6.124–6.126], as the effective mass of holes in the heavy hole sub-band of quantum well layers can be significantly reduced in the plan of the wells by compressive strain. This should result in lower threshold currents and other improvements of the laser parameters.

The wavelength tuning properties of strained InGaAs quantum well lasers using an external grating for feedback has been reported recently [6.127]. Tunable laser oscillation has been observed over a range of 170 nm, between 840 and 1010 nm, under pulsed current excitation.

The increased reliability of InGaAs lasers has been attributed to both lattice hardening and strain accommodation [6.129] in strained layer material system. $In_xGa_{1-x}As$ quantum well in an AlGaAs/GaAs material system experience mismatch strain which increases with higher indium composition. As long as $In_xGa_{1-x}As$ layers are kept below some compositionally dependent critical thickness [6.130] the strain can be accommodated elastically. Although superior reliability has been routinely demonstrated for shorter wavelength InGaAs lasers, increased reliability at wavelengths beyond 1.0 μm is not a foregone conclusion in view of the high strains. A strained layer $In_{0.37}Ga_{0.63}As$ SQW lasers at 1.01 μm with over 20 000 h CW operation was reported recently [6.131].

The structure and properties of low threshold current semi-insulating-blocked planar buried heterostructure lasers with thin strained layer MQW active regions. The laser structure was grown by atmospheric pressure metal–organic vapour phase epitaxy (MOVPE) using t-butylarsine (TBA) [6.132]. It is similar to the separate confinement heterostructure laser. It employs four 25 Å thick $In_{0.77}Ga_{0.23}As$ QWs with 1.53% compression, with 90 Å thick InGaAsP (lattice matched, 1.25 μm composition) barriers.

Er-doped fiber optical amplifiers (EDFA) pumped by a 0.98 μm InGaAs/GaAs strained-layer quantum well laser are essentially attractive because they yield a lower noise figure, high gain coefficient, lower threshold and higher slope efficiency. EDFA require high pump power because the gain saturation level increases as the pump power increases. This means that high output power as well as high power conversion efficiency are desirable attributes for the InGaAs/GaAs strained-layer laser.

High power strained-layer InGaAs/GaAs graded-index separate confinement heterostructure (GRIN-SCH) single quantum well (SQW) lasers at an emission wavelength of 0.98 μm have been fabricated. A light power as high as 270 mW and a maximum front power conversion efficiency of 51.5% have been obtained for the antireflective and highly-reflective coated laser with a 9 μm wide ridge and a 600 μm long cavity [6.128].

SURFACE EMITTING QUANTUM WELL LASERS

Vertical-cavity, surface-emitting lasers have great potential because of their inherent two-dimensional nature and very small gain medium volumes which are essential to low threshold currents. Possible applications are optical switching/computing, photonic interconnection, high/low power laser sources, image processing, neural networks etc. There have been numerous reports on vertical-cavity, surface-emitting lasers using InGaAs/GaAs/AlAs or GaAs/AlGaAs structures [6.133–6.138].

A top–surface-emitting quantum well laser structure is a vertical p–i–n junction where the electrical current is injected through the bottom and top mirrors (see Fig. 6.25) [6.139, 6.140]. The whole structure is grown on a Si-doped n^+-GaAs substrate by molecular beam epitaxy. The Si-doped bottom mirror and has 27.5 pairs of AlAs/$Al_{0.15}Ga_{0.85}As$ quarter-wave stack designed at 850 nm. The undoped active region consists of four 100 Å thick GaAs quantum wells separated by three 70 Å thick $Al_{0.3}Ga_{0.7}As$ barriers. The Be-doped top mirror consists of 20 periods of

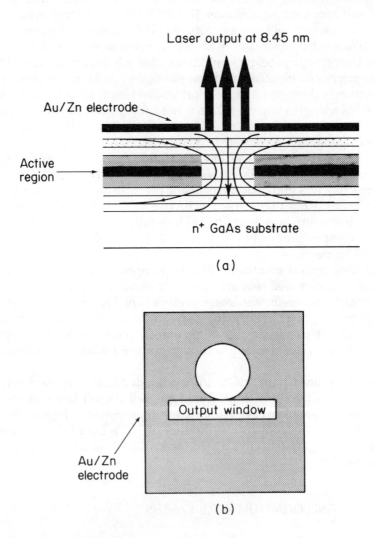

Laser output at 8.45 nm

Au/Zn electrode

Active region

n^+ GaAs substrate

(a)

Output window

Au/Zn electrode

(b)

Fig. 6.25 Top-surface-emitting quantum well laser structures. (From [6.139] Lee *et al.* (1990) *Electron. Lett.* **26**, 710–711. Reproduced by permission. ©1990 IEE.)

$AlAs/Al_{0.58}Ga_{0.42}As/Al_{0.16}Ga_{0.84}As/Al_{0.58}Ga_{0.42}As$. The two intermediate $Al_{0.58}Ga_{0.42}As$ layers are introduced to reduce p-mirror resistivities.

CW threshold currents are 3.5–8.0 mA for 10–30 μm diameter lasers, increasing in value with the linear dimension of the lasers. Threshold voltages are 3.7–4.2 V, depending on the implant condition. 10 μm diameter top-surface-emitting lasers show an initial CW slope efficiency of 1.2 mW/mA up to 0.6 mW CW output power, operating in the fundamental transverse mode (TEM_{00}) at room temperature. 78% differential quantum efficiency is obtained. The reasons for high quantum efficiencies are partly because of the high crystal quality and the optimized laser structure, but more extensive analysis will be published elsewhere. The top-surface-emitting lasers show stable CW characteristics with lasing wavelengths between 845 and 848 nm. Also measured spectral linewidth is 0.023 Å, using a scanning Fabry–Perot etalon of finesse 100.

6.5.3 Quantum well optical modulators

There is considerable interest in optical modulators for fiber transmission systems, as well as for interconnect and signal processing systems. Especially in high-bit-rate fiber transmission systems, the frequency chirp associated high-speed direct laser modulation causes the performance limitations which have led to intensive research on external optical modulators.

The quantum well structure has attracted much interest in optical modulators because of its large field-induced change in index n associated with the excitonic quantum-confined Stark effect (QCSE) being much larger than in bulk material.

PRINCIPLES OF QUANTUM WELL OPTICAL MODULATORS

A quantum well is a thin heterostructure of a low bandgap material (such as GaAs) sandwiched between two layers of a high bandgap material (such as AlGaAs). For most of the work in this field, the thickness of the low bandgap material has been roughly 100 Å, or about 30 atomic layers. Wells this thin manifest quantum mechanical effects in macroscopic ways, which is the reason they are termed 'quantum well'. The left-hand side of Fig. 6.26 [6.141] shows the band diagram of such a quantum well. Electrons and holes in the material tend to become localized in the region of the low bandgap material where their potential energies are lowest. Figure 6.26(a) shows the energy of the valence band maximum and conduction band minimum as a function of position for a quantum well. The energy levels of the ground states for electrons and holes are shown, along with wavefunction envelopes. Figure 6.26(b) shows these items when an electric field is applied perpendicular to the quantum well layers.

For such thin wells, the electrons and holes behave in a two-dimensional fashion. This alters their density-of-states functions, which sharpens the absorption edge [6.142]. In addition to this two-dimensional effect, the motion of electrons and holes in the quantum well is affected by their confinement. In accordance with the laws of quantum mechanics [6.143], the energies of the particles cannot be equal to the minimum energies of their respective wells. Because their position is localized in the region of the well, the Heisenberg uncertainty principle requires that the particles

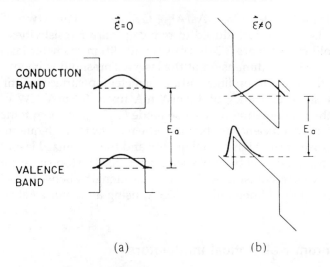

Fig. 6.26 Electroabsorption effect in MQW. (From [6.141] Wood (1988) *J. Lightwave Technol.* **LT-6**, 743–757. Reproduced by permission. ©1988 IEEE.)

have a non-zero momentum uncertainty. This translates into a 'zero-point energy', which displaces the ground state energy of a particle from the bottom of the well. These zero-point energies are very important, because it is their modulation with electric field that makes efficient MQW modulators possible.

Light is absorbed by the quantum well when a photon of sufficient energy excites an electron from the valence band into the conduction band, making an electron–hole pair. The lowest energy absorption occurs when the photon creates an electron–hole pair bound together in a state called an 'exciton' [6.144]. In an exciton, the electron and hole are bound in a hydrogen-like state due to their electrostatic attraction.

By looking at Fig. 6.26(a), one sees that the photon energy that creates this exciton is just the sum of the bulk bandgap of the semiconductor making up the well and the zero-point energies of the electron and hole, with a correction for the exciton binding energy. That is, the energy needed to create the exciton E_a is given by

$$E_a = E_g + E_{e1} + E_{h1} - B \tag{6.88}$$

where E_g is the bulk band gap of the well material, E_{e1} and E_{h1} are the zero-point energies of the electron and hole, respectively, and B is the binding energy of the exciton [6.157–6.159].

Because of their small binding energy, the creation of excitons is generally observed in bulk semiconductors only at low temperature. An important early observation was the discovery that, because of the two-dimensional nature of the electron–hole gas in quantum wells, the binding energy is increased and the excitons are more stable in MQWs than in bulk GaAs, which means that they are observable at room temperature in the absorption spectrum of MQWs. The narrow exciton energy peak sharpens the absorption edge, which improve the performance of MQW modulators. However, it is important to recognize that the presence of these room-temperature excitons is not central to the operation of MQW modulators.

When an electric field is applied perpendicular to the quantum well layers, the potential energy well that the electrons and holes see is changed from that shown in Fig. 6.26(a) to that shown in part (b). The primary effect is to modify the zero-point energies of the particles. This effect, dubbed the 'quantum confined Stark effect (QCSE)' [6.145–6.149], arises because the potential wells seen by the particles in the two halves of Fig. 6.26 are very different. The effect can be understood by viewing the electric field as a perturbation. As the field is increased, the electron and hole wavefunctions relax into the displaced wavefunctions shown in Fig. 6.26(b). These modified wavefunctions are lower in energy than the original wavefunctions, so the zero-point energies for the electron and hole are reduced.

As [6.88] shows, this change in zero-point energies results in a decrease in the effective bandgap of the quantum well. As the electric field applied to the wells increases, the absorption peaks shift toward lower energy. The magnitude of the shifts of these peaks with electric field is well understood theoretically, and can be fitted with no adjustable parameters [6.146–6.148].

Figure 6.27 [6.141] shows the schematic structure of the sample used to take this data. This device is called a 'transverse' modulator because the light propagates perpendicular to the MQW layers. In the MQW layers. In the centre of the sample is a region labelled 'MQW active' which contains a set of 50 GaAs quantum wells, each 95 Å thick, separated from each other by $Al_{0.32}Ga_{0.68}As$ barriers, 98 Å thick. In order to apply the electric field to the wells, they are fabricated in the undoped region of a p–i–n diode. The back-biased p–i–n diode makes it possible to apply fairly large fields to the wells without significant leakage current, which might cause sample heating. The electric field profile at two applied voltages is shown at bottom of Fig. 6.27. At 0 V applied, the field is close to zero for the MQWs, but at an applied back-bias voltage of 8 V, the field increases to approximately 7×10^4 V/cm. Note that there is some inhomogeneity in the electric field across the wells. This is due to a residual doping of approximately 5×10^{15} cm^{-3} in the i region of the diode. To minimize this inhomogeneity, it is important to minimize the residual doping [6.151].

SOME CHARACTERISTICS OF MQW OPTICAL MODULATORS

An important chacteristic of an intensity modulator is its on/off ratio. For a transverse device, the on/off ratio R is given by

$$R = \exp(\Delta\alpha L) \qquad (6.89)$$

where $\Delta\alpha$ is the maximum achievable change in the absorption coefficient, and L is the interaction length between the light and the electroabsorptive material. Clearly, to make a high quality short device, one would like $\Delta\alpha$ to be large as possible.

A more novel and useful way to increase the effective interaction length, and thus improve device performance, is the waveguide type modulators. Here, the direction of propagation of the light has been changed to lie in the plane of the MQW layers, and a slab optical waveguide has been added to the structure to confine the light along the length of the device. In this structure, there is no prescribed geometric

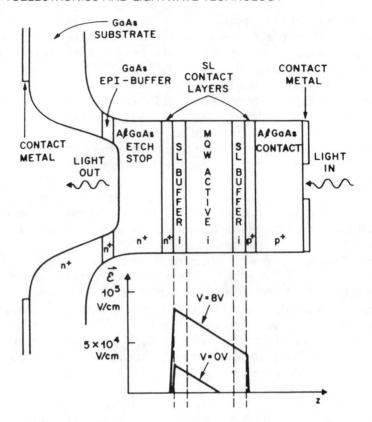

Fig. 6.27 Absorption experiments in MQW modulators (From [6.141] Wood (1988) *J. Lightwave Technol.* **LT-6**, 743–757. Reproduced by permission. ©1988 IEEE.)

relationship between the thickness of the MQW region and the optical interaction length, so the interaction length can be as long as necessary to produce the desired on/off ratio. In addition, while the previous geometry presented only one possible polarization (the electric field of the light parallel to the plane of MQW layers), the waveguide geometry presents both this and the orthogonal polarization (the electric field of the light perpendicular to the layers). These polarizations (termed TE and TM, respectively) have different absorption spectra, and thus present the opportunity for polarization-dependent devices.

Insertion loss is another key parameter for any modulator. For the purposes of the optical power at the output of the device to the power at its input, when the device is biased for maximum transmission. Both these powers are measured in air. For applications which require coupling to optical fibers, there would presumably be additional losses associated with fiber coupling.

For most of the applications envisioned for these MQW devices, speed of response is a critical parameter. Like any high-speed device, MQW modulators can be characterized in the time or frequency domain. In time-domain measurements, an electrical pulse with short rise and fall times is applied to the modulator, and the speed of the generated optical pulse is observed. In frequency-domain measurements, the device is sinusoidally modulated to perhaps a ±10% change in transmission. In principle, both time- and frequency-domain techniques provide accurate measurements of the speed of response of a device; both characterizations can be found in the literature.

A time-domain measurement of the waveguide modulator took place when an electrical impulse was applied to the device. This pulse had a full width at half maximum (FWHM) of 60 ps. The optical signal, which was detected by a high-speed photodetector, passed through a preamplifier, and recorded on a sampling oscilloscope. The observed FWHM of this pulse is 128 ps; however, after deconvolving the effect of the limited speed of response of the detection system, it was estimated [6.152] that the optical pulse itself had a FWHM of 97 ps.

HIGH-SPEED MULTIPLE QUANTUM WELL OPTICAL MODULATORS

High-speed optical waveguide modulators operating at a wavelength of 1.5 μm are important optoelectronic devices which can be used to extend the bandwidth–distance capacity of optical fiber transmission systems. These devices are also potential candidates for monolithic optoelectronic integrated circuits where lasers, modulators, detectors, and other electronic circuit components are incorporated. Multiple quantum well (MQW) electro-absorption waveguide modulators operating near 1.5 μm wavelength have been demonstrated using the InGaAs/InP [6.153], InGaAs/InAlAs [6.154], InGaAlAs/InAlAs [6.155], InGaAsP/InP [6.156], and GaSb/AlGaDb [6.157] material systems. The first four of these material systems were based on InP substrates and all of them exhibited relatively large electro-absorption effects due to the quantum confined Stark effect (QCSE).

As an example, a high-speed and high on/off ratio InGaAlAs/InAlAs MQW optical modulator is discussed here [6.158]. The MQW structures were fabricated on an InP substrate by molecular-beam epitaxy. The diodes consist of 0.40 μm thick undoped MQW layers sandwiched between a Si-doped InAlAs layer and a Be-doped InAlAs layer. This p–i–n structure gives rise to a strong electric field in the intrinsic region when reverse bias is applied. This leads to a change in absorption close to the fundamental absorption gap (quantum confined Stark effect). The MQW layers consist of 30 periods of an 86 Å InGaAl/As well and a 50 Å InAlAs barrier. The device is processed into a high-mesa-type geometry with a width of 4 μm and a length of 90–120 μm. To minimize stray capacitance, polyimide was spin-coated under the bonding pad over a 50 μm-square area. The capacitance of the individual diodes, including the submount stray capacitance (about 0.07 pF), is about 0.2 pF. Over 25 dB on/off ratio has been obtained at low-operating voltages of 5–8 V. The mean insertion loss is 18 dB. The reflection loss is 3 dB. The insertion loss is mainly coupling loss, which is estimated by extrapolating from zero length of the MQW guide and is found to be 14 dB. This large coupling loss is due to mode mismatch between fiber and guide. The residual absorption loss is also estimated from this figure and is determined to be 5–10 dB/mm.

The frequency response of the MQW modulator is reported in [6.158]. The modulator was driven by a 5 dBm RF signal at a dc bias of 4 V, where an on/off ratio of 10 dB was obtained. The measured electrical 3 dB bandwidth was over 22 GHz and agreed with calculations based on the RC time constant and stray inductance of the modulator using the parameters of 50 Ω 0.2 pF, and 0.3 nH. This indicates that the present modulator bandwidth is limited by the RC time constant and stray inductance.

REFERENCES

[6.1]   T. P. Lee and C. Zah (1989) "Wavelength-tunable and single-frequency semiconductor lasers for photonic communications networks" *IEEE Commun. Mag.* **27**(10), 42–52.

[6.2]   H. Kogelnik and C. V. Shank (1971) "Stimulated emission in a periodic structure" *Appl. Phys. Lett.* **18**, 152–154.

[6.3]   D. R. Scifres *et al.* [1974] "Distributed feedback single heterostructure laser" *Appl. Phys. Lett.* **25**, 203.

[6.4]   T. Okoshi (1986) "Ultimate performance of heterodyne/coherent optical fiber communications" *J. Lightwave Technol.* **LT-4**, 1556–1562.

[6.5]   R. E. Wagner *et al.* (1987) "Coherent lightwave systems for interoffice and loop-feeder applications" *J. Lightwave Technol.* **LT-5**, 429–438.

[6.6]   L. G. Kazovsky (1986) "Coherent optical receivers: performance analysis and laser linewidth requirements" *Opt. Eng.* **24**(4), 575–579.

[6.7]   A. L. Schawlow and C. H. Townes (1958) "Infrared and optical lasers" *Phys. Rev.* **112**, 1940–1949.

[6.8]   M. Lax (1966) "Quantum noise V: Phase noise in a homogeneously broadened maser" in *Physics and Quantum Electronics*, ed P. L. Kelley, B. Lax, and P. E. Tannenwald (New York: McGraw-Hill), 735–747.

[6.9]   R. D. Hempstead and M. Lax (1967) "Classical noise V1: Noise in self sustained oscillators near threshold" *Phys. Rev.* **161**, 350–366.

[6.10] H. Gerhardt *et al.* (1972) "Measurements of the laser line width due to quantum phase and quantum amplitude noise above and below threshold" *Z. Phys.* **253**, 113–126.

[6.11] M. W. Fleming and A. Mooradian (1981) "Fundamental line broadening of single-mode (GaAlAs) diode lasers" *Appl. Phys. Lett.* **38**, 511.

[6.12] C. H. Henry (1982) "Theory of the linewidth of semiconductor laser" *IEEE J. Quantum Electron.* **QE-18**, 259–264.

[6.13] C. H. Henry, *et al.* (1981) "Spectral dependence of the change in the refractive index due to carrier injection in GaAs lasers" *J. Appl. Phys.* **52**, 4457–4461.

[6.14] C. H. Harder *et al.* (1983) "Measurement of the linewidth enhancement factor α of semiconductor lasers" *Appl. Phys. Lett.* **42**, 328–330.

[6.15] I. D. Henning and J. V. Collins (1983) "Measurement of the semiconductor laser linewidth broadening factor" *Electron. Lett.* **19**, 927–929.

[6.16] R. Schimpe *et al.* (1984) "FM noise of index-guided GaAlAs diode lasers" *Electron. Lett.* **20**, 206–208.

[6.17] C. H. Henry *et al.* (1985) "Locking range and stability of injection locked 1.54-μm InGaAsP semiconductor lasers" *IEEE J. Quantum Electron.* **QE-21**, 1152–1156.

[6.18] B. Daino *et al.* (1983) "Phase noise and spectral line shape in semiconductor lasers" *IEEE J. Quantum Electron.* **QE-19**, 266–269.

[6.19] M. Takahashi *et al.* (1989) "Narrow spectral linewidth 1.5 μm GaInAsP/InP distributed Bragg Reflector (DBR) lasers" *IEEE J. Quantum Electron.* **QE-25**, 1280–1287.

[6.20] F. Koyama *et al.* (1983) "1.5–1.6 μm GaInAsP/InP dynamic-single-mode (DSM) lasers with distributed Bragg reflector" *IEEE J. Quantum Electron.* **QE-19**, 1042–1051.

[6.21] M. Yamada and Y. Suematsu (1979) "A condition of single longitudinal mode operation in injection lasers with index-guiding structure" *IEEE J. Quantum Electron.* **QE-15**, 743–749.

[6.22] Y. Suematsu and K. Furuya (1977) "Theoretical spontaneous emission factor of injection lasers" *Trans. Inst. Electron. Commun. Eng. Japan* **E60**(9), 467–477.

[6.23] K. Utaka *et al.* (1981) "Lasing characteristic of 1.5–1.6 μm GaInAsP/InP integrated twin-guide lasers with first order distributed Bragg reflectors" *IEEE J. Quantum Electron.* **QE-17**, 651–658.

[6.24] S. Akiba *et al.* (1987) "1.5-μm λ/4-shifted InGaAsP/InP DFB lasers" *J. Lightwave Technol.* **LT-5**, 1564–1573.

[6.25] Y. Suematsu *et al.* (1983) "Dynamic single-mode semiconductor lasers with a distributed reflector" *J. Lightwave Technol.* **LT-1**, 161–176.

[6.26] H. Kogelnik and C. V. Shank (1972) "Coupled-wave theory of distributed feedback lasers" *J. Appl. Phys.* **43**, 2327–2335.

[6.27] H. A. Haus and C. V. Shank (1976) "Antisymmetric taper of distributed feedback lasers" *IEEE J. Quantum Electron.* **QE-12**, 532–539.

[6.28] K. Utaka *et al.* (1984) "Analysis of quarter-wave-shifted DFB laser" *Electron. Lett.* **20**, 326–327.

[6.29] C. H. Henry (1985) "Performance of distributed feedback lasers designed to favor the energy gap mode" *IEEE J. Quantum Electron.* **QE-21**, 1913–1918.

[6.30] K. Utaka *et al.* (1986) "λ/4-shifted InGaAsP/InP DFB laser" *IEEE J. Quantum Electron.* **QE-22**, 1042–1051.

[6.31] S. Sasaki *et al.* (1988) "Effects of dynamic mode behavior and mode partition of DFB lasers on GB/s transmission systems" *Electron. Lett.* **24**, 26.

[6.32] C. E. Zah *et al.* (1989) "Performance of 1.5 μm quarter-wavelength-shifted DFB-SIBPH laser diodes with electron-beam defined and reactive ion etched gratings" *OFC'89*, Houston, TX, WB5.

[6.33] T. Ikegami *et al.* (1983) "Stress tests on 1.3-μm buried heterostructure laser diodes" *Electron. Lett.* **19**, 282–283.

[6.34] K. Iga and K. Kawabata (1985) "Active Bragg reflector and its application to semiconductor laser" *Japan. J. Appl. Phys.* **14**, 427–428.

[6.35] K. Sakai *et al.* (1982) "1.5-μm range InGaAsP/InP distributed feedback lasers" *IEEE J. Quantum Electron.* **QE-18**, 1272–1278.

[6.36] F. Koyama *et al.* (1983) "1.5–1.6 μm GaInAsP/InP dynamic-single-mode (DSM) lasers with Bragg reflectors" *IEEE J. Quantum Electron.* **QE-19**, 1042–1051.

[6.37] M. Asada *et al.* (1981) "The temperature dependence of the threshold current of GaInAsP/InP DH lasers" *IEEE J. Quantum Electron.* **QE-17**, 611–619

[6.38] H. Soda *et al.* (1986) "Mode analysis of λ/4-shifted GaInAsP/InP DFB lasers considering a refractive index change due to a spatial hole burning along a laser axis" in *Proc. IECE Japan Tech. Meet. on Opt. Quantum Electron.*, 49–56.

[6.39] K. Kobayashi and I. Mito (1985) "High light output-power single longitudinal mode semiconductor laser diodes" *J. Lightwave Technol.* **LT-3**, 1202–1210.

[6.40] G. P. Agrawal and N. K. Dutta (1985) "Polarization characteristics of distributed feedback semiconductor lasers" *Appl. Phys. Lett.* **46**, 213–215.

[6.41] T. Ikegami (1972) "Reflectivity of mode at facet and oscillation mode on double-heterostructure injection lasers" *IEEE J. Quantum Electron.* **QE-8**, 470–476.

[6.42] W. Streifer *et al.* (1975) "Coupling coefficients for distributed feedback single- and double-heterostructure diode lasers" *IEEE J. Quantum Electron.* **QE-11**, 867–873.

[6.43] W. Streifer *et al.* (1976) "TM-mode coupling coefficients in guided-wave distributed feedback lasers" *IEEE J. Quantum Electron.* **QE-12**, 74–78.

[6.44] K. Utaka *et al.* (1984) "λ/4-shifted InGaAsP/InP DFB lasers by simultaneous holographic exposure of positive and negative photoresists" *Electron. Lett.* **20**, 1008–1010.

[6.45] I. M. Ito *et al.* (1983) "InGaAsP double-channel-planar-buried- heterostructure laser diode (DC-PBH) with effective current confinement" *J. Lightwave Technol.* **LT-1**, 195–202.

[6.46] K. Utaka *et al.* (1984) "Effect of mirror facets on lasing characteristics of distributed feedback InGaAsP/InP laser diodes at 1.5 μm range" *IEEE J. Quantum Electron.* **QE-20**, 236–245.

[6.47] Y. Kotaki *et al.* (1990) "Narrow linewidth and wavelength tunable multiple quantum well λ/4 shifted distributed feedback laser" *Presented at OFC, San Francisco, CA, 1990* paper THE3.

[6.48] H. Yamazaki *et al.* (1990) "250 kHz linewidth operation in long cavity 1.5 µm multiple quantum well DFB-LD's with reduced linewidth enhancement factor" *Presented at OFC, San Francisco, CA, 1990* paper PD33–1.

[6.49] H. Soda *et al.* (1987) "Stability in single longitudinal mode operation in GaInAs/InP phase adjusted DFB lasers" *IEEE J. Quantum Electron.* **QE-23**, 804–814.

[6.50] M. Okai *et al.* (1990) "Corrugation-pitch-modulated MQW-DFB laser with narrow spectral linewidth (170 kHz)" *IEEE Photon. Technol. Lett.* **PTL-2**, 529–530.

[6.51] M. Okai *et al.* (1989) "Corrugation-pitch-modulated phase-shifted DFB laser" *IEEE Photon. Technol. Lett.* **1**, 200–201.

[6.52] M. Okai *et al.* (1989) "Novel method to fabricate corrugation for λ/4-shifted distributed feedback laser using a grating photomask" *Appl. Phys. Lett.* **55**, 415–417.

[6.53] T. Okoshi *et al.* (1980) "Novel method for high resolution measurement of laser output spectrum" *Electron. Lett.* **16**, 630–631.

[6.54] M. Kuznetsov (1988) "Theory of wavelength tuning in two-segment distributed feedback lasers" *IEEE J. Quantum Electron.* **QE-24**, 1837–1844.

[6.55] Y. Suematsu *et al.* (1985) "Dynamic single-mode semiconductor lasers with a distributed reflector" in *Semiconductors and Semimetals* **22**, part B, ed W. T. Tsang (Orlando, FL: Academic).

[6.56] W. T. Tsang (1985) "The cleaved-coupled-cavity (C^3) laser" in *Semiconductors and Semimetals* **22**, part B, ed W. T. Tsang (Orlando, FL: Academic).

[6.57] Y. Yoshikuni *et al.* (1986) "Broad wavelength tuning under single-mode oscillation with a multi-electrode distributed feedback laser" *Electron. Lett.* **22**, 1153–1154.

[6.58] S. Murata *et al.* (1987) "Over 720 GHz (5.8 nm) frequency tuning by a 1.5 µm DBR laser with phase and Bragg wavelength control regions" *Electron. Lett.* **23**, 403–405.

[6.59] T. L. Koch *et al.* (1988) "Continuously-tunable 1.5µm multiple-quantum-well GaInAs/GaInAsP distributed-Bragg-reflector lasers" *Electron. Lett.* **24**, 1431.

[6.60] L. A. Coldren and S. W. Corzine (1987) "Continuously-tunable single-frequency semiconductor lasers" *IEEE J. Quantum Electron.* **QE-23**, 903–908.

[6.61] Y. Tohmori *et al.* (1983) "Wavelength tuning of GaInAsP/InP integrated laser with butt joint built-in distributed Bragg reflector" *Electron. Lett.* **19**, 656–657.

[6.62] M. Kitamura *et al.* (1983) "High-power single-longitudinal-mode operation of 1.3 µm DFB-DC-PBH LD" *Electron. Lett.* **19**, 840–841.

[6.63] H. A. Haus and C. V. Shank (1976) "Antisymmetric taper of distributed feedback lasers" *IEEE J. Quantum Electron.* **QE-12**, 532–539.

[6.64] N. Eda *et al.* (1985) "Axial mode selectivity in active distributed-reflector for dynamic-single-mode lasers" *J. Lightwave Technol.* **LT-3**, 400–407.

[6.65] T. L. Koch and L. A. Coldren (1985) "Optimum coupling junction and cavity lengths for coupled-cavity semiconductor lasers" *J. Appl. Phys.* **57**, 740–754.

[6.66] R. Wyatt and W. J. Delvin (1983) "10 kHz linewidth 1.5 µm InGaAsP external cavity laser with 55 nm tuning range" *Electron. Lett.* **19**, 110–112.

[6.67] F. Heismann *et al.* (1987) "Narrow-linewidth, electro-optically tunable InGaAsP-Ti:LiNbO$_3$ extended cavity laser" *Appl. Phys. Lett.* **51**, 164–164.

[6.68] G. Coguin *et al.* (1988) "Single- and multiple- wavelength operation of acousto-optically tuned semiconductor lasers at 1.3 µm" *Proc. 11th IEEE Int Semiconductor Laser Conf., Boston, MA, 1988*, 130–131.

[6.69] M. W. Fleming and A. Mooradian (1981) "Fundamental line broadening of single-mode GaAlAs diode lasers" *Appl. Phys. Lett.* **38**, 511–513.

[6.70] D. Mehuys *et al.* (1989) "Optimized Fabry–Perot AlGaAs quantum-well lasers tunable over 105 nm" *Electron. Lett.* **25**, 143–145.

[6.71] M. Okai *et al.* (1989) "Wide-Range continuous tunable double-sectioned distributed feedback lasers" *15th Euro. Cong. on Opt. Commun., Gothenburg, Sweden, 1989*.

[6.72] H. A. Haus (1984) *Waves and Fields in Optoelectronics* (Englewood Cliffs, NJ: Prentice-Hall).

[6.73] S. L. McCall and P. M. Platzman (1985) "An optimized π/2 distributed feedback laser" *IEEE J. Quantum Electron.* **QE-21**, 1899–1904.

[6.74] C. H. Henry (1982) "Theory of the linewidth of semiconductor lasers" *IEEE J. Quantum Electron.* **QE-18**, 259–264.

[6.75] L. D. Westbrook (1985) "Dispersion of linewidth-broadening factor in 1.5 μm laser diode" *Electron. Lett.* **21**, 1018–1019.

[6.76] L. D. Westbrook (1986) "Measurements of dg/dN and dn/dN and their dependence of photon energy in λ = 1.5 μm InGaAsP laser diodes" *Proc. IEE* **133**(J), 135–142.

[6.77] M. Yamaguchi *et al.* (1985) "Wide range wavelength tuning in 1.3 μm DBR-DC-PBH LDs by current injection into the DBR region" *Electron. Lett.* **21**, 63–65.

[6.78] L. D. Westbrook *et al.* (1985) "Monolithic 1.5 μm hybrid DFB/DBR lasers with 5 nm tuning range" *Electron. Lett.* **20**, 957–959.

[6.79] S. Murata *et al.* (1987) "Frequency modulation and spectral charactersitics for a 1.5 μ phase-tunable DFB laser" *Electron. Lett.* **23**, 12–14.

[6.80] Y. Tohmori *et al.* (1983) "Wavelength tuning of GaInAsP/InP integrated laser with butt-jointed built-in distributed Bragg reflector" *Electron. Lett.* **19**, 656–657.

[6.81] Y. Kotaki *et al.* (1987) "1.55 μm wavelength tunable FBH-DBR laser" *Electron. Lett.* **23**, 325–327.

[6.82] Y. Kotaki *et al.* (1988) "Tunable DBR laser with wide tuning range" *Electron. Lett.* **24**, 503–505.

[6.83] S. Murata *et al.* (1988) "Tuning ranges for 1.5 μm wavelength tunable lasers" *Electron. Lett.* **24**, 577–579.

[6.84] K. Iwashita and T. Matsumoto (1987) "Modulation and detection characteristics of optical continuous phase FSK transmission system" *J. Lightwave Technol.* **LT-5**, 452–460.

[6.85] S. Murata *et al.* (1987) "Over 720 GHz (5.8 nm) frequency tuning by a 1.5 μm DBR laser with phase and Bragg wavelength control regions" *Electron. Lett.* **23**, 403–405.

[6.86] K. Kobayashi and I. Mito (1988) "Single frequency and tunable laser diodes" *IEEE J. Lightwave Technol.* **6**, 1623–1633.

[6.87] Y. Kotaki *et al.* (1988) "Tunable DBR laser with wide tuning range" *Electron. Lett.* **24**, 503–505.

[6.88] T. L. Koch *et al.* (1988) "Continuously tunable 1.5 μm multiple-quantum-well GaInAs/GaInAsP distributed-Bragg-Reflector laser" *Electron. Lett.* **24**, 23.

[6.89] Y. Kotaki and H. Ishikawa (1989) "Spectral characteristics of a three-section wavelength-tunable DBR laser" *IEEE J. Quantum Electron.* **QE-25**, 1340–1345.

[6.90] W. Streifer *et al.* (1975) "Effect of external reflectors on longitudinal modes of distributed feedback lasers" *IEEE J. Quantum Electron.* **QE-11**, 154–161.

[6.91] B. Tromborg *et al.* (1987) "Transmission line description of optical feedback and injection locking for Fabry–Perot and DFB lasers" *IEEE J. Quantum Electron.* **QE-23**, 1875–1889.

[6.92] J. E. Bowers *et al.* (1986) "High-speed InGaAsP constricted-mesa lasers" *IEEE J. Quantum Electron.* **QE-22**, 833–843.

[6.93] R. Olshansky *et al.* (1986) "Frequency response of an InGaAsP vapor phase regrown buried heterostructure laser with 18 GHz bandwidth" *Appl. Phys. Lett.* **49**, 128–130.

[6.94] J. C. Cambell *et al.* (1985) "A high-speed InP/InGaAsP/InGaAs avalanche photodiode exhibiting a gain-bandwidth product of 60 GHz" *ECOC Venice, Italy, 1985*, post-deadline paper.

[6.95] K. Taguchi *et al.* (1986) "High-sensitivity planar InGaAs avalanche photodiodes with preferential lateral extended guard ring for Gbit/s-range communication" *OFC, Atlanta, GA, 1986*, paper WC2.

[6.96] A. H. Gnauck et al., (1985) "4 Gbit/s transmission over 103 km of optical fiber using a novel electronic multiplexer/demultiplexer" *J. Lightwave Technol.* **LT-3**, 1032–1035.

[6.97] S. Fujita *et al.* (1986) "4-Gbit/s long-span transmission experiments employing high-speed DFB-LD's and InGaAs APD's" *Proc. ECOC, Barcelona, Spain, 1986*, 507.

[6.98] S. Kobayashi *et al.* (1982) "Direct frequency modulation in AlGaAs semiconductor lasers" *IEEE J. Quantum Electron* **QE-18**, 582–595.

[6.99] K. Y. Lau and A. Yariv (1985) "Ultra-high speed semiconductor lasers" *IEEE J. Quantum Electron.* **QE-21**, 21–138.

[6.100] H. Imai and T. Kaneda (1988) "High-speed distributed feedback lasers and InGaAs avalanche photodiodes" *IEEE J. Lightwave Technol.* **6**, 1634–1642.

[6.101] K. Kamite *et al.* (1989) "14 GHz single-mode picosecond optical pulse train generation in Zn-doped distributed-feedback lasers" *Appl. Phys. Lett.* **54**(3), 208–209.

[6.102] T. Cella *et al.* (1988) "High-speed 1.3 micron InGaAsP distributed feedback lasers" *Proc. 11th IEEE Int. Semiconductor Laser Conf. Boston, MA, 1988*, 50–51.

[6.103] Y. Hirayama *et al.* (1988) "High-speed (13 GHz) 1.5 μm self- aligned constricted mesa DFB lasers grown entirely by MOCVD" *Proc., 11th IEEE Int Semiconductor Laser Conf, Boston, MA, 1988*, 46–47.

[6.104] K. Uomi *et al.* (1989) "Ultra-High-Speed 1.55 μm/4 shifted DFB lasers with bandwidth of 17 GHz" *Electron. Lett.* **25**, 668–669.

[6.105] K. Uomi *et al.* (1987) "Ultra-High relaxation oscillation frequency (up to 30 GHz of highly P-doped GaAs/GaAlAs multiple Quantum well lasers" *Appl. Phys. Lett.* **51**, 78–80.

[6.106] Y. Hirayama *et al.* (1989) "High-speed 1.5 μm self-aligned constricted mesa DFB lasers grown by MOCVD" *IEEE J. Quantum Electron.* **25**, 1320–1323.

[6.107] Y. Hirayama *et al.* (1988) "High-speed 1.5 μm self-aligned constricted mesa lasers grown entirely by MOCVD" *Electron. Lett.* **24**, 452–454.

[6.108] L. Piccari and P. Spano (1982) "New method for measuring ultrawide frequency response of optical detectors" *Electron. Lett.* **18**, 116–118.

[6.109] V. N. Lutskii (1970) "Quantum size effect-present state and perspectives of experimental investigations" *Phys. Status Solidi* **1**, 199–220.

[6.110] M. I. Elinson *et al.* (1972) "Quantum size effect and perspectives of its practical application" *Thin Solid Films* **12**, 383–397.

[6.111] A. Ya. Shik (1975) "Superlattices-periodic semiconductor structures" *Sov. Phys. Semicond.* **8**, 1195–1209.

[6.112] R. Dingle (1975) "Confined carrier quantum states in ultrathin semiconductor heterostructures" in *Festkörper Probleme XV (Advances in Solid State Physics)* ed H. J. Quisser (New York: Pergamon), 21–28.

[6.113] J. R. Schrieffer (1955) "Effective carrier mobility in surface-space charge layers" *Phys. Rev.* **97**, 641–646.

[6.114] F. Stern and W. E. Howard (1967) "Properties of semiconductor surface inversion layers in the electric quantum limit" *Phys. Rev.* **163**, 816–835.

[6.115] N. Holonyak *et al.* (1980) "Quantum-well heterostructure lasers" *IEEE J. Quantum Electron.* **QE-16**, 170–185.

[6.116] E. A. Rezek *et al.* (1978) "Single and multiple thin layer ($L_z < 400$ Å) $In_{1-x}Ga_xP_{1-z}As_z$/InP heterostructure light emitters and lasers ($\lambda \approx 1.1 \mu$m, 77 K)" *J. Appl. Phys.* **49**, 69–74.

[6.117] R. Fivaz (1967) "Theory of layer structures" *J. Phys. Chem. Solids* **28**, 839–845.

[6.118] R. J. Elliott and A. F. Gibson (1974) *Solid State Physics and Its Applications* (New York: Harper and Row) 44.

[6.119] N. Holonyak *et al.* (1980) "Phonon-assisted recombination and stimulated emission in quantum well $Al_xGa_{1-x}As$/GaAs heterostructures" *J. Appl. Phys.* **51**.

[6.120] Y. Matsushima *et al.* (1989) "Narrow spectral linewidth of MBE- grown GaInAs/AlInAs MQW lasers in the 1.55 μm range" *IEEE J. Quantum Electron.* **QE-25**, 1376–1380.

[6.121] T. P. Lee (1985) "Linewidth of single-frequency semiconductor lasers for coherent lightwave communications" in *Tech. Dig. IOOC/ECOC'85, Venezia, Italy* **2**, 189–196.

[6.122] Y. Arakawa and A. Yariv (1985) "Theory of gain, modulation response, and spectral linewidth in AlGaAs quantum well lasers" *IEEE. J. Quantum Electron.* **QE-21**, 1674.

[6.123] H. D. Wolf *et al.* (1989) "High-speed AlGaAs/GaAs multiple quantum well ridge waveguide lasers" *Electron. Lett.* **25**, 1245–1246.

[6.124] A. R. Adams (1986) "Band structure engineering for low threshold high efficiency semiconductor lasers" *Electron. Lett.* **22**, 249–250.

[6.125] E. Yablonvitch and E. O. Kane (1988) "Band structure engineering of semiconductor lasers for optical communications" *J. Lightwave Technol.* **6**, 1292–1299.

[6.126] T. Ohtoshi and N. Chinone (1989) "Linewidth enhancement factor in strained QW lasers" *Photonic Tech. Lett.* **11**, 117–119.

[6.127] L. E. Eng (1990) "Broadband tuning (170 nm) of InGaAs quantum well lasers" *Electron. Lett.* **26**, 1675–1677.

[6.128] T. Takeshita *et al.* (1990) "High-power operation in 0.98-μm strained-layer InGaAs/GaAs single-quantum-well ridge waveguide lasers" *IEEE Photon. Technol. Lett.* **2**(12), 849–851.

[6.129] R. M. Kolbas *et al.* (1988) "Strained-Layer InGaAs-AlGaAs photopumped and current injection lasers" *IEEE J. Quantum Electron.* **QE-24**, 1605–1613.

[6.130] K. J. Beernink *et al.* (1989) "Characterization of InGaAs-GaAS strained layer lasers with quantum wells near the critical thickness" *Appl. Phys. Lett.* **55**, 2167–2169.

[6.131] S. L. Yellen (1990) "20000 h InGaAs quantum well lasers" *Electron. Lett.* **26**, 2083–2084.

[6.132] U. Koren *et al.* (1990) "Low threshold highly efficient strained quantum well lasers at 1.5 micrometre wavelength" *Electron. Lett.* **26**, 465–467.

[6.133] K. Iga *et al.* (1987) "Microcavity GaAlAs/GaAs surface emitting laser with $1_{th} = 6$ mA" *Electron. Lett.* **23**, 134–136.

[6.134] A. Ibaraki *et al.* (1989) "Buried heterostructure GaAs/GaAlAs distributed Bragg reflector surface emitting laser with very low threshold (5.2 mA) under room temperature CW conditions" *Japan J. Appl. Phys.* **28**, L667–668.

[6.135] J. L. Jewell *et al.* (1989) "Low-threshold vertical-cavity surface emitting microlasers" *Electron. Lett.* **25**, 1123–1124.

[6.136] D. Botez *et al.* (1989) "Low-threshold-current-density vertical cavity surface emitting AlGaAs/GaAs lasers" *IEEE Photon. Technol. Lett.* **8**, 205–208.

[6.137] J. L. Jewell *et al.* (1990) "Surface emitting microlasers for photonic switching and interchip connections" *Opt. Eng.* **29**, 210–214.

[6.138] Y. H. Lee *et al.* (1990) "Effects of etch depth and ion implantation on surface emitting microlasers" *Electron. Lett.* **26**, 225–227.

[6.139] Y. H. Lee *et al.* (1990) "Top-surface-emitting GaAs four-quantum-well lasers emitting at 0.85 μm" *Electron. Lett.* **26**, 710–711.

[6.140] Y. H. Lee *et al.* (1990) "High efficiency (1.2 mW/mA) top-surface-emitting GaAs quantum well lasers" *Electron Lett.* **26**, 1308–1310.

[6.141] T. H. Wood (1988) "Multiple quantum well (MQW) waveguide modulators" *J. Lightwave Technol.* **LT-6**, 743–757.

[6.142] R. Dingle (1975) "Confined carrier quantum states in ultrathin semiconductor heterostructures" in *Feskorperprobleme: (Advances in Solid State Physics)* ed H. J. Quesser, **21** (New York: Pergamon), 15.

[6.143] M. E. Robert (1961) *Fundamentals of Modern Physics* (New York: Wiley), 239–256.

[6.144] R. C. Miller and D. A. Kleinman (1985) "Excitons in GaAs quantum wells" *J. Luminescence* **30**, 520.

[6.145] T. H. Wood *et al.* (1984) "High-speed optical modulation with GaAs/GaAlAs quantum wells in a p-i-n diode structure" *Appl. Phys. Lett.* **44**, 16.

[6.146] D. A. B. Miller *et al.* (1984) "Band-edge electroabsorption in quantum well structure: the quantum-confined Stark effect" *Phys. Rev. Lett.* **53**, 2173.

[6.147] D. A. B. Miller *et al.* (1985) "Electric field dependence of optical absorption near the bandgap of quantum-well structures" *Phys. Rev. B.* **32**, 1043.

[6.148] C. Alibert *et al.* (1985) "Measurement of electric-field-induced energy-level shifts in GaAs single-quantum-wells using electroreflectance" *Solid State Commun.* **53**, 457.

[6.149] D. A. B. Miller *et al.* (1986) "Relation between electroabsorption in bulk semiconductors and in quantum wells: The quantum-confined Franz–Keldysh effect" *Phys. Rev. B.* **33**, 6976.

[6.150] G. Bastard (1985) "Electronic energy levels in semiconductor quantum wells and superlattices" *Superlattices and Microstructures* **1**, 265.

[6.151] D. J. Newson and A. Kurobe (1987) "Effect of residual doping on optimum structure of multiquantum-well optical modulators" *Electron. Lett.* **23**, 439.

[6.152] T. H. Wood *et al.* (1985) "100-ps waveguide multiple quantum well (MQW) optical modulator with 10:1 on/off ratio" *Electron. Lett.* **21**, 693.

[6.153] U. Koren *et al.* (1987) "Low loss InGaAs/InP multiple quantum well optical electroabsorption waveguide modulator" *Appl. Phys. Lett.* **51**, 1132–1134.

[6.154] K. Wakita *et al.* (1986) "Long-wavelength waveguide multiple-quantum-well (MQW) optical modulator with 30:1 on/off ratio" *Electron. Lett.* **22**, 907–908.

[6.155] I. Kotaka *et al.* (1989) "High-speed InGaAlAs/InAlAs multiple quantum well optical modulators with bandwidths in excess of 20 GHz at 1.55 μm" *IEEE photon. Technol. Lett.* **1**, 100–102.

[6.156] H. Temkin *et al.* (1987) "InGaAsP/InP quantum well modulators grown by gas source molecular beam epitaxy" *Appl. Phys. Lett.* **50**, 1776–1778.

[6.157] T. H. Wood *et al.* (1987) "High-speed waveguide optical modulator made from GaSb/AlGaSb multiple quantum wells (MQW's)" *Electron. Lett.* **23**, 540–542.

[6.158] K. Wakita *et al.* (1990) "High-speed InGaAlAs/InAlAs multiple quantum well optical modulators" *J. Lightwave Technol.* **8**, 1027–1031.

7

GUIDED-WAVE
OPTICAL DEVICES

7.1 THEORY OF GUIDED-WAVE OPTICS

The theory of mode coupling was first developed by J. R. Pierce [7.1] to analyse coupling of electron-beam wave and waves on electromagnetic structures in electron-beam tubes. Later, coupled-mode theory was developed for the analysis of optical waveguides [7.2–7.5]. Coupled-mode theory is approximate, with approximations that are not always self-evident. For this reason, it was worth looking for a formal theoretical framework from which coupled mode theory could emerge in an unequivocal way.

Papers published by Hardy and Streifer [7.6–7.8] give propagation constants and coupling coefficients which are more accurate than the previously published. The results for slab waveguides computed from this H-S theory indeed are better when compared with the exact theory than those in [7.2–7.5]. One of the H-S theories is to show how power non-orthogonality modifies the coupling behaviour when the coupling is strong. Some results in [7.9], derived from the variational principle for the propagation constant of a waveguide–wave solution using a superposition of the uncoupled modes as a trial field, obtained similar equations to [7.6–7.8].

7.1.1 Coupled mode theory of parallel waveguides

Coupling between parallel waveguides has long been a subject of interest in the fields of integrated optics [7.2–7.5] and semiconductor diode lasers [7.10–7.13]. Although coupled mode theory utilizes the interaction of the individual waveguide modes, and is thereby inherently limited, it is possible to develop the theory in a self-consistent manner by retaining the radiation modes in the derivation and neglecting them only after their influence is determined. In this way H-S theory obtained unambiguous formulae for propagation constants and coupling coefficients which are more accurate than those previously published. The increased accuracy is particularly significant for coupling between non-identical waveguides and H-S theory appears to extend coupled mode theory to more strongly coupled waveguide geometries than in previous theories; it becomes inaccurate in the very strong coupling limit.

THE COUPLED MODE EQUATIONS OF PARALLEL WAVEGUIDES [7.6]

The two parallel planar optical waveguides condition has been discussed already [7.6]. We select it as an example of coupled mode analysis. Fig. 7.1 [7.6] illustrates the coupling between waveguides in which the guided mode of waveguide 'b' couples efficiently into waveguide 'a', but not conversely, i.e. $|k_{ab}|^2 \gg |k_{ba}|^2$. Power is conserved in this process.

The dimensions and refractive indexes are such that only one TE mode propagates in each guide individually. With $\lambda = 0.8\,\mu m$, a numerical solution for the two array modes of the two-waveguide system yields propagation constants $\sigma_1 = 27.201\,\mu m^{-1}$ and $\sigma_2 = 26.931\,\mu m^{-1}$, which compare favourably with the H-S results, i.e. $\sigma_1 = 27.200\,\mu m^{-1}$ and $\sigma_2 = 26.926\,\mu m^{-1}$. According to the previous theory $\sigma_1 = 27.210\,\mu m^{-1}$ and $\sigma_2 = 26.953\,\mu m^{-1}$. Thus the beat length between the two array modes, i.e. $2\pi/(\sigma_1 - \sigma_2)$, as calculated by H-S theory, differs from the exact value by $\approx 1.5\%$, whereas the prior results are inaccurate by $\approx 5\%$.

Consider two parallel dielectric waveguides denoted 'a' and 'b', which may differ in shape, size, and/or refractive index [7.6]. If the refractive indexes are real, then clearly the waveguide index must exceed that of the surroundings. The modal vectorial fields of the individual guides are designated by superscripts (a) and (b), respectively. Now following [7.14], we have an electromagnetic field {$\mathbf{E}(x,y,z)$, $\mathbf{H}(x,y,z)$} which satisfies Maxwell's equations plus the boundary conditions of the entire structure. Such fields include the guided modes of the entire waveguide structure.

For discussion, each individual waveguide supports only one guided mode is assumed so that the transverse field may be written

$$\mathbf{E}_t(x,y,z) = U(z)\mathbf{E}_t^{(a)}(x,y) + V(z)\mathbf{E}_t^{(b)}(x,y) + \mathbf{E}_r(x,y,z) \qquad (7.1a)$$

and

$$\mathbf{H}_t(x,y,z) = U(z)\mathbf{H}_t^{(a)}(x,y) + V(z)\mathbf{H}_t^{(b)}(x,y) + \mathbf{H}_r(x,y,z) \qquad (7.1b)$$

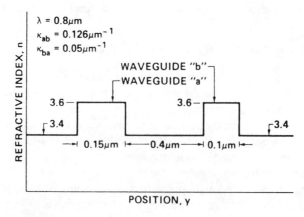

Fig. 7.1 The coupling between two parallel waveguides. (From [7.6] Hardy and Streifer (1985) *J. Lightwave Technol.* **LT-3**, 1135–1146. Reproduced by permission. ©1985 IEEE.)

where $U(z)$ and $V(z)$ express the z-dependence of the individual guided modes whose transverse fields are $\{\mathbf{E}_t^{(a)}, \mathbf{H}_t^{(a)}\}$, $\{\mathbf{E}_t^{(b)}, \mathbf{h}_t^{(b)}\}$, and $\mathbf{E}_r$ and $\mathbf{H}_r$ are the residual fields. Since the system modes are not exactly linear combinations of the individual waveguide modes, E_r and H_r are required for these to be equations. In this derivation the residue field was assumed to be orthogonal to both $\mathbf{E}_t^{(a)}$ and $\mathbf{E}_t^{(b)}$, and that requirement leads to the more accurate expressions derived herein. The residue field may or may not be completely guided by the system of two waveguides depending on whether or not $\{\mathbf{E}_t, \mathbf{H}_t\}$ is a guided mode of the system. In both cases the residue field is expressed formally in terms of the radiation modes of either waveguide, but it is necessary to calculate it explicitly.

The objective here is to derive differential equations relating $U(z)$ and $V(z)$. The derivation is quite mathematical and tedious; it appears in its entirety in [7.6, Appendix B]. Following substantial manipulation, one could obtain

$$\frac{dU}{dz} = i\gamma^{(a)}U + i\varkappa_{ab}V \tag{7.2a}$$

$$\frac{dV}{dz} = i\gamma^{(b)}V + i\varkappa_{ba}U \tag{7.2b}$$

where the effects of the residual fields have been neglected only after deriving expressions for the coefficients in (7.2a) and (7.2b). The propagation constants $\gamma^{(a)}$ and $\gamma^{(b)}$ are given by

$$\gamma^{(a)} = \beta^{(a)} + [\tilde{\varkappa}_{aa} - C_{ab}\tilde{\varkappa}_{ba} + C_{ab}C_{ba}(\beta^{(a)} - \beta^{(b)})]/(1 - C_{ab}C_{ba}) \tag{7.3a}$$

and

$$\gamma^{(b)} = \beta^{(b)} + [\tilde{\varkappa}_{bb} - C_{ba}\tilde{\varkappa}_{ab} + C_{ab}C_{ba}(\beta^{(b)} - \beta^{(a)})]/(1 - C_{ab}C_{ba}) \tag{7.3b}$$

where the corrections to the individual waveguide mode propagation constants. The coupling coefficients are

$$\varkappa_{ab} = \{\tilde{\varkappa}_{ab} + C_{ab}[\beta^{(a)} - \beta^{(b)} - \tilde{\varkappa}_{bb}]\}/(1 - c_{ab}C_{ba}) \tag{7.4a}$$

and

$$\varkappa_{ba} = \{\tilde{\varkappa}_{ba} + C_{ba}[\beta^{(b)} - \beta^{(a)} - \tilde{\varkappa}_{aa}]\}/(1 - C_{ab}C_{ba}). \tag{7.4b}$$

In (7.3a)–(7.4b), the coefficients C_{ab} and C_{ba} describe the individual waveguide mode overlap, i.e.

$$C_{pq} = 2\hat{z} \cdot \int_{-\infty}^{\infty} \int [\mathbf{E}_t^{(q)} \times \mathbf{H}_t^{(p)}] \, dx\, dy \qquad p, q = a, b \tag{7.5}$$

and C_{pq} will appear in power equations describing the joint interactions of the individual waveguide modes. The individual modes are normalized [7.6, Appendix A] such that

$$C_{aa} = C_{bb} = 1. \tag{7.6}$$

The constants $\tilde{\varkappa}_{pq}$ depend on the perturbations to the individual waveguides and are defined by

$$\tilde{\varkappa}_{pq} = \omega \int_{-\infty}^{\infty} \int \Delta\epsilon^{(p)} \left[\mathbf{E}_t^{(p)} \cdot \mathbf{E}_t^{(q)} - \frac{\epsilon^{(q)}}{\epsilon_0 n^2} E_z^{(p)} E_z^{(q)} \right] dxdy \qquad p,q = a,b \tag{7.7}$$

where $\epsilon^{(p)}$, $p = a,b$ refer to the individual waveguides and $\Delta\epsilon^{(p)}$, $p = a,b$ are the perturbations to the respective guide. Note that the integrals in (7) extend only over the regions in which the perturbations occur. For example, in a simple case, waveguide 'b' is the perturbation to waveguide 'a', and conversely. In contrast, the integrals in (7.5) extend over all space.

More complicated waveguiding structures are decomposed into two individual guides as follows:

$$\epsilon^{(a)}(x,y) + \Delta\epsilon^{(a)}(x,y) = \epsilon_0 n^2(x,y) \tag{7.8a}$$

and

$$\epsilon^{(b)}(x,y) + \Delta\epsilon^{(b)}(x,y) = \epsilon_0 n^2(x,y) \tag{7.8b}$$

where $\epsilon^{(a)}(x,y)$ and $\epsilon^{(b)}(x,y)$ each represent the individual waveguides singly embedded in the surrounding medium. Note that $\epsilon^{(a)}(x,y) + \epsilon^{(b)}(x,y) \neq \epsilon(x,y) = \epsilon_0 n^2(x,y)$, since each function contains the same surroundings. Moreover, most generally $\epsilon_0 n^2(x,y)$ need not even equal the mathematical union of $^{(a)}(x,y)$ and $\epsilon^{(b)}(x,y)$, because an additional index perturbation not encompassed by either waveguide could be included.

In [7.5], quantities labelled $M_{(a)}$ and $M_{(b)}$ were derived (see [7.5, equation (52)]) as corrections to the propagation constants, i.e. $\gamma^{(p)} = \beta_1^{(p)} - iM_{(p)}$, $p = a,b$. The expressions for $M_{(p)}$, aside from the multiplicative factor $-i$, differ from $\tilde{\varkappa}_{pp}$ in that the latter include integrals which contain the z-components of the individual waveguide modes. Those integrals vanish for TE modes, but even then (7.3a) and (7.3b) for $\gamma^{(p)}$, $p = a,b$, are more accurate in that they incorporate $\tilde{\varkappa}_{ab}$, $\tilde{\varkappa}_{ba}$, C_{ab} and C_{ba}. An expression for the coupling coefficient $\varkappa_{ab}$ is also derived in [7.5] but is immediately specialized to the case of identical waveguides. Completing the derivation of [7.5] for dissimilar waveguides yields

$$\tilde{\varkappa}_{ab} = \omega \int_{-\infty}^{\infty} \int \Delta\epsilon^{(b)} \mathbf{E}_{t1}^{(a)} \cdot \mathbf{E}_{t1}^{(b)} dxdy \tag{7.9a}$$

and

$$\tilde{\varkappa}_{ba} = \omega \int_{-\infty}^{\infty} \int \Delta\epsilon^{(a)} \mathbf{E}_{t1}^{(b)} \cdot \mathbf{E}_{t1}^{(a)} dxdy. \tag{7.9b}$$

Thus $\tilde{\varkappa}_{ba}$ resembles $\tilde{\varkappa}_{ab}$ in (7.7) without the second integral and similarly for $\tilde{\varkappa}_{ab}$ and $\varkappa_{ba}$; however, although $\varkappa_{ab}$ and $\tilde{\varkappa}_{ab}$ do not have similar forms, the latter does approximate the former numerically just as $\tilde{\varkappa}_{ba}$ approximates $\varkappa_{ba}$.

It is important to note that neither $\varkappa_{ab}$ and $\varkappa_{ba}$ as defined by (7.4a) and (7.4b) nor $\bar{\varkappa}_{ab}$ and $\bar{\varkappa}_{ba}$ in (7.9a) satisfy complex conjugate relationships for lossless waveguides unless the guides are identical. The usual derivation of that result relies on an approximate application of power conservation. Consequently, the coupling coefficients are equal and may be determined approximately from the exact modal propagation constants according to [7.6, Appendix F]

$$\varkappa_{ab} = \varkappa_{ba} = (\sigma_1 - \sigma_2)/2. \tag{7.10}$$

POWER FLOW AND POWER TRANSFER BETWEEN PARALLEL WAVEGUIDES

The time-averaged power flow in the z-direction is given by

$$P_z = \tfrac{1}{2}\hat{z}\int_{-\infty}^{\infty}\int \mathrm{Re}(\mathbf{E}_t \times \mathbf{H}_t^*)dxdy. \tag{7.11}$$

Now if the unperturbed waveguide system has neither loss nor gain (hereafter referred to as lossless) we then obtain

$$P_z = \tfrac{1}{4}[\,|U|^2 + |V|^2 + (C_{ab} + C_{ba})\mathrm{Re}(UV^*)\,] + \tfrac{1}{2}\hat{z}\int_{-\infty}^{\infty}\int \mathrm{Re}(\mathbf{E}_r \times \mathbf{H}_r^*)dxdy \tag{7.12}$$

upon substituting (7.1a) and (7.1b) in (7.11). The usual derivations showing that $\varkappa_{ab} \equiv \varkappa_{ba}^*$ in a lossless system employs only the first two terms of (7.12). In the following discussion only the last term is neglected.

Let us consider two parallel lossless guide with $U \equiv 0$ and $V = V_0$ at $z = 0$. It does not need to address the question of how that initial connection is established, but that issue must be considered in any analysis regardless of the equations used to describe the propagation and coupling along the length where the guides are parallel. Upon solving (7.2a) and (7.2b) subject to these initial conditions, we then obtain

$$V(z) = V_0\left[\cos(\psi z) + \frac{i\Delta}{\psi}\sin(\psi z)\right]e^{i\phi z} \tag{7.13a}$$

and

$$U(z) = V_0\frac{i\varkappa_{ab}}{\psi}\sin(\psi z)e^{i\phi z} \tag{7.13b}$$

where

$$\phi = [\gamma^{(a)} + \gamma^{(b)}]/2 \tag{7.14a}$$

$$\Delta = [\gamma^{(b)} - \gamma^{(a)}]/2 \tag{7.14b}$$

and

$$\psi = \surd(\Delta^2 + \varkappa_{ab}\varkappa_{ba}). \tag{7.14c}$$

We observe that, when $\psi z = \pi/2$, $|U|$ is a maximum. If waveguide 'b' could be terminated at that point, the modal amplitude and power in 'a' could be estimated. We multiply the entire field by $H_t^{(a)}$ and integrate to obtain $U = C_{ab}V$. Based on the estimate the power remaining in waveguide 'a' is given by

$$P^{(a)} = \frac{1}{4}|U + C_{ab}V|^2 = \frac{1}{4} \frac{|V_0|^2}{\psi^2}|\varkappa_{ab} + C_{ab}\Delta|^2 \qquad (7.15)$$

and this result is independent of the formula used to evaluate $\varkappa_{ab}$.

Instead, however, let the coupling process begin with $U = V_0$ and $V = 0$ at $z = 0$. If, after propagating $\psi z = \pi/2$, waveguide 'a' can be terminated, then as above the modal amplitude in waveguide 'b' would be $V + C_{ba}U$. Consequently, the power remaining in waveguide 'b' is estimated to be

$$P^{(b)} = \frac{1}{4}|V + C_{ba}U|^2 = \frac{1}{4} \frac{|V_0|^2}{\psi^2}|\varkappa_{ba} - C_{ba}\Delta|^2. \qquad (7.16)$$

In order to satisfy reciprocity in a lossless waveguide system, $P^{(a)}$ and $P^{(b)}$ in (7.15) and (7.16) differ negligibly.

Finally, in order that power be conserved when propagating along z, P_z must be independent of z. Equations (7.13a) and (7.13b) are substituted in (7.12), with the residue fields neglected, to obtain

$$P_z = \frac{1}{4}|V_0|^2 \left\{ 1 + \frac{\varkappa_{ab}}{\psi^2}[(\varkappa_{ab} - \varkappa_{ba}) + \Delta(C_{ab} + C_{ba})]\sin^2(\psi z) \right\}. \qquad (7.17)$$

Now for $P_z = $ constant, the coefficient of the $\sin^2(\psi z)$ term must be zero, but except for identical guides that result could be demonstrated analytically. This deficiency should be expected since the last term of (7.12) has been neglected fortuitously.

7.1.2   COUPLED MODE ANALYSIS OF MULTIWAVEGUIDES [7.7]

The phenomenon of coupling among several waveguides arises in many situations including high power phased array diode lasers [7.15, 7.16], integrated optical multiport switches [7.17–7.19] and fiber optical couplers [7.20, 7.21]. The theory currently available to describe the coupling process [7.5] is thought to be generally applicable, but as shown here it is neither quantitatively nor qualitatively accurate, even in some cases of weak coupling between neighbouring waveguides.

The analysis of coupling among parallel multiwaveguides is presented. The development is based on the two parallel waveguide theory in the above section. Here we consider N such guides which in general differ in shape, size, and/or refractive index and are located in specific, but arbitrary, not necessarily collinear, relation to each other. For simplicity, each waveguide is presumed to support only one guided mode, but that restriction is not a fundamental limitation. Together with the continuum of radiation modes they form a complete set. Depending on the waveguide shape, the guided mode(s) may be TE, TM, or hybrid; the theory is of general utility in that applies to any such mode. Moreover, loss and/or gain are encompassed.

In the course of the derivation, a matrix of constants representing the overlap of the individual waveguide modes and a matrix of coefficients which are related to the individual waveguide propagation constants and coupling among the guides are all introduced. Finally, an inversion and matrix multiplications are needed to obtain a matrix of coupling coefficients. That matrix is then available to solve for the power distribution among the waveguides as a function of the propagation distance z given an initial field distribution. Additionally, eigenfunctions of the matrix are the system supermodes [7.16] and the eigenvalues are their propagation constants. The field patterns of the supermodes are invariant with z except for exponential growth or decay. Thus the supermodes are linear superpositions of the individual waveguide modes, which are only approximations to the exact array modes. Propagation constants for the latter are not identical to those of the supermodes but when the theory presented in section is employed, the supermode patterns and propagation constants closely approach the exact values.

FORMULATION OF COUPLED MODES IN MULTIWAVEGUIDES

Coupled mode equations are derived for a system of N waveguides which is invariant in z and is characterized by a dielectric distribution $\epsilon(x,y) = \epsilon_0 n^2(x,y)$. The guides are in general not identical and their relative locations are specified, but arbitrary and not necessarily colinear. Each guide, numbered $m = 1, 2, \ldots, N$, has a dielectric constant distribution $\epsilon^{(m)}(x,y) = \epsilon_0 n_m^2(x,y)$, which includes the guide plus the surrounding medium, excluding the other guides. For each waveguide a perturbation is defined by

$$\Delta\epsilon^{(m)}(x,y) = \epsilon_0 \Delta n_m^2 = \epsilon(x,y) - \epsilon^{(m)}(x,y) \qquad (7.18)$$

where, as noted above, $\epsilon(x,y)$ describes the entire waveguide system. If it is inconvenient or otherwise inappropriate to separate $\epsilon(x,y)$ exactly into N individual waveguides, then that part of $\epsilon(x,y)$ not included in some waveguide will be common to all the functions $\Delta\epsilon^{(m)}$, $m = 1, 2, \ldots, N$ [7.6].

When the waveguides are isolated, the dielectric distribution $\epsilon^{(m)}(x,y)$ is assumed to support a single guided mode with transverse field components $\{\mathbf{E}_t^{(m)}, \mathbf{H}_t^{(m)}\}$ and a propagation constant $\beta^{(m)}$. In terms of these field components overlap integrals are defined as

$$c_{pm} = 2\hat{z} \int_{-\infty}^{\infty} \int \mathbf{E}_t^{(m)} \times \mathbf{H}_t^{(p)} dx dy \qquad \forall p, m \qquad (7.19)$$

and the mode normalization is such that

$$c_{pp} = 1 \qquad \forall p. \qquad (7.20)$$

Considering next any arbitrary electromagnetic field propagating in the $+z$ direction, $\{E_t(x,y,z), H_t(x,y,z)\}$ could be written as

$$E_t(x,y,z) = \sum_{m=1}^{N} u_m(z) E_t^{(m)}(x,y) + Q_t(x,y,z) \tag{7.21a}$$

$$H_t(x,y,z) = \sum_{m=1}^{N} u_m(z) H_t^{(m)}(x,y) + R_t(x,y,z) \tag{7.21b}$$

where $Q_t(x,y,z)$ is that part of E_t which cannot be represented as a linear sum of the individual waveguide modes and similarly for R_t, H_t. As shown in [7.7, Appendix A], when the function Q_t (or R_t) is negligible, the functions $u_m(z)$ written as the elements of the column matrix $U(z)$ are found to satisfy the matrix differential equation

$$C\frac{dU}{dz} = i(BC + K)U \tag{7.22}$$

where C, B, and K are $N \times N$ matrices. The elements of C are c_{pm}, B is a diagonal matrix of the propagation constants $\beta^{(m)}$, and the elements of K are denoted $\tilde{\varkappa}_{pm}$. These quantities are defined by

$$\tilde{\varkappa}_{pm} = \omega \int_{-\infty}^{\infty} \int \Delta\epsilon^{(p)} \left[E_t^{(p)} \cdot E_t^{(m)} - \frac{\epsilon^{(m)}}{\epsilon_0 n^2} E_z^{(p)} E_z^{(m)} \right] dx dy. \tag{7.23}$$

The diagonal elements $\tilde{\varkappa}_{mm}$ are associated with corrections of the propagation constants $\beta^{(m)}$ caused by the perturbations $\Delta\epsilon^{(p)}$. The off-diagonal constants are related to the coupling between waveguides 'p' and 'm', but the true coupling coefficients are only obtained when (7.22) is solved for dU/dz. It is important to recognize that the integrals which define the elements of C are substantially larger than those appearing in the definitions of $\tilde{\varkappa}_{mm}$ because the latter are localized to the regions where $\Delta\epsilon^{(p)}$ is non-zero, whereas the former extend over all space. The formal solution of (7.22) requires inverting C and, provided det $(C) \neq 0$,

$$\frac{dU}{dz} = i[C^{-1}BC + C^{-1}K]U. \tag{7.24}$$

By analogy with the earlier theory [7.15, 7.16], the diagonal elements of $C^1BC + C^{-1}K$ are the propagation constants whereas the off-diagonal elements are the coupling coefficients.

Since the matrix $C^{-1}BC + C^{-1}K$ is constant, (7.24) is easily solved for $U(z)$ given an initial condition $U(0)$; in formal mathematical terms [7.22]

$$U(z) = \exp[i(C^{-1}BC + C^{-1}K)z]U(0). \tag{7.25}$$

The solution for $U(z)$ determines the distribution of power among the waveguides as a function of z. The time-averaged power flow in the z direction is defined by

$$P_z = \frac{1}{2} \int_{-\infty}^{\infty} \int \text{Re}\{\hat{z} \cdot \mathbf{E}_t \times \mathbf{H}^*_t\} dx dy \qquad (7.26)$$

which is substantially simplified when $\mathbf{Q}_t(x,y,z)$ [7.6] and its counterpart for the $\mathbf{H}_t(x,y,z)$ field, namely $\mathbf{R}_t(x,y,z)$ are neglected. In particular, for a system of lossless waveguides or if all gain and/or loss are treated as perturbations,

$$P_z = \frac{1}{4} \text{Re}(\tilde{U} C U^*) \qquad (7.27)$$

where the tilde denotes transpose, the expansion (7.21b) of the $\mathbf{H}_t$-field has been employed, and the definition of C invoked.

Modes of the entire waveguide system, i.e. supermodes, are patterns which propagate without change (other than exponential gain or loss). In general, there exist N such modes, which are represented as

$$e_t^{(\nu)} = \sum_{m=1}^{N} u_m^{(\nu)}(z) \mathbf{E}_t^{(m)} \qquad (7.28)$$

and the propagation condition requires the $U^{(\nu)}(z)$ be in the form

$$U^{(\nu)}(z) = U_0^{(\nu)} e^{i\sigma_\nu z}. \qquad (7.29)$$

The supermode propagation constants are determined by substituting (7.28) in (7.24) and requiring that the determinant be zero for the existence of a non-trivial solution, namely

$$\det[C^{-1}BC + C^{-1}K - \sigma_\nu I] = 0 \qquad (7.30)$$

where I is the identity matrix. Associated with each σ_ν, $\nu = 1, 2, \ldots, N$ is an eigenvector of coefficients for the supermode amplitudes $U_0^{(\nu)}$, i.e. the supermode is given by (7.28) with $u_m^{(\nu)}$, $m = 1, 2, \ldots, N$ being the elements of $U^{(\nu)}$.

In carrying out the above derivation, a new set of differential equations is obtained for describing coupling among an array of waveguides. Superficially, the formulation appears more involved than those utilized previously in that one must evaluate additional integrals and several matrix manipulations are required. Relative to the first point, the C-matrix elements are needed, but these are not difficult to compute since they depend only on the unperturbed mode patterns and not the dielectric perturbations. Elements of the K-matrix differ somewhat from those specified by previous theories. Although the principal diagonal elements of K have been defined previously [7.5], they are often ignored [7.15, 7.16]. Furthermore, in many analyses all elements of K, except those on the two principal off-diagonals, are neglected and these elements are evaluated assuming only nearest-neighbour coupling. If those assumptions are justified, they may also be exploited in simplifying the K matrix of this paper. But in all cases it is necessary to include the principal diagonal elements of K and non-negligible elements of C, which at a minimum include the two principal off-diagonals.

7.2 INTEGRATED OPTICS IN LITHIUM NIOBATE WAVEGUIDES

The design of guided-wave devices or systems depends strictly on the possibility of obtaining the best electromagnetic wave confinement and propagation conditions by controlling the waveguide optical properties and geometrical configuration.

Although most of the past work in integrated optoelectronics was devoted to solving electromagnetic problems and to developing individual devices, a considerable amount of work has been done recently on understanding the basic physics of guided-wave devices to lead to improved performances and higher reliability.

The essential element of a device is the optical waveguide which consists of a thin layer on a material substrate. The refractive index of the thin layer is greater than that of the substrate.

A wide range of materials has been used for thin-film optical devices and circuits. Active devices such as switches, deflectors and modulators have been formed in electro-optical or acousto-optical materials, while passive components such as filters and lenses have been formed in glass substrates. In the development of active devices lithium niobate is the favoured substrate because of its high optical quality an electro-optical and piezoelectrical properties, and because it provides easy fabrication of good quality optical waveguides.

Planar and stripe dielectric waveguides have been formed on lithium niobate using various technological processes. Techniques for device fabrication have been derived largely from the microelectronics area in the sense that they have had to be properly developed because of different materials and device requirements. Waveguide fabrication techniques aim to create a local variation of refractive index with a graded- or step-profile, involving conventional techniques such as vacuum thin-film deposition, thermal diffusion, proton exchange, ion exchange and ion implantation.

Lithium niobate is available as large, transparent, single crystals with excellent electro-optical and acousto-optical properties. Since the first fabrication of indiffused waveguides in $LiNbO_3$ in 1974 [7.24], this material has been extensively used in integrated optics research and it became possible to simply fabricate low loss waveguides. Consequently, high performance devices (modulators, switches, etc) were envisaged. The emergence of the single mode fiber as the dominating optical transmission medium in the early 1980s further increased the interest in $LiNbO_3$ integrated optics devices since they, like most other waveguide devices, basically require single mode operation to be efficient.

The $LiNbO_3$ device research has encompassed material development of the $LiNbO_3$ crystal (specification, impurity contents, doping, etc.), improved understanding of the Ti indiffusion process as well as the development of new fabrication processes (e.g. proton exchange [7.25], ion implantation [7.26]). Parallel to this development, progress was made in understanding complex devices and device modelling [7.27, 7.28]. As a result of this, a number of highly complex high-performance devices have been developed during the past few years. This has made the $LiNbO_3$ devices far more mature than the corresponding semiconductor devices (in GaAs/GaAlAs or InP/InGaAsP) with their complex fabrication technology. Development regarding pigtailing with fibers and packaging of $LiNbO_3$ chips has been considerably slower which is one of the reasons for the (at this stage) rather scarce commercial availability of $LiNbO_3$ integrated optics components as well as the few reported systems demonstrations in the telecommunications area.

7.2.1 Lithium niobate guided-wave devices [7.30]

$LiNbO_3$ integrated optics devices incorporate light guiding channels in the $LiNbO_3$ crystal with higher refractive index than the surrounding medium. The manipulation of light is based on the linear electro-optic effect [7.29]:

$$\Delta B_{ij} = r_{ijk}E_k \qquad (7.31)$$

where r_{ijk} is the electro-optic tensor and E_k is the electric field.
  Further:

$$B_{ij}K_{jk} = \delta_{ik} \qquad (7.32)$$

where B and K are the dielectric impermeability and permeability tensors, respectively; the latter is generally referred to as the dielectric constant tensor. The change in refractive index is derived from the change in the dielectric constant [7.29]. Thus an applied voltage perturbs the indicatrix, and hence the refractive index, seen by the optical fields [7.29]. In a common orientation in $LiNbO_3$ (electric field along the z-axis) the index change (for polarization along the z-axis) is:

$$\Delta n = -\tfrac{1}{2}n_{33}^3 r_{33}E_3. \qquad (7.33)$$

Using numerical values at $\lambda = 1.3\,\mu m$ ($r_{33} \sim 30 \times 10^{-12}m/V, n_{33} = 2.15$) and an E field of $1\,V/\mu m$, it is seen that the ensuing index change is only 1.5×10^{-4}. This is one of the important boundary conditions for $LiNbO_3$ and sets a limit to the level of integration obtainable, since device operation relies on this index change and requires (order of magnitude)

$$k_0 L \Delta n = \pi \qquad (7.34)$$

where $k_0 = 2\pi/\varkappa_0$ and L is the device length.
  The electrically controlled index change can be used for phase modulation, and the phase modulation can, in turn, be used to generate amplitude modulation in several ways:

(1) interferometric: Mach–Zehnder–modular, balanced bridge switch,
(2) phase match control: directional coupler.

However, the index perturbations can be used to create optical field changes, which are employed in cutoff modulators and directional couplers.

The most common integrated optical switch is the coupled waveguide structure of Fig. 7.2 [7.30], the electro-optic directional coupler [7.31]. Here two phase-matched optical waveguides are arranged at such a small separation that light is periodically coupled back and forth between the waveguides in the direction of light propagation. By arranging electrodes at the waveguides, the refractive index can be changed via an applied electric field, employing the electro-optic effect. Hence, the original phase-matching between the waveguides is destroyed and switching of light between the output ports can be accomplished.

In order to ease fabrication tolerances (the coupler otherwise has to be exactly an odd multiple of coupling or crossover lengths in order to implement the cross state) the stepped-$\Delta\beta$ configuration [7.32] was devised and has been extensively used. However, other four-port switching devices exist, and Fig. 7.3 [7.30] shows, in addition to the directional coupler, an X-switch [7.33, 7.34] and a 'BOA' [7.35]. The electro-optically induced index change has different effects in the latter two (a 'BOA' is essentially an elongated X-switch) and this can be analysed as follows [7.36].

The applied voltage will, depending on the electrode configuration, give rise to an even (Δn_e) or odd (Δn_o) index perturbation, referring to the switch lateral coordinate. The even perturbation will change the β's of the modes to the composite two-channel structure (the even and odd 'supermodes'):

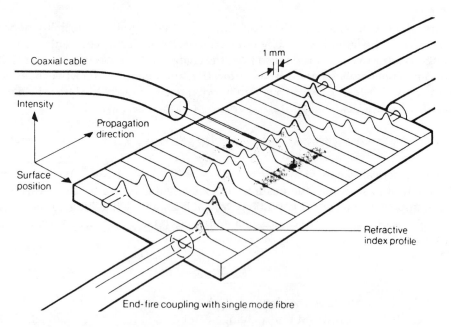

Fig. 7.2 LiNbO$_3$ waveguide electro-optic directional coupler. (From [7.30] Thylen (1988) *J. Lightwave Technol.* **LT-6**, 847–861. Reproduced by permission. ©1988 IEEE.)

$$\Delta\beta_e \propto (\psi_e, \Delta n_e \psi_e) \tag{7.35}$$

$$\Delta\beta_0 \propto (\psi_0, \Delta n_e \psi_0) \tag{7.36}$$

where the brackets denote scalar products involving suitably normalized even and odd supermodes ψ_e, ψ_0. It is assumed that the unperturbed coupler is symmetric; similarly, the coupling between the even and odd modes, $\varkappa_{e,o}$ is given by

$$\varkappa_{e,0} \propto (\psi_0, \Delta n_e \psi_e) = 0 \tag{7.37}$$

for symmetry reasons.

An odd perturbation will, complementarily, leave β_e and β_0 unchanged (of first order) but couple ψ_e, ψ_0.

Hence, the BOA and X with their strong coupling and large mismatch between β_e and β_0 are preferably operated by applying an even index perturbation, changing β_e and β_0 by different amounts, without coupling of the supermodes. This also gives a periodic transfer function (light power versus voltage) since the output power will be proportional to $\cos^2((\beta_e - \beta_0)L/2)$ [7.34], and this makes electric tuning possible without relying on more complex structure, as in the directional coupler.

The BOA could be operated with an odd perturbation; since strong coupling of the supermodes is then needed large voltages are required; however, these only reflect the fact that the device length taken as the coupling length

$$L_c = \pi/(\beta_e - \beta_0) \tag{7.38}$$

is short and hence the switch voltage obeys the usual length scaling law, in this case the non-periodic directional coupler type response. The directional coupler structure, on the other hand, has a larger coupling length (smaller $\beta_e - \beta_0$) and should be operated with an odd perturbation Δn_0, to couple the supermodes to one another. Here an even perturbation is less efficient for switching.

Fig. 7.3 shows the required electrode structure on z-cut LiNbO$_3$ for achieving an odd perturbation for the directional coupler Fig. 7.3(a) and an even perturbation for the BOA and X (Fig. 7.3(b) and (c)).

It should be obvious from the above that it is the electrode arrangement that determines whether a device is operated as a directional coupler, or as a BOA, with a periodic light output versus voltage response.

The switches described above form the basic building blocks for time, space, and wavelength switches and are used to generate more complex devices.

LITHIUM NIOBATE WAVEGUIDE SWITCH ARRAYS

One of the most interesting aspects of integrated optics space switches is their ability to route information practically irrespective of its bandwidth and coding format ('frequency and code transparency') in marked contrast to electronic space switches. While it may be argued that this is also a property of mechanical switches, which are in addition polarization independent without resorting to elaborate construction,

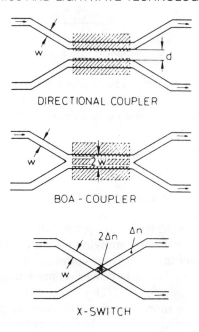

Fig. 7.3 Different types of waveguide optical switching devices. (From [7.30] Thylen (1988) *J. Lightwave Technol.* **LT-6**, 847–861. Reproduced by permission. ©1988 IEEE.)

integrated optics offers all the advantages of integration and in addition the possibility for high-speed switching. However, these two last features are in conflict, since large-scale integration entails short switches, giving high voltages and impracticality for high-speed operation.

The requirement on the switch arrays concerning blocking properties (strictly non-blocking, rearrangeably non-blocking, blocking), broadcast requirements, etc, will strongly influence their size and characteristics. By strictly non-blocking, we mean that any unoccupied input/output port can be connected without disturbing existing signal paths, in rearrangeably non-blocking structures, existing paths might have to be re-routed to yield the non-blocking feature, temporarily interrupting the connection and causing crosstalk. In a synchronous time-switched network this is obviously of less concern. Blocking arrays, with fewer switches, can of course be made with a higher number of input/outputs on a given substrate area. However, the number of switches is not the most important parameter in LiNbO$_3$ integrated optics, but rather the switch array depth, i.e. the number of switches in cascade, since the size of LiNbO$_3$ wafers is limited. Fig. 7.4 [7.30] summarizes these features for three prevalent types: the crossbar, lattice, and tree structure.

The crossbar type [7.37–7.39] is strictly non-blocking, but requires a large 'depth'. The lattice structure [7.40, 7.41] is shorter (in the switch region) but can only be used where the system is not sensitive to rearrangement transients. It should, however, be noted that the length advantage of the lattice structure is somewhat superficial, since all inputs/outputs have to be brought together over a shorter distance than is the case for the crossbar structure. The 'tree' structure [7.42, 7.43] is interesting in that it trades length for width, which is advantageous in integrated

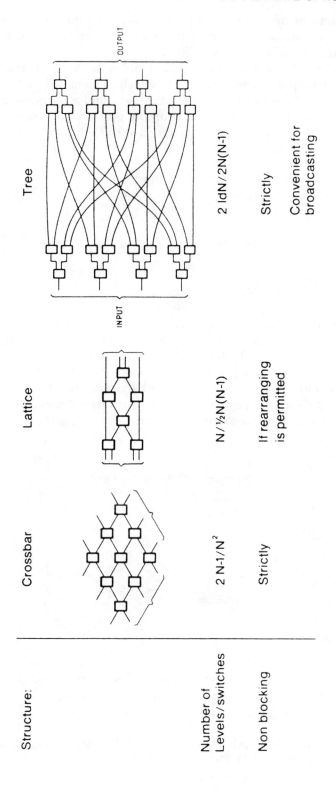

Structure:	Crossbar	Lattice	Tree
Number of Levels/switches	2 N-1/N²	N/½N(N-1)	2 ldN/2N(N-1)
Non blocking	Strictly	If rearranging is permitted	Strictly
			Convenient for broadcasting

Fig. 7.4 Comparison of different switch array architectures. (From [7.30] Thylen (1988) *J. Lightwave Technol.* **LT-6**, 847–861. Reproduced by permission. ©1988 IEEE.)

optics. Thus, the length or depth grows logarithmically with the switch size. It requires more switches than the crossbar arrays, but has the attractive feature of allowing broadcasting, as can be seen from Fig. 7.4.

The largest switch arrays fabricated to date [7.37] employ the crossbar structure, and have 8 inputs and 8 outputs, giving a total of 64 switch elements. Fig. 7.5 [7.30] shows each switch element can be individually addressed, and any non-occupied input port can be connected to any non-occupied output port, without rearranging previously established signal paths. The matrix of [7.37] has the following characteristics:

Crosstalk:	-30 ± 4.1 dB
Bar state voltage:	18.6 ± 3.6 V
Cross state voltage:	26.4 ± 1.6 V.

The tolerances refer to standard deviation for all switches in a matrix.

LITHIUM NIOBATE WAVEGUIDE MODULATORS

One of the most obvious applications is external modulation. In fact, the highest length–bit rate product has been achieved using an Ti:LiNbO$_3$ external modulator in an 8 Gbit/s experimental system [7.44]. The advantage of external modulation over direct modulation is the absence (ideally) of 'chirping': in the laser diode under direct modulation, the carrier dynamics bring about a dynamic sweep of the laser frequency which gives a dispersion penalty. Certain external modulators (push–pull operated Mach–Zehnder, directional coupler in the cross channel) give zero 'chirp'

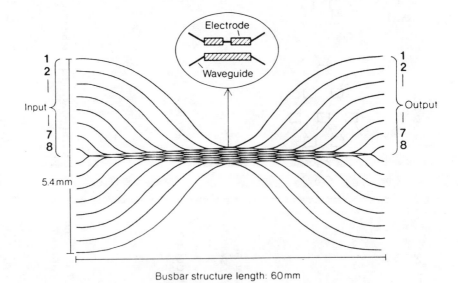

Fig. 7.5 Structure of 8×8 crossbar switch matrix. (From [7.30] Thylen (1988) *J. Lightwave Technol.* **LT-6**, 847–861. Reproduced by permission. ©1988 IEEE.)

[7.45], but only at frequencies lower than their bandwidth. A time domain analysis of travelling wave devices does, however, imply that this is only approximately correct, but without altering the basic advantage of the external modulation over direct modulation. This advantage has to be traded against the insertion loss of the modulator and in turn compared to the possible output power advantages of a laser with relaxed high speed modulation requirements.

In general, travelling wave devices [7.47–7.52] are preferred. In these devices, as opposed to the lumped RC-type devices (Fig. 7.2) the guided light interacts with the controlling microwave signal, which propagates on a microwave transmission line (see Fig. 7.6) [7.30], and the bandwidth is determined by a combination of the velocity mismatch between the optical and electrical signals and the microwave losses of the transmission line, rather than by RC constants. In Fig. 7.6, incident CW laser light (left) is split up in a 3 dB Y-junction and propagates under a microwave transmission line, interacting with the propagating microwave, which produces a relative dephasing of the light in the two arms. By optical interference in the next Y-junction this phase modulation is converted to amplitude modulation. The velocity mismatch between light and microwave limits the bandwidth. An appropriate figure of merit for modulators is the 3 dB bandwidth divided by drive voltage for 100% modulation/switching.

The quotient of the figures of merit for travelling wave devices and lumped type RC-devices

$$an_\mu/(n_\mu - n_o) \tag{7.39}$$

where $a = 1.9$ for a directional coupler switch employing uniform electrodes and $a = 1.4$ for a $\Delta\beta$-reversal switch and for a Mach–Zehnder modulator. n_μ is the microwave effective index, and n_o is the optical effective index.

Attempts to increase bandwidth (such as thicker buffer layers, complex electrode structures) tend to also increase the drive voltage, although in the small signal regime certain improvements can be made [7.49–7.51].

An integrated-optic device in Ti:LiNbO$_3$ which implements such an optical bypass function as well as high-frequency modulation in a 2.4 Gbit/s system has been developed [7.46].

A schematic of the device and its systems environment is depicted in [7.30]. In normal operation, the incident data on the fiber bus are coupled to the detector via switch S1, detected and processed. Data to be transmitted (whether repeated or terminal-generated) modulate the directional coupler travelling-wave modulator XM, the output of which is coupled to the output fiber via switch S2. The two bypass switches S1 and S2, are designed so that the chip becomes optically transparent to the fiber if the corresponding terminal goes down.

7.2.2 Fabrication techniques of lithium niobate waveguides [7.53]

To date, the best way to produce good quality optical waveguides in lithium niobate seems to be Ti in-diffusion in a dry O$_2$ atmosphere at temperatures above 1000°C, and for long diffusion times (>10 h). However, despite the large amount of research

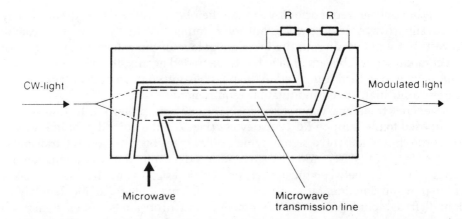

Fig. 7.6 Travelling wave Mach–Zehnder electro-optic modulator. (From [7.30] Thylen (1988) *J. Lightwave Technol.* **LT-6**, 847–861. Reproduced by permission. ©1988 IEEE.)

done in this field, further microscopical investigation are required to characterize better the guiding layer. Correlation between optical properties (mode number, mode index, field distribution, scattering levels, total attenuation, and optical stability) and structural characteristics (homogeneity, defects, compound formation, crystal quality) is clearly needed in order to define a precise model of the fabrication process. Moreover, influence of crystal composition on the kinetics of the waveguide fabrication process should be more clearly determined in order to eliminate existing confusion on the values of some fabrication parameters.

DIFFUSION PROCESSES IN LITHIUM NIOBATE

Diffusion processes may be used in integrated optics or optoelectronics to form an optical waveguide, i.e. a surface layer several μm thick, with a higher refractive index than the substrate, where light confinement, and then the electro-optical interaction takes place.

Several diffusion processes for producing optical waveguides in lithium niobate have been experimented with.

Out-diffusion

$LiNbO_3$ can crystallize in the slightly non-stoichiometric form $(Li_2O)_y(Nb_2O_5)_{1-y}$ [7.54]. Although the ordinary index n_0 is not affected as Li_2O is removed from the crystal surface, the extraordinary index n_e increases approximately linearly as y decreases in the $0.48 < y < 0.50$ range, with $dn_e/dy = -1.6$. So, because of the high mobility of lithium ions, a decrease of y may be easily achieved by out-diffusion of Li_2O from the surface. Thus, out-diffusion gives rise to an extraordinary index gradient, with the maximum at the surface and a gradual in-depth decrease approaching the bulk index. A vaporizing surface flux which was constant with time was demonstrated using a diffusion model. The activation energy (≈ 68 kcal/mole) does not vary with crystal orientation, but the refractive index change at the surface is larger for perpendicular diffusion than for diffusion parallel to the optical axis [7.55].

The refractive index profile depends on the parameters of the diffusion and vaporization processes at the out-diffusion temperature. However, the extraordinary index profile in the out-diffused layer assumes, in general, a form very close to the complementary error function:

$$\Delta n_e = \Delta n_e(0)\text{erfc}(x/2\sqrt{Dt}) \qquad (7.40)$$

where D is the diffusion coefficient ($D = 1.5 \times 10^{-9}$ cm^2/s, and 4.2×10^{-9} cm^2/s for diffusion parallel, and diffusion perpendicular to the optic c-axis at 1100°C, respectively), t is the diffusion time, x is the depth, and $\Delta n_e(0)$ is the index change at the surface ($\Delta n_e(0) = 10^{-3}$–10^{-2}).

The out-diffusion process can produce optical waveguides of thickness of the order of 10 μm, with a large effective width which may limit the operation of integrated optoelectronic devices which, in most application, require stripe waveguides. However, strip or ridge waveguides may be also formed by etching [7.56].

In-diffusion of metals

Several species of metal atom, such as Mg, Ni, Zn, Fe, Co, Cr, V, and Ti, have been used as diffusants into LiNbO$_3$, producing some modifications of the optical properties and, in particular, of refractive index [7.57, 7.58]. Thin films (10–100 nm) of metal are first deposited on the crystal surface by vacuum evaporation or sputtering. The diffusion process occurs at high temperature (850–1150°C) in inert or reactive atmosphere, with activation energy typically in the 1.0–3.0 eV range. The diffusion coefficients are anisotropic, and range between 10^{-4} to 10^{-6} cm^2/s.

The diffusion of divalent metallic atoms such as Ni, Zn, and Mg causes a reduction of the extraordinary index n_e. The ordinary index n_o also decreases when Mg is utilized as diffusant. The index n_0 increases when Ni or Zn diffuses. The study of titanium in-diffusion into lithium niobate has received greater attention because Ti yields a good light confinement, with a relatively large increase of both ordinary and extraordinary refractive indices ($\Delta n_e < 0.04$; $\Delta n_0 < 0.02$). Consequently, Ti in-diffusion allows the fabrication of optical waveguides supporting both TE and TM modes.

Diffusion of Ti thin layers is carried out at a temperature from 900 to 1150°C in argon, nitrogen, oxygen or air ambient, for diffusion times ranging from 0.5 to 30 h. Diffusion anisotropy has been found to be marked mainly in the y-cut and z-cut samples. In [7.61] the values of some fabrication parameters are reported.

Some attention has been paid to the mechanism of refractive index modification by titanium in-diffusion. It is suggested [7.59] that the refractive index change is due to the increase of polarizability, and to the photoelastic effect caused by the different size of Ti and Nb ions. The dependence of the effective index of both TE and TM waves on diffusion time t_d, for samples annealed in dry O$_2$ consists of a decrease of the index with t_d. The relationships between the surface index change, diffusion depth and fabrication conditions. A linear variation of the extraordinary index change with Ti concentration has been determined in [7.62] in y-cut samples. Ti concentrations and index profiles in z-cut Ti:LiNbO$_3$ slab waveguides under several diffusion conditions. A non-linear relation between titanium concentration and refractive index change are also determined.

To fabricate single mode waveguides, the problem of the out-diffusion of lithium oxide from the sample surface occurring during Ti in-diffusion must be solved. Excitation of a single wave into the waveguide is, in fact prevented with the refractive index changes, owing to out-diffusion, superimposed onto the Ti in-diffusion index profile. Considerable attention has been devoted to the suppression of out-diffusion of lithium oxide from the crystal surface. Out-diffusion can affect significantly the performances of stripe waveguide devices owing to the formation of an unwanted surface guiding layer for the extraordinary wave. Several methods of suppressing lithium oxide out-diffusion have been successfully tried. They have been reviewed in [7.64]. The characteristic features of Ti:LiNbO$_3$ waveguides with suppressed out-diffusion of lithium oxide was studied in [7.65]. However, a very effective and simple method of suppressing out-diffusion appears to be Ti in-fusion in the presence of water vapour (pure O$_2$ bubbling through a fixed volume of pure water at 80–90°C during diffusion), which inhibits and decomposes the LiNbO$_3$O$_8$ phase that affects the optical properties of slab waveguides increasing the in-plane scattering levels [7.66].

Optical losses in Ti:LiNbO$_3$ slab waveguides are owing mainly to in-plane scattering losses. The dependence of in-plane scattering levels on the diffusion time is investigated [7.66–7.67], by both microanalytical techniques and optical measurements. The dependence of scattering levels on the waveguide fabrication parameters was observed in [7.68]. It inferred that the major sources of light scattering in the guiding layer are the refractive index inhomogeneity and surface roughness. Other possible sources of scattering are strain and dislocations induced in the crystal by the thermal process, as confirmed also in [7.69]. However, good quality Ti:LiNbO$_3$ waveguides with losses < 1 dB/cm and high reproducibility, having taken particular care in avoiding contamination during the Ti film deposition process.

Ion exchange

Ion exchange, widely used to produce optical waveguides in glasses, was first used experimentally in [7.70] to fabricate lithium niobate waveguides. After immersing x-cut LiNbO$_3$ crystals in silver nitrate at 360°C for several hours (> 3 h) a silver/lithium ion exchange occurs. The ion exchange produces an increase of the extraordinary index with maximum change ≈ 0.12, and a step-like index profile. Unfortunately poor quality waveguides were produced with optical attenuation around 6 dB/cm.

A significant improvement was obtained in [7.71]: waveguides with lower optical losses (< 2 dB/cm) and $\Delta n_e \approx 0.12$–0.13, by thallium/lithium exchange.

Proton exchange

Proton exchange in lithium niobate consists of the replacement of Li with H. It was first used to fabricate optical waveguides in [7.72, 7.73]. An accurate study of proton exchanged waveguides has been carried out in [7.74].

Proton exchanges takes place when the $LiNbO_3$ crystal is submerged for the desired time in an acid or hydrate melt which constitutes the proton source. In [7.75], a list of some proton sources is reported. A commonly used source is benzoic acid at 120–250°C. A complete or partial exchange may be generated depending on the acidity of the medium. During waveguide fabrication, a partial exchange (65–75%) is carried out because complete exchange cause structural modifications and cracks in the substrate. To avoid face etching in y-cut samples, a small amount of lithium benzoate (typically 1%) can be added to benzoic acid. A low hydrogen concentration at the crystal surface results in a reduction or elimination of surface etching.

Proton exchange increases the extraordinary index ($\Delta n_e \approx 0.1$) and decreases the ordinary index ($\Delta n_0 \approx -0.05$) [7.73, 7.76, 7.77]. The extraordinary refractive index assumes a very nearly step distribution, as it results from the effects of diffusion rate and hydrogen content.

ION IMPLANTATION IN LITHIUM NIOBATE

Impurity additions to produce optical waveguides may also be introduced into lithium niobate by ion implantation. This technique consists of bombarding the substrate with ions (from atoms such as N, O, He, Ne, Ar, H, and Ti) with energies from 7 keV to 2 MeV. The impinging ions penetrate the substrate surface, losing their energy through electronic excitation and nuclear collisions. It results in a crystalline surface layer damaged slightly, and separated from the substrate by a low index amorphous region. Using He ions with energies in the 1–2 MeV range, the low-index region is placed from 2 to 4 μm below the surface [7.78, 7.79]. Implantation of O^+ at 170 keV showed an index change increasing with ion does up to the saturation value ($>5 \times 10^{15}$ cm^{-2}). In y-cut samples the maximum index changes are $\Delta n_o = -0.165$ and $\Delta n_e = -0.105$.

The main advantages of the ion implantation technique can be identified as follows:

(1) possibility of producing well reproducible guides
(2) homogeneity of optical parameter change in a well defined region
(3) large index changes
(4) reasonably high electro-optic coefficients.

For these reasons, embedded strip waveguides have been fabricated by combined Ti in-diffusion and ion implantation [7.80].

7.3 OPTICAL WAVEGUIDES IN SEMICONDUCTORS

Integrated optoelectronic devices made of semiconductors, particularly semiconductor compounds, have attracted considerable interest in recent years. Such components may form the basis for integrated optoelectronic circuits in which light sources and detectors, as well as optical waveguide devices and electronic circuitry, are monolithically integrated on the same substrate.

GaAs (or InP) semiconductors, due to their direct bandgap properties, are one of the most promising materials for the monolithic integration of guided-wave optical components including electro-optic [7.81–7.86] and acousto-optic deflector/modulators [7.87, 7.88], and integrated optoelectronic devices [7.89–7.91] in a common substrate. In this section, we will discuss some considerations about guided-wave optical components in GaAs.

One of the problems, associated with such guided-wave devices, however, has been the relatively high propagation losses in single mode channel waveguides made from semiconductors [7.92–7.95]. So, there is a strong motivation to develop low loss waveguides in GaAs/GaAlAs and InP/InGaAsP semiconductors as this system is generally more suitable for long distance, high bit rate optical fiber transmission.

7.3.1 GaAs and InP optical waveguides

LOW-LOSS SINGLE MODE GaAs/AlGaAs OPTICAL WAVEGUIDES

A GaAs/AlGaAs single mode waveguide grown by organometallic vapour phase epitaxy (OMVPE) will be the example discussed here. The waveguide propagation losses were minimized by using compensated epitaxial layers to reduce the free-carrier absorption and by employing a thick lower cladding layer in order to reduce the substrate radiation losses. Average propagation losses as low as 0.15 dB/cm were measured at 1.52 μm wavelength for single mode waveguides.

The schematic cross section of the GaAs/AlGaAs ridge waveguides is shown in Fig. 7.7 [7.96]. The slab waveguide layers consisted of a 8 μm thick 30 Å GaAs/50 Å Al$_{0.1}$Ga$_{0.9}$As superlattice (SL) lower cladding layer, a 2 μm thick GaAs guiding layer, and a $\approx$2.5 μm thick 30 Å GaAs/50 Å Al$_{0.1}$Ga$_{0.9}$As SL upper cladding layer. These layers were grown in an atmospheric pressure, horizontal, rf heated OMVPE reactor using trimethylgallium (TMG), trimethylaluminum (TMA), and a 2% mixture of arsine (AsH$_3$) in hydrogen. The semi-insulating GaAs substrates were oriented 6° off the (100) plane toward the (111)A. The growth temperature (625°C) and the input mole ratio AsH$_3$/(TMG + TMA) were adjusted to achieve compensated layers with free-carrier concentration less than 10^{14} cm^{-3}.

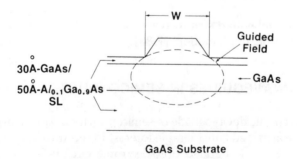

Fig. 7.7 Cross section of the GaAs/AlGaAs ridge waveguide. (From [7.96] Kapon and Bhat (1987) *Appl. Phys. Lett.* **50**, 1628–1630. Reproduced by permission of AIP.)

Ridge waveguides of various widths were fabricated on the slab waveguide wafers by using conventional photolithography and wet chemical etching. The etching mask consisted of photoresist (Shipley AZ-1350J) stripes of widths ranging from 2 to 10 μm, and the etching solution used was $H_2SO_4{:}H_2O_2{:}H_2O$ 1:8:40 by volume. The etching depth was 2.3 μm and the waveguide were ≈ 1 cm long.

The propagation losses of the waveguides were evaluated by measuring the finesse of the Fabry–Perot (FP) waveguide resonators formed by cleaving two opposite ends of the waveguides [7.95, 7.97]. The relative transmission of such a FP waveguide is given by

$$T(\phi) = \frac{(1-R)^2 e^{-\alpha l}}{(1-\tilde{R})^2 + 4\tilde{R}\sin^2\phi} \tag{7.41}$$

where L is the waveguide length, α is the propagation loss coefficient, R is the facet reflectivity, $\tilde{R} = R \exp(-\alpha L)$ and $\phi = 2\pi n_{\text{eff}} L/\lambda_0$, n_{eff} being the effective mode index. The transmission function T is periodic in ϕ, and the contrast of the transmission oscillations is given by

$$K \equiv \frac{T_{\text{max}} - T_{\text{min}}}{T_{\text{max}} + T_{\text{min}}} = \frac{2\tilde{R}}{1 + \tilde{R}^2}. \tag{7.42}$$

By using (7.42), it follows that

$$\ln[(1 - \sqrt{1 - K^2})/K] = \ln R - \alpha L. \tag{7.43}$$

Equation (7.43) shows that a measurement of the transmission fringe contrast K can yield the loss coefficient α, provided that R and L are known. It is important to note that, in general, the reflectivity R for a guided mode can be significantly different than the plane-wave Fresnel reflectivity (for normal incidence) $R_{\text{pw}} = (n_{\text{eff}} - 1)^2/(n_{\text{eff}} + 1)^2$ [7.92].

SEMICONDUCTOR WAVEGUIDE OPTICAL SWITCHES

A GaAs carrier injection type optical waveguide switch was discussed as an example of semiconductor switches [7.97]. Its switching ability was achieved at all wavelengths within the $\lambda = 1.06{-}1.55$ μm range with 90 mA injected current. This broad wavelength bandwidth is very useful for optical switching in WDM (FDM) transmission systems.

A schematic view of the optical switch having X-shaped waveguide configuration is shown in Fig. 7.8(a) [7.97]. The refractive index decreases because of electrons and holes injected through the electrodes that cut across the region where the waveguides intersect, due to the free carrier plasma dispersion effect. Therefore, total reflection occurs. Injected free carriers are confined within the waveguide core region by the double-heterojunction. The reflectional coefficient of incident optical field is related to the effective refractive index change which is given by [7.98]:

$$\Delta n_{\text{eff}} = \frac{1}{n_{\text{eff}}} \frac{\int_{-\infty}^{\infty} n(x)\, \Delta n(x)\, \psi^2(x) dx}{\int_{-\infty}^{\infty} \psi^2(x) dx}. \tag{7.44}$$

Here, n_{eff} is effective refractive index of the waveguide; $n(x)$ is the refractive index profile of the waveguide in the X direction in Fig. 7.8(a); $\Delta n(x)$ is the refractive index change due to the plasma dispersion effect; and $\psi(x)$ is the mode field profile of the slab waveguide consisting of GaAs core and AlGaAs cladding. The $\Delta n(x)$ by plasma dispersion effect is given by

$$\Delta n(x) = \begin{cases} \dfrac{\lambda^2 q^2}{8\pi^2 c^2 n\epsilon_0} \left(\dfrac{N_e}{m_e} + \dfrac{N_h}{m_h} \right) & |X| \leqslant \dfrac{d}{2} \\[2ex] 0 & |X| > \dfrac{d}{2} \end{cases} \tag{7.45}$$

where λ is wavelength in a vacuum; q is electron charge; c is light velocity; n is refractive index, and ϵ_0 is permeability in a vacuum. N_e and N_h are electron and hole densities, while m_e and m_h are electron and hole effective masses, respectively. Injected carrier density is given by

$$N = J\tau/qd \tag{7.46}$$

where J is injected current density, τ is carrier lifetime, and d is core thickness. Here, Δn_{eff} is calculated by (7.44)–(7.46) under conditions of constant current density.

Based on the above analysis, optical switches were fabricated. A cross-sectional view of the current injection area is shown in Fig. 7.8(b). The fabricational process is as follows: n-$Al_{0.4}Ga_{0.6}As$ and GaAs layers were grown on n-GaAs substrate by a metal organic chemical vapour deposition (MOCVD) system. Crossing waveguide patterns were fabricated by the conventional photo-lithography technique and chemical etching. The p-$Al_{0.4}Ga_{0.6}As$, n-$Al_{0.4}Ga_{0.6}As$, and n-GaAs cap layers were then grown. Waveguide cross-angles are 3, 4, and 5°. Switching properties were investigated with the 3° waveguide cross angle at 1.06, 1.3, and 1.55 μm wavelengths. The optical power of the lowest-order mode was maximized during measurement. Incident light is switched to another port with injected current of 90 mA (about 4.8 kA/cm^2). The crosstalk value was -8.9 dB without current injection and -10.1 dB with 100 mA injected current at $\lambda = 1.3$ μm. Switched power dependence on the injected current was similar for all in range wavelengths. The wavelength range of the switch can at least reach 1.06–1.55 μm.

7.3.2 Some calculations of semiconductor directional couplers [7.99]

A design of a double-heterostructure (DH) directional coupler [7.99] in order to obtain low loss and low driving voltage (by optimizing the composition, thickness, and doping level of the various heterostructure layers) was selected here as an example for the design of semiconductor optical waveguide components. The

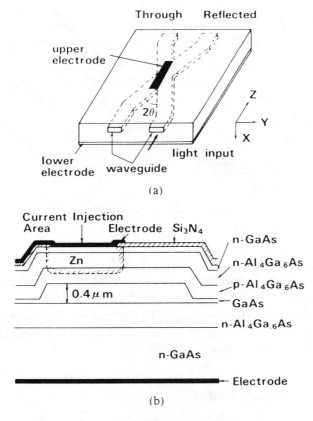

Fig. 7.8 A GaAs carrier injection type optical waveguide switch. (From [7.97] Ito *et al.* (1989) *IEEE J. Quantum Electron.* **QE-25**, 1677–1681. Reproduced by permission. ©1989 IEEE.)

complex effective index method is used to calculate the transverse refractive index method is used to calculate the transverse refractive index distribution. The electrostatic field and electrons and holes distributions are calculated by solving a non-linear Poisson equation by means of a finite difference method. The resulting effective index change is calculated with the help of a perturbation method. To calculate the optical propagation along the modulator, the beam propagation method (BPM), which is very useful in integrated optics, since the field can be simply obtained along complicated structures without any modal formalism.

The schematic of the device is shown in Fig. 7.9 [7.99] (in this case, it is a Schottky-type modulator, but the situation would be identical with a p^+ layer instead of a Schottky barrier). The n^+ substrate is grounded and a pair of electrodes overlays each waveguide in order to obtain either a uniform or a reversed $\Delta\beta$ structure.

EFFECTIVE INDEX METHOD

The ridge optical waveguides are analysed by the effective index method [7.100]. The transverse coupler effective index distribution is calculated using a complex

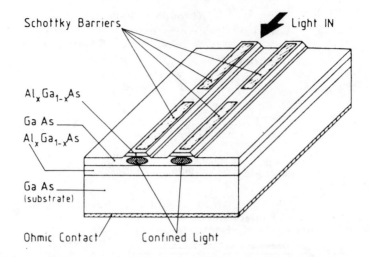

Fig. 7.9 A Schottky diode type electro-optic directional coupler. (From [7.99] Sansonetti *et al.* (1989) *J. Lightwave Technol.* **LT-7**, 385–389. Reproduced by permission. ©1989 IEEE.)

multilayer model. Once the transverse effective index distribution is found, we look for the imaginary part maximum, and only consider waveguides with values corresponding to a loss below 0.1 dB/cm.

ELECTROSTATIC FIELD COMPUTATION

In order to compute the electrostatic field in the device, the following coupled equations must be solved with appropriate boundary conditions.

Poisson equation:

$$\text{div}(\epsilon_r \nabla \phi) = q(n - p + \text{dop})/\epsilon_0 \tag{7.47}$$

transport equation:

$$qn\mu_n \nabla \phi_n = J_n$$

$$qp\mu_p \nabla \phi_p = J_p \tag{7.48}$$

conservation equation

$$q\partial n/\partial t = -\text{div}J_n$$

$$q\partial p/\partial t = \text{div}J_p \tag{7.49}$$

where ϕ is the electrostatic potential, ϕ_n and ϕ_p the quasi-Fermi potentials which can be deduced from the electron and hole concentrations (n and p) with Boltzmann statistics, q is the electron charge, ϵ_0 and ϵ_r the vacuum and relative electric permitivities, μ_n(respectively, μ_p) is the electron (respectively, hole) mobilities, dop is the ionized doping level, J_n and J_p are the electron and hole current densities. In the electro-optic directional coupler described above, no current is flowing because of the reverse biased Schottky contacts. Consequently, J_n and J_p are assumed to be equal to zero and the quasi-Fermi potentials are constant inside the device. With Boltzmann statistics, n (respectively, p) is then depending on ϕ only:

$$n = K \exp(q\phi/kT)$$

$$p = K'\exp(-q\phi/kT) \tag{7.50}$$

with K and K' constant.

The set of equations (7.47)–(4.49) is then reduced to a single equation, called the 'non-linear Poisson equation':

$$\text{div}(\epsilon_r \nabla \phi) = q/\epsilon_0 [K \exp(q\phi/kT) - K' \exp(-q\phi/kT) + \text{dop}]. \tag{7.51}$$

This equation can be solved easily with numerical techniques [7.101]: the non-linear equation is first linearized by using a Newton–Raphson method; then the successive Dirichlet problems which are deduced from this iterative procedure are solved with the use of a spatial discretization in either a two-dimensional cross section of the device with a finite element method or in a one-dimensional cross section with a finite difference method.

ELECTRO-OPTIC PERTURBATION

Denoting by $\psi(x)$ the optical field depth distribution, the optical wave equation is multiplied by $\psi^*(x)$, integrating it over the whole depth and differentiating versus the refractive index difference due to the linear electro-optic effect. For a TE mode propagating along the (-110) direction and taking notation of [7.102]:

$$\Delta n_{\text{eff}}/V = G \times F \tag{7.52}$$

with

$$G = r_{41}n_s^3/2 \tag{7.53}$$

and

$$F = \frac{\int_{-\infty}^{\infty} (n^4(x)/n_{\text{eff}}n_s^3)E(x)|\psi(x)|^2 dx}{\int_{-\infty}^{\infty} |\psi(x)|^2 dx \int_{-\infty}^{\infty} E(x)dx} \tag{7.54}$$

where Δn_{eff} is the effective index variation in the ridge where the voltage V is applied, n_s the substrate refractive index, n_{eff} the real part of the effective index, $n(x)$ and $E(x)$ are, respectively, the real part of the refractive index and the electrostatic field depth distributions, r_{41} the linear electro-optic coefficient; G is a material parameter and F represents an overlap factor between the optical and electrostatic fields. Since the operating wavelength is far from the bandgap, the quadratic effect can be neglected [7.103].

BEAM PROPAGATION METHOD

The BPM is then used to calculate the optical field along the coupler. A fast Fourier transform algorithm is used; 256 grid points are used in the transverse direction; the field is multiplied by a graded absorber at the boundaries of the computational window, in order to correctly compute the electric field in case an important fraction of the power escapes from the computational window [7.104].

A Gaussian beam (at its waist), corresponding to the image of a laser diode output onto the device, is launched into a single waveguide structure and propagates in it for 1 mm before entering the interaction region where both waveguides are present; this ensures that the propagation eigenmode of a single waveguide is launched into the coupler interaction region.

MODEL COMPARED WITH EXPERIMENTAL RESULTS

This model is compared with experimental results obtained for TE waves at 1.56 μm in a DH containing 10% Al in the confining layers, whose thicknesses confirm to the design requirement for the propagation loss. The 1.7 μm thickness of the guiding layer is sufficient to position the second-order mode slightly above cutoff, but due to the launching conditions and the higher propagation loss of the second-order mode, only the fundamental mode is observed.

The near-field pattern at the coupler output is measured with a scanning mirror and a slit in front of the detector. The device interaction length is 9 mm, the whole device length being cleaved at 9 mm [7.99]. It can be seen that the agreement between the experimental results and the model is good.

REFERENCES

[7.1]   J. R. Pierce (1954) "Coupling of modes of propagation" *J. Appl. Phys.* **25**, 179–183.

[7.2]   E. A. J. Marcatili (1969) "Dielectric rectangular waveguide and directional coupler for integrated optics" *Bell Syst. Tech. J.* **48**, 2071–2101.

[7.3]   D. Marcuse (1971) "The coupling of degenerate modes in two parallel dielectric waveguides" *Bell Syst. Tech. J.* **50**, 1791–1816.

[7.4]   H. Kogelnik and R. V. Schmit (1976) "Switched directional couplers with alternation $\Delta\beta$" *IEEE J. Quantum Electron.* **QE-12**, 396–401

[7.5]   H. F. Taylor and A. Yariv (1974) "Guided wave optics" *Proc. IEEE* **62**, 1044–1060.

[7.6]   A. Hardy and W. Streifer (1985) "Coupled mode theory of parallel waveguides" *J. Lightwave Technol.* **LT-3**, 1135–1146.

[7.7]   A. Hardy and W. Streifer (1986) "Coupled modes of multiwaveguide systems and phased arrays" *J. Lightwave Technol.* **LT-4**, 90–96.

[7.8]   W. Streifer *et al.* (1987) "Coupled-mode theory of optical waveguides" *J. Lightwave Technol.* **LT-5**, 1–4.

[7.9]   H. A. Haus *et al.* (1987) "Coupled-mode theory of optical waveguides" *J. Lightwave Technol.* **LT-5**, 16–23.

[7.10] D. R. Scifres *et al.* (1979) "High-power coupled-multiple-stripe phase-locked injection laser" *Appl. Phys. Lett.* **34**, 259–261.

[7.11] D. E. Ackliy and R. W. H. Engelmann (1981) "Twin-stripe injection laser with leak-mode coupling" *Appl. Phys. Lett.* **37**, 866–868.

[7.12] J. Katz *et al.* (1983) "Far-field distributions of semi-conductor phase-locked arrays with multiple contacts" *Electron. Lett.* **19**, 660–662.

[7.13] J. P. van der Zeil *et al.* (1984) "High-power picosecond pulse generation in GaAs multiquantum well phased-locked laser arrays using pulsed current injection" *IEEE J. Quantum Electron.* **QE-20**, 1236–1242.

[7.14] H. Kogelnik (1975) "Theory of dielectric waveguides" in *Integrated Optics* ed T. Tamir (New York: Springer), ch. 2.

[7.15] J. K. Butler *et al.* (1984) "Coupled mode analysis of phased-locked injection laser array" *Appl. Phys. Lett.* **44**, 293–295.

[7.16] E. Kapon *et al.* (1984) "Supermode analysis of phase-locked arrays of semiconductor lasers" *Opt. Lett.* **9**, 125–127.

[7.17] L. A. Molter-Orr and H. A. Haus (1984) "$N \times N$ coupled waveguide switch" *Opt. Lett.* **9**, 466–467.

[7.18] S. Ruschin and E. Marom (1984) "Coupling effects in symmetrical three-guide structures" *J. Opt. Soc. Am. A* **1**, 1120–1128.

[7.19] M. O. Vassell (1976) "A theoretical analysis of mode-mixing optical waveguides with nearest-neighbor mode coupling" *Opt. Quantum Electron.* **8**, 23–30.

[7.20] E. G. Rawson and M. D. Bailey (1979) "Bitaper star couplers with up to 100 fiber channels" *Electron. Lett.* **15**, 432–433.

[7.21] C. C. Wang *et al.* (1985) "9×9 single-mode fiber-optic star couplers" *Opt. Lett.* **10**, 49–51.

[7.22] W. Kaplan (1962) *Operational Methods for Linear Systems* (Reading, MA: Addison-Wesley), p. 490.

[7.23] W. Streifer and E. Kapon (1979) "Application of the equivalent-index method to DH diode lasers" *Appl. Opt.* **18**, 3724–3725.

[7.24] R. V. Schmidt and I. P. Kaminow (1974) "Metal diffused optical waveguides in $LiNbO_3$" *Appl. Phys. Lett.* **25**, 458–460.

[7.25] J. L. Jackel, *et al.* (1982) "Proton exchange for high index waveguides in $LiNbO_3$" *Appl. Phys. Lett.* **41**, 607.

[7.26] G. L. Destefanis *et al.* (1978) "The formation of waveguides and modulators in $LiNbO_3$ by ion implantation" *J. Appl. Phys.* **50**, 7898.

[7.27] L. Thylen *et al.* (1983) "Computer analysis and design of $Ti:LiNbO_3$ integrated optics devices and comparison with experiments" *Proc. ECOC*, 425–428.

[7.28] M. D. Feit *et al.* (1983) "Comparison of calculated and measured performance of diffused channel-waveguide couplers" *J. Opt. Soc Am.* **73**, 1296–1304.

[7.29] I. P. Kaminow (1974) *An Introduction to Electrooptic Devices* (New York and London: Academic).

[7.30] L. Thylen (1988) "Integrated optics in $LiNbO_3$: Recent developments in devices for telecommunications" *J. Lightwave Technol.* **LT-6**, 847–861.

[7.31] T. Tamir, Ed. (1985) *Integrated Optics* (Berlin: Springer).

[7.32] R. V. Schmidt and H. Kogelnik (1975) "Electrooptically switched coupler with stepped $\Delta\beta$ reversal using Ti-diffused $LiNbO_3$ waveguides" *Appl. Phys. Lett.* **28**, 503–506.

[7.33] C. S. Tsai *et al.* (1978) "Optical channel waveguide switch and coupler using total internal reflection" *IEEE J. Quantum Electron.* **QE-14**, 513–517.

[7.34] A. Neyer (1983) "Electrooptic x-switch using single mode $Ti:LiNbO_3$ channel waveguides" *Electron. Lett.* **19**, 553.

[7.35] M. Papuchon and A. Roy (1977) "Electric active bifurcation: BOA" *Appl. Phys. Lett.* **31**, 266–267.

[7.36] A. Neyer *et al.* (1985) "A beam propagation method analysis of active and passive waveguide crossing" *J. Lightwave Technol.* **LT-3**, 635–642.

[7.37] P. Granestrand *et al.* (1986) "Strictly nonblocking 8×8 integrated optical switch matrix" *Electron. Lett.* **22**, 816–818.

[7.38] L. McCaugan and G. A. Bogert (1985) "4× Ti:LiNbO$_3$ integrated optical crossbar switch array" *Appl. Phys. Lett.* **47**, 348–350.

[7.39] A. Neyer *et al.* (1986) "Nonblocking 4×4 switch array with 16 switches in Ti:LiNbO$_3$" *Proc. OSA Top. Meet. Integrated and Guided Wave Optics*, paper WAA2.

[7.40] H. F. Taylor (1974) "Optical wave-guide connecting networks" *Electron. Lett.* **10**, 41–43.

[7.41] D. Hoffman *et al.* (1986) "Low-loss rearrangeably nonblocking 4×4 switch matrix module" *Proc. ECOC*, 167–170.

[7.42] K. Habara and K. Kikuchi (1985) "Optical time division space switches tree-structured directional couplers" *Electron. Lett.* **21**, 631–632.

[7.43] G. A. Bogert (1987) "4×4 Ti:LiNbO$_3$ switch array with full broadcast capability" *Proc. OSA Top. Meet. Photonic Switching*, paper ThD3.

[7.44] A. H. Gnauck *et al.* (1986) "Information bandwidth limited transmission at 8 Gbit/s over 68.3 km of optical fiber" *Proc. OFC-86*, paper PD-9.

[7.45] F. Koyama and K. Iga (1987) "Frequency chirping in some types of external modulators" *Proc. OFC/IOOC-87*, paper W04.

[7.46] L. Thylen *et al.* (1985) "Integrated optic device for high-speed databuses" *Electron. Lett.* **21**, 491–493.

[7.47] M. Isutsu *et al.* (1977) "Broad-band traveling-wave modulator using a LiNbO$_3$ optical waveguide" *IEEE J. Quantum Electron.* **QE-13**, 287–290.

[7.48] R. A. Becker (1984) "Traveling-wave electrooptic modulator with maximum bandwidth-length product" *Appl. Phys. Lett.* **45**, 1168–1170.

[7.49] L. Thylen and A. Djupsjöbacka (1985) "Bandpass response traveling-wave modulator with a transit time difference compensation scheme" *J. Lightwave Technol.* **LT-3**, 47–51.

[7.50] A. Djupsjöbacka (1985) "Novel type of broad-band traveling wave integrated optic modulator" *Electron. Lett.* **21**, 908–909.

[7.51] M. Nasarathy and D. Dolfi (1987) "Velocity mismatch compensated spread spectrum traveling-wave modulators using pseudorandom switched electrode patterns" *Proc. OFC/IOOC'87*, paper TUQ37.

[7.52] T. Sawano *et al.* (1987) "Integrated-optic 2×4 traveling-wave switch matrix and its use in optical time demultiplexing" *Proc. OFC/IOOC'87*, paper WB3.

[7.53] M. N. Armenise (1988) "Fabrication techniques of lithium niobate waveguides" *IEE Proc.* **135**, 85–91.

[7.54] I. P. Kaminow and J. R. Carruthers (1973) "Optical waveguiding layer in LiNbO$_3$ and LiNbO$_3$" *Appl. Phys. Lett.* **22**, 326–328

[7.55] J. R. Carruthers *et al.* (1974) "Diffusion kinetics and optical waveguiding properties of outdiffused layers in lithium niobate and lithium tanatlate" *Appl. Opt.* **13**, 2333–2342.

[7.56] I. P. Kaminow *et al.* (1974) "Lithium niobate ridge waveguide modulator" *Appl. Phys. Lett.* **24**, 622–624.

[7.57] R. V. Schmidt and I. P. Kaminow (1974) "Metal-diffused optical waveguides in lithium niobate" *Appl. Phys. Lett.* **25**, 458–460.

[7.58] J. Noda and H. Iwasaki (1979) "Impurity diffusion into LiNbO$_3$ and LiTaO$_3$" *Proc. 2nd Meeting on Ferroelectric Materials and Their Applications*, 149–154.

[7.59] K. Sugii *et al.* (1978) "A study on titanium diffusion into LiNbO$_3$ waveguides by electron probe analysis and X-ray diffraction methods" *J. Mat. Sci.* **13**, 523–533.

[7.60] H. Naitoh *et al.* (1977) "Mode control of Ti-diffused LiNbO$_3$ slab optical waveguide" *Appl. Opt.* **16**, 2546–2549.

[7.61] G. J. Griffiths and R. J. Esdaile (1984) "Analysis of titanium diffused planar optical waveguides in lithium niobate" *IEEE J. Quantum Electron.* **QE-20**, 149–159.

[7.62] M. Minakata *et al.* (1978) "Precise determination of refractive index changes in Ti diffused LiNbO$_3$ optical waveguides" *J. Appl. Phys.* **49**, 4677–4682.

[7.63] S. Fouchet *et al.* (1987) "Wavelength dispersion of Ti induced refractive index change in LiNbO$_3$ as a function of diffusion parameters" *J. Lightwave Tech.* **LT-5**, 700–708.

[7.64] J. L. Jackel (1982) "Suppression of out diffusion in Ti diffused LiNbO$_3$: a review" *Opt. Commun.* **3**, 82–85.

[7.65] K. S. Buritskii and V. A. Chernykh (1986) "Properties of Ti:LiNbO$_3$ waveguides with suppressed outdiffusion of Li$_2$O" *Sov. J. Quantum Electron.* **16**, 1424–1426.

[7.66] M. N. Armenise *et al.* (1984) "The dependence of in-plane scattering levels in Ti:LiNbO$_3$ optical waveguides on diffusion time" *IEE Proc. H* **131**, 295–298.

[7.67] M. N. Armenise *et al.* (1984) "In-plane scattering in tithium-diffused LiNbO$_3$ optical waveguides" *Appl. Phys. Lett.* **45**, 326–328.

[7.68] J. Singh *et al.* (1985) "An experimental study of in-plane light scattering in titanium diffused y-cut LiNbO$_3$ optical waveguides" *J. Lightwave Technol.* **3**, 67–76.

[7.69] M. E. Twigg (1987) "Study of structural faults in Ti-diffused lithium niobate" *Appl. Phys. Lett.* **50**, 501–503.

[7.70] M. L. Shah (1975) "Optical waveguides in LiNbO$_3$ by ion exchange technique "*Appl. Phys. Lett.* **26**, 652–653.

[7.71] J. L. Jackel (1980) "Ion exchange for optical waveguides in lithium niobate and lithium tantalate" *Proc. Topical Meeting on Integrated Optics and Guide-wave Optics*, paper WB4-1–4.

[7.72] J. L. Jackel and C. E. Rice (1981) "Topotactic LiNbO$_3$ and LiTaO$_3$" *Ferroelectrics* **38**, 801–804.

[7.73] J. L. Jackel *et al.* (1982) "Proton exchange for high-index waveguides in LiNbO$_3$" *Appl. Phys. Lett.* **41**, 607–608.

[7.74] A. C. G Nutt (1985) "Techniques for fabricating integrated optical components on lithium niobate" *PhD Thesis*, University of Glasgow.

[7.75] J. L. Jackel and R. J. Holmes (1983) "Recent advances in LiNbO$_3$ processing" *Proc. 2nd European Conf. on Integrated Optics*, 38–41.

[7.76] M. De Micheli *et al.* (1982) "Fabrication and characterization of titanium indiffused proton exchanged (TIPE) waveguides in lithium niobate" *Opt. Commun.* **42**, 101–103.

[7.77] J. L. Jackel *et al.* (1983) "Composition control in proton exchanged LiNbO$_3$" *Electron. Lett.* **19**, 387–388.

[7.78] G. M. Destefanis *et al.* (1979) "The formation of waveguides and modulators in LiNbO$_3$ by ion implantation" *J. Appl. Phys.* **50**, 7898–7905.

[7.79] G. M. Destefanis *et al.* (1980) "Optical waveguides in LiNbO$_3$ formed by ion implantation" *Radiat. Effects* **48**, 63–68.

[7.80] J. Heibei and E. Voges (1982) "Strip waveguides in LiNbO$_3$ fabricated by combined metal diffusion and ion implantation" *IEEE J. Quantum Electron.* **QE-18**, 820–825.

[7.81] J. C. Campbell *et al.* (1975) "GaAs electrooptic directional-coupler switch" *Appl. Phys. Lett.* **27**, 212.

[7.82] F. J. Leonberger *et al.* (1976) "GaAs p$^+$n–n$^+$ directional-coupler switch" *Appl. Phys. Lett.* **29**, 652.

[7.83] K. Tada and H. Yanagawa (1978) "New coupled-waveguide optical modulators with Schottky contacts" *J. Appl. Phys.* **49**, 5404.

[7.84] A. Alping *et al.* (1986) "Highly efficient waveguide phase modulator for integrated optoelectronics" *Appl. Phys. Lett.* **48**, 1243.

[7.85] S. Y. Wang *et al.* (1987) "GaAs traveling-wave polarization electrooptic waveguide modulator with bandwidth in excess of 20 GHz at 1.3 μm" *Appl. Phys. Lett.* **51**, 83

[7.86] X. Cheng and C. S. Tsai (1988) "Electrooptic Bragg-diffraction modulators in GaAs/AlGaAs heterostructure waveguides" *J. Lightwave Technol.* **6**, 809.

[7.87] C. S. Tsai (1979) "Guided-wave acoustooptic Bragg modulators for wideband integrated optical communications and signal processing" *IEEE Trans. Circuits Syst.* **CAS-26**, 1072.

[7.88] C. J. Lii *et al.* (1986) "Wide-band guided-wave acoustooptic Bragg cells in GaAs-GaAlAs waveguide" *IEEE J. Quantum Electron.* **QE-22**, 868.

[7.89] A. Yariv (1984) "The beginning of integrated optoelectronic circuits" *IEEE Trans. Electron. Devices* **ED-3**, 1656

[7.90] I. Hayshi (1986) "Research aiming for future optoelectronic integration. The Optoelectronics Joint research Laboratory" *IEE Proc. J.* **133**, 237.

[7.91] O. Wada *et al.* (1986) "Recent progress in optoelectronic integrated circuits" *IEEE J. Quantum Electron.* **QE-22**, 805.

[7.92] F. J. Leonberger *et al.* (1981) "Low loss GaAs optical waveguides formed by lateral kepitaxial growth over oxide" *Appl. Phys. Lett.* **38**(5), 313–315.

[7.93] M. Erman *et al.* (1983) "Low loss waveguide grown on GaAs using localized vapor phase epitaxy" *Appl. Phys. Lett.* **43**(10), 894–895.

[7.94] P. Buchman *et al.* (1984) "Reactive ion etched GaAs optical waveguide modulators with low loss and high speed" *Electron. Lett.* **20**, 295–297.

[7.95] R. G. Walker (1985) "Simple and accurate loss measurement technique for semiconductor optical waveguides" *Electron. Lett.* **21**, 581–583.

[7.96] E. Kapon and R. Bhat (1987) "Low-loss single-mode GaAs/AlGaAs optical waveguides grown by organometallic vapor phase epitaxy" *Appl. Phys. Lett.* **50**, 1628–1630.

[7.97] F. Ito *et al.* (1989) "A carrier injection type optical switch in GaAs using free carrier plasma dispersion with wavelength range from 1.06 to 1.55 μm" *IEEE J. Quantum Electron.* **QE-25**, 1677–1681.

[7.98] H. A. Haus (1984) *Waves and Fields in Optoelectronics* (Englewood Cliffs, NJ: Prentice-Hall), p. 189.

[7.99] P. Sansonetti *et al.* (1989) "Design of semiconductor electrooptic directional couple with the beam propagation method" *J. Lightwave Technol.* **LT-7**, 385–389.

[7.100] G. B. Hocker and W. K. Burns (1977) "Mode dispersion in diffused channel waveguides by the effective index method" *Appl. Opt.* **16**, 113–118.

[7.101] E. Caquot *et al.* (1985) "ETHER: Software simulating the electrical behavior of heterostructure devices" *Physica* **129B**, 356–360, 1985.

[7.102] M. J. Akams *et al.* (1986) "Optimum overlap of electric and optical fields in semiconductor waveguide devices" *Appl. Phys. Lett.* **48**, 820–822.

[7.103] H. G. Bach *et al.* (1983) "Electrooptical light modulation in InGaAsP/InP double heterostructure diodes" *Appl. Phys. Lett.* **42**, 692–694.

[7.104] B. Hermansson and D. Yevick (1983) "A two-dimensional proagating beam method analysis of the coupling of semiconductor laser light into tapers" *Proc. ECOC'83, Geneva, Switzerland*, 459–462.

INDEX